CHEMICAL SIGNALS IN VERTEBRATES 3

CHEMICAL SIGNALS IN VERTEBRATES 3

Edited by

Dietland Müller-Schwarze

and

Robert M. Silverstein

State University of New York
Syracuse, New York

Springer Science+Business Media, LLC

Library of Congress Cataloging in Publication Data

International Symposium on Chemical Signals in Vertebrates (3rd: 1982: Sarasota, Fla.)

Chemical signals in vertebrates 3.

"Proceedings of the Third International Symposium on Chemical Signals in Vertebrates, held April 11–13, 1982, in Sarasota, Florida"—T.p. verso.

Includes bibliographical references and index.

1. Vertebrates—Behavior—Congresses. 2. Chemical senses—Congresses. 3. Pheromones—Congresses. 4. Animal communication—Congresses. I. Müller-Schwarze, Dietland. II. Silverstein, Robert Milton, 1916– . III. Title.
[DNLM: 1. Animal communication—Congresses. 2. Chemoreceptors—Congresses. 3. Pheromones—Congresses. 4. Vertebrates—Physiology—Congresses. W3 SY5O635 3rd 1982c / WL 102 I6592 1982c]

QL750.I48 1982 596′.0159 83-2151

ISBN 978-1-4757-9654-4 ISBN 978-1-4757-9652-0 (eBook)
DOI 10.1007/978-1-4757-9652-0

Proceedings of the Third International Symposium on Chemical Signals in Vertebrates, held April 11–13, 1982, in Sarasota, Florida

Originally published by Plenum Press, New York in 1983
Softcover reprint of the hardcover 1st edition 1983

PREFACE

The first volume in this series appeared in 1977, the second in 1980. From these volumes and the present one, some research trends in chemical communication can be perceived.

In the 1977 volume, studies on 13 animal taxa were reported. In the present volume, the number is 25. This taxonomic diversification of research since the first volume of this series demonstrates the wide variety of ecological adaptions, although no new general principles of chemical communication have emerged. Furthermore, divergences in chemical communication below the species level have become more apparent. In general, more sophisticated observations and techniques have led to greater awareness of the complexities in chemical communication. As such awareness has also developed in the field of insect chemical communication, there has been a corresponding increase in the identification of the chemical compounds involved. However, in the vertebrates, no such correlation exists; in the present volume, conclusive chemical identifications of semiochemicals are remarkable by their paucity.

Continuing the comparison with the field of insect communication, we note also a disparity in efforts to reduce findings to practice. Traps baited with semiochemicals are used on a routine basis for detection and survey of insect pest populations on a world-wide basis; population control by mass trapping and by permeation has been reported for several species; although many difficulties remain, insect pest control, at least in part, by semiochemicals can be safely projected. There is no indication in the present volume that corresponding progress in the vertebrates has been made or is in prospect in the near future. Vertebrate pheromones, however, present different problems and opportunities. While it is unlikely that behavior can be manipulated to any significant extent by chemical stimuli alone, limited and selective effects may be possible, especially when chemical signals are combined with other stimuli.

Basic studies, in contrast, remain a rewarding exercise. Notable are the advances made in the role of priming pheromones in reproduction, in the rapidly developing investigations of vomeronasal reception, and in neuroendocrine control of semiochemical use and reception.

Syracuse, New York
March 1983

D. Müller-Schwarze
R. M. Silverstein

CONTENTS

Part One: Reptiles

Strike-Induced Chemosensory Searching by Rattlesnakes: The Role of Envenomation-Related Chemical Cues in the Post-Strike Environment 1
David Chiszar, Charles W. Radcliffe
Kent M. Scudder, and David Duvall

Fossil and Comparative Evidence for Possible Chemical Signaling in the Mammal-Like Reptiles 25
David Duvall, Michael B. King, and
Brent M. Graves

Part Two: Vomeronasal Organ

Snake Tongue Flicking Behavior: Clues to Vomeronasal System Functions 45
Mimi Halpern, and John L. Kubie

The Accessory Olfactory System: Role in Maintenance of Chemoinvestigatory Behavior 73
Gary K. Beauchamp, Irwin G. Martin,
Judith L. Wellington, and Charles J.
Wysocki

Flehmen Behavior and Vomeronasal Organ Function 87
Benjamin L. Hart

Part Three: Neuroendocrinology of Olfaction

Olfaction in Central Neural and Neuroendocrine Systems: Integrative Review of Olfactory Representations and Interrelations 105
W. B. Quay

The Neuroendocrinology of Scent Marking 119
Pauline Yahr and Deborah Commins

Part Four: Chemical Signals and Endocrines

Priming Pheromones in Mice 135
Anna Marchlewska-Koj

Pheromonal Control of the Bovine Ovarian Cycle 153
J. G. Vandenbergh and M. K. Izard

Synchronizing Ovarian and Birth Cycles by Female Pheromones 159
Martha K. McClintock

Volatile and Nonvolatile Chemosignals of Female Rodents: Differences in Hormonal Regulation 179
John Nyby

Part Five: Chemical Signals and Genetics

Communication Disparities Between Genetically-Diverging Populations of Deermice 195
Glenn Perrigo and F. H. Bronson

Part Six: Field Studies (Pheromone Ecology)

The Ecological Importance of the Anal Gland Secretion of Yellow Voles (Lagurus luteus) 211
Fan Zhiqin

Odor as a Component of Trap Entry Behavior in Small Rodents 223
D. Michael Stoddart

Experimental Modulation of Behavior of Free-Ranging Mammals by Semiochemicals 235
Dietland Müller-Schwarze

Part Seven: Social Odors: Discrimination and Recognition

Mechanisms of Individual Discrimination in Hamsters 245
Robert E. Johnston

Human Olfactory Communications 259
Michael J. Russell

Part Eight: Chemistry

Studies of the Chemical Composition of Secretions from Skin Glands of the Rabbit *Oryctolagus cuniculus* 275
B. S. Goodrich

Part Nine: Pheromones and Other Physiological Functions

Thermal and Osmolarity Properties of Pheromonal Communication in the Gerbil, *Meriones unguiculatus* 291
Del D. Thiessen and Arthur E. Harriman

Part Ten: Abstracts

The Evolution of Alarm Pheromones 309
Paul J. Weldon

Investigation into the Origin(s) of the Freshwater Attractant(s) of the American Eel 313
Peter W. Sorensen

A Pregnancy Block Resulting from Multiple-Male Copulation or Exposure at the Time of Mating in Deer Mice (*Peromyscus maniculatus*) 317
Donald A. Dewsbury

Olfactory Communication in Kangaroo Rats (*D. merriami*) 321
Jan A. Randall

Odor Preferences of Young Rats: Production of an Attractive Odor by Males 325
Richard E. Brown

Rate and Location of Scent Marking by Pikas During the Breeding Season 329
Carron Meaney

Effects of Urine on the Response to Carrot-Bait in the European Wild Rabbit 333
Diana Bell, Stephen Moore, and David Cowan

An Investigation into the 'Bruce Effect' in Domesticated Rabbits 339
Diana Bell and Cindy Reece

Individual Discrimination on the Basis of Urine in Dogs and Wolves 343
Donna S. Brown and Robert E. Johnston

Throat-Rubbing in Red Howler Monkeys (*Alouatta seniculus*) 347
Ranka Sekulic and John F. Eisenberg

Author Index 351

Subject Index 361

STRIKE-INDUCED CHEMOSENSORY SEARCHING BY RATTLESNAKES: THE ROLE OF ENVENOMATION-RELATED CHEMICAL CUES IN THE POST-STRIKE ENVIRONMENT

David Chiszar, Charles W. Radcliffe,
and Kent M. Scudder

University of Colorado
Boulder, Co.

David Duvall

University of Wyoming
Laramie, Wy.

INTRODUCTION: STRIKE-INDUCED CHEMOSENSORY SEARCHING

Rattlesnakes and many other viperids typically strike and release adult rodent prey (Gans, 1966; O'Connell et al., 1982; Radcliffe et al., 1980), allowing the envenomated rodent to wander up to 600 cm before succumbing to the venom (Estep et al., 1981). The snakes then follow the chemical trail left by the envenomated prey. Although this predatory strategy risks losing the prey, it avoids tissue damage that could result from rodent teeth and claws if the snake attempted to hold the struggling prey after the strike. Even some of the deadliest elapids exhibit this strategy when they prey upon rodents (Chiszar et al., under review; Radcliffe et al., 1982; Shine & Covacevich, 1982), indicating that rodents are formidable prey and that the strike-release-trail system probably appeared very early in the evolution of venomous snakes (see Marx & Rabb, 1965, for a discussion of viperid evolution).

Of course, this system places considerable importance on the snake's ability to find the envenomated prey, especially when the rodent has traveled far enough to get out of the snake's visual and/or thermodetection fields. Accordingly, many researchers have focused on the chemosensory abilities of snakes which exhibit the strike-release-trail strategy (see Burghardt, 1970, for a thorough review). In particular, Dullemeijer (1961) showed that rattlesnakes (Crotalus ruber) were able to follow rodent trails on the basis of chemical cues alone, and that vomeronasal chemoreception was the most important sensory system for this behavior. It was, therefore, surprising when we failed to observe increases in rate of tongue flicking (RTF) when rattlesnakes were presented with odors derived from rodent prey (Chiszar & Radcliffe, 1977; Chiszar et al., 1976). These data led to many subsequent experiments which revealed that rattlesnakes become interested in such chemical cues after they have delivered a successful predatory strike (see Chiszar & Scudder, 1980, for a review).

Two recently completed studies will illustrate our general methodology as well as the robustness of this phenomenon, called strike-induced chemosensory searching (SICS). Both studies were conducted by Kathryn Stimac. In the first experiment, six adult specimens of the banded rock rattlesnake (Crotalus lepidus klauberi) were subjects. They were collected by Charles Radcliffe and Barbara O'Connell from the Chiricahua Mountains near Portal, Arizona. After the animals had acclimated to the laboratory and to the weekly feeding schedule, each snake was observed in two conditions: NS, a live mouse (Mus musculus, about 16 gm) was suspended via forceps into the snake's cage (about 30 cm from the snake's head) for 3 sec and was then removed; and S, same as condition NS except that the snake was allowed to strike the mouse at the end of the 3 sec presentation (the mouse was removed immediately after being released by the snake). Intertrial interval (ITI) was one week, and three snakes experienced NS followed by S whereas the remaining snakes received the reverse order. Fig. 1 shows the mean RTF (per min) for 10 min prior to mouse presentations and for 20 min following them. It is clear that RTF increased consequent to mouse presentations and that RTF was much higher after S than NS presentations. It is especially noteworthy that mice never touched the floor or walls of snake cages during these observations; hence, it is unlikely that snakes were responding to odors present in the post-strike environment. Rather, we believe that stimulation encountered during the strike is responsible for the elevation of RTF which would, under normal circumstances, permit the snake to locate the envenomated mouse's trail. Although the strike is released by visual and/or thermal cues arising from prey (Scudder, 1982), the strike then releases chemosensory searching. A final point about this study is that these C. l. klauberi probably fed almost entirely on lizards in nature. It is also probable that the snakes strike and hold lizards. Nonetheless, they exhibited the strike-release strategy in dealing with rodents in this experiment. In fact, they did so beginning with the first post-neonatal mice we offered them. Accordingly, SICS occurs even in these montane rattlesnakes which (at most) are facultative rodent feeders. Therefore, SICS is not restricted only to those rattlesnakes which specialize on rodent prey.

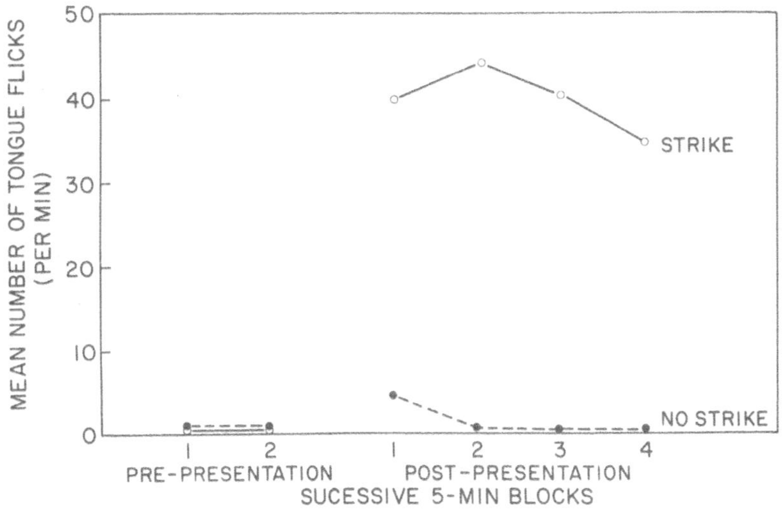

Fig. 1. Mean number of tongue flicks per min by 6 specimens of *Crotalus lepidus klauberi*. Each snake was observed for 10 min prior to presentation of a mouse, and for 20 min thereafter. In the NO STRIKE condition the mouse was suspended via forceps into the snake's cage but held out of striking range for 3 sec. In the STRIKE condition the snakes were allowed to deliver a predatory strike at the end of 3 sec. Mice were always removed after presentation.

The second study that will be mentioned by way of introduction involved a single specimen of *C. triseriatus*. This animal was capable of very limited movement of the most anterior and posterior extremes of its body (5-10 cm). Fig. 2 reveals extensive ossification of the spinal column. Furthermore, this snake never ingested food during the last six months of its life; indeed, it rarely even *attempted* to swallow. Nonetheless, the snake would strike rodent prey. So, Ms. Stimac observed the animal in 5 replications of the NS and S conditions described above. Fig. 3. shows the mean RTF's, and it is clear that SICS occurred. We conclude that SICS in this *C. triseriatus* was probably an obligate consequence of striking prey, even though other aspects of the predatory chain (e.g., yawning, locating the dead prey's head, swallowing) were not (Chiszar et al., 1980a, 1981a).

DURATION OF SICS

Typically, our experiments record SICS for 10-15 min post-strike; and, RTF usually accelerates to an asymptotic level during

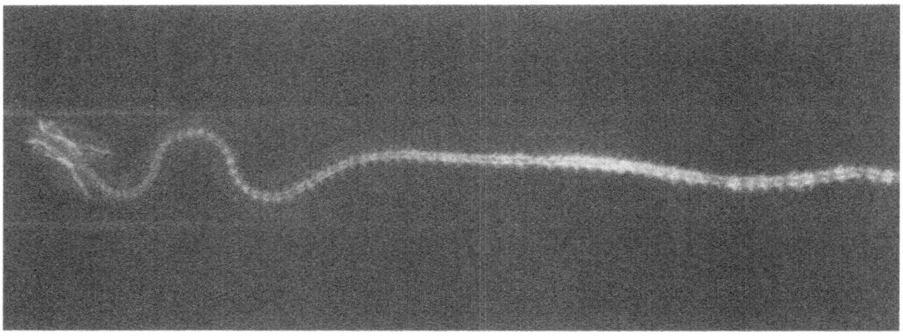

Fig. 2a Radiograph of a specimen of Crotalus triseriatus. Vertebrae in anterior 1/3 of the body appear normal, but high density of the vertebrae in the midsection is indicative of bony proliferation which is characteristic of arthritis and osteomyelitis.

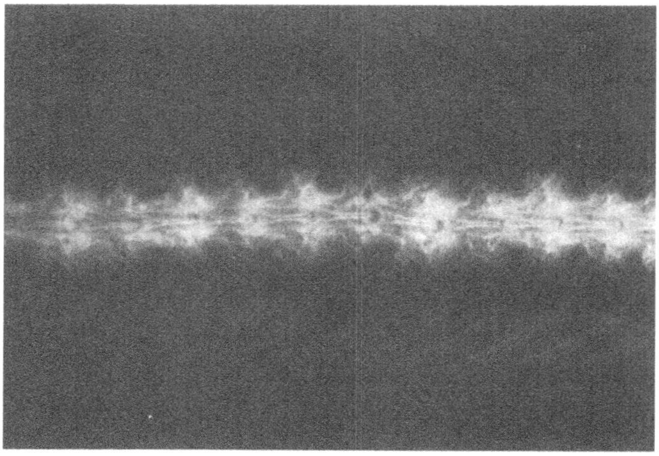

Fig. 2b Enlargement of the midsection of Fig. 2a showing extensive ossification of joints. The snake was unable to coil or even to bend this section of its trunk. In fact the animal was never seen to crawl during the last six months of life. Radiographs were made and interpreted by Kevin Fitzgerald and Barbara Shor, Colorado State University, School of Veterinary Medicine.

min 1-5 and remains at that level until observation is discontinued. It is interesting that molecules acquired by the vomeronasal organs (VMO) during the strike probably remain available to those organs for 10-15 min (Kubie, 1977; Meredith,

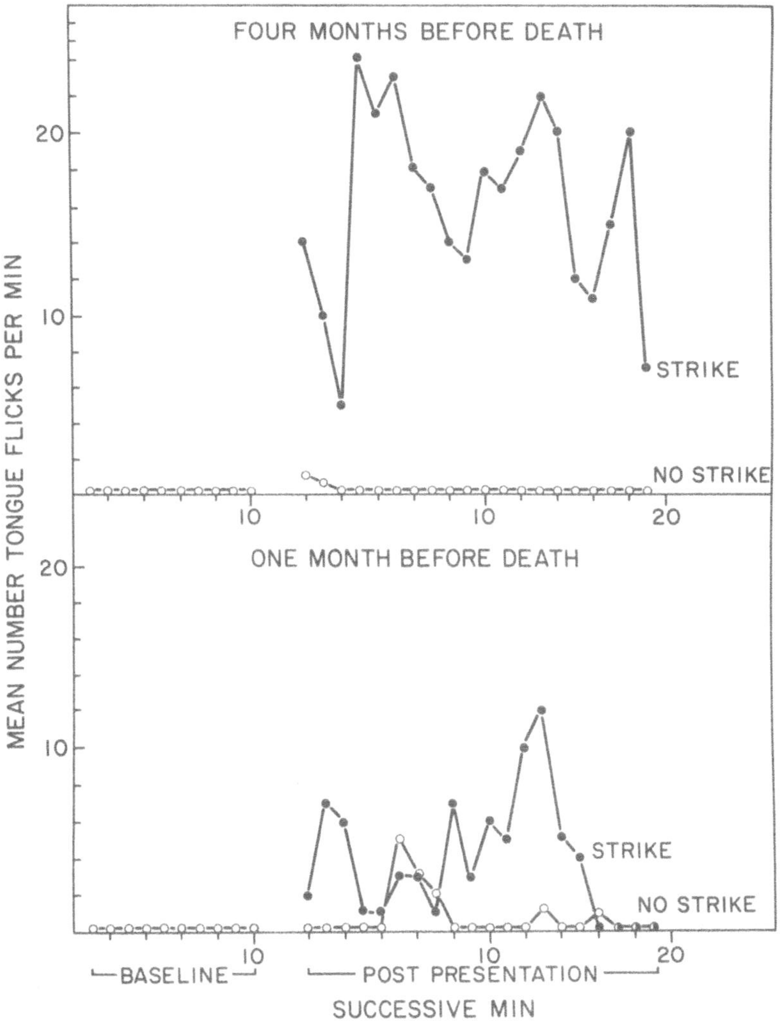

Fig. 3. Mean number of tongue flicks per min by the specimen of <u>C. triseriatus</u> shown in Fig. 2. Conditions (NO STRIKE and STRIKE) were identical to those described in Fig. 1. Note that SICS was apparent during the last four months (4 replications) and even during the last month (1 replication) of this animal's life, even though no ingestive behavior occurred. This is evidence of the obligate nature of SICS.

1980; Meredith & Burghardt, 1978). Hence, we wanted to determine if SICS would continue beyond this time with no rodent-derived chemical cues available in the post-strike environment. If RTF suddenly returned to baseline after 15 min, it would be reasonable to conclude that SICS was supported entirely by direct chemical stimulation of VMO.

Scudder et al. (under review) observed *C. v. viridis* and *C. durissus terrificus* for 30 min after S and NS presentations. In both taxa, SICS remained at the asymptotic level (reached after 5 min) until the end of the observation period. Chiszar et al. (1982) observed *C. v. viridis* and *C. e. enyo* for 300 min after S and NS presentations. Both taxa exhibited asymptotic RTF for about 60 min and then gradually returned to baseline. However, RTF remained significantly above baseline for about 150 min. Hence, SICS lasts longer than would be expected if molecules acquired by VMO during the strike were entirely responsible for maintaining this behavior. For this reason, we have suggested that striking prey results in the formation of an enduring neural representation of chemosensory properties of prey (we have also used the term "search image"; O'Connell et al., 1981; Chiszar et al., 1980b)). This is presumably a CNS phenomenon, and it serves as a basis for comparison. That is, we hypothesize that snakes attempt to locate stimuli in the post-strike environment which match elements of this representation (Chiszar et al., 1981b, 1982).

It would be interesting to determine the specificity of this neural representation. If a snake strikes a mouse, will a dead rat or a dead lizard be as acceptable as a dead mouse to the searching predator? Does *envenomation* influence the representation (e.g., can the snake *distinguish* between envenomated vs non-envenomated mice? Are they equally acceptable?)? Other fundamental questions concern the ontogeny of this representation. For example, does a snake form a new image after each predatory strike, or is the image innate and simply retrieved from some CNS storage location as a consequence of stimulation arising from the strike? Do neonatal rattlesnakes exhibit SICS and related aspects of adult predatory behavior upon their first encounter with rodents? We have data on many of these issues; in fact, these questions have guided our research over the past two years.

ACCESSIBILITY OF CHEMOSENSORY SEARCHING

We have concentrated on SICS, but striking is not the only factor that will trigger relatively high RTFs. Cowles and Phelan

(1959) implied that prolonged exposure of hungry rattlesnakes (C. ruber) to several strong odors (esp. putrified flesh) would cause RTF increases. Gillingham and Baker (1982) showed that C. atrox would locate and ingest putrified rodent carcasses, and this almost certainly involved high RTFs. Gillingham and Clark (1981) observed RTF increases in C. atrox after prolonged exposure to visual and/or thermal cues arising from mice. Chiszar et al. (1981c) reported similar behavior in C. v. viridis, C. e. enyo, and S. catenatus tergeminus. Also, Chiszar et al. (1981b) found that increased hunger was associated with increased sensitivity to rodent odors in C. v. viridis and C. e. enyo, even though none of the snakes had an opportunity to strike rodents prior to being exposed to odors.

At the level of behavioral description, these studies suggest that rattlesnakes are facultatively polyethic with respect to predatory strategies. Relatively well fed specimens are probably ambushers, and probably exhibit the strike-release-trail strategy most of the time (Klauber, 1956). With mild increases in hunger the animals may shift ambushing sites, and they might even become foragers. This, however, requires sensitivity to prey odors in advance of striking. Carrion feeding may emerge when rattlesnakes are very hungry, or when a foraging specimen encounters odors arising from putrified flesh. Indeed, some workers consider that carrion may be a relatively frequent, even modal, component of rattlesnake diets (Gillingham & Baker, 1982).

The studies cited in this section have interesting implications for the idea that rattlesnakes have CNS representations of chemosensory properties of prey. If the RTF elevations reported by Gillingham and Clark (1981) and Chiszar et al. (1981b & c) are functionally analogous to SICS, they suggest that chemosensory searching and the representational system can be accessed in ways other than by striking prey. That prolonged visual and/or chemical stimulation can activate chemosensory searching is not particularly surprising. That increased hunger causes rattlesnakes to become sensitive to prey odors may mean that a CNS representation of prey attributes is being made available to the predators on the basis of purely interoceptive events in order to guide foraging behavior. This, in turn, suggests that when snakes are in the ambushing mode striking does not result in the formation of an image; rather, striking results in the retrieval of an image which is already present in the CNS but which needs to be brought into foreground (to use a computer metaphor). That newborn rattlesnakes exhibit SICS and trailing behavior is additional support for the hypothesis that some form of prey-related CNS representation is innate and merely needs to be "activated" by a predatory strike (Scudder et al, in prep.; see also Burghardt, 1970).

ENVENOMATED VS NONENVENOMATED MICE

Given that a rattlesnake has struck a mouse and that SICS has been activated, will that snake distinguish between envenomated (E) and nonenvenomated (but dead; NE) mice? Duvall et al. (1978) allowed rattlesnakes to choose between such mice, and the E mouse was grasped and swallowed on a majority of the trials. Then Duvall et al. (1980) presented rattlesnakes with such mice wrapped in hardware mesh bags. The amount of time and the number of tongue flicks directed to each bag were recorded. Both *C. v. viridis* and *C. d. terrificus* showed significantly more interest in the bag containing the E rodent. Finally, Chiszar et al. (1980) recorded the same measures when rattlesnakes were successively presented with E and NE mice wrapped in mesh bags. Once again the E mice received more tongue flicks and were examined for a greater length of time than NE mice. This latter study indicates that the ability of rattlesnakes to distinguish between E and NE mice does not depend upon a contrast effect deriving from simultaneous availability of both sets of cues. Such a result is consistent with the idea that rattlesnakes are comparing cues from the environment with features of a CNS representation of chemosensory properties of prey.

Although considerably more work is necessary, enough data exist to support the conclusion that rattlesnakes can distinguish E from NE mice. It is not yet possible to say what the critical difference(s) between these mice may be. Perhaps *venom* alters mice in ways detectable to rattlesnakes. Perhaps *venom* has little to do with the phenomenon: The snakes may be responding to alarm pheromones released by mice as a consequence of being *attacked*.

TRAILING

It is essential to recognize that SICS and trailing are not synonymous. In most of the previously cited studies of SICS great care was taken to eliminate or minimize rodent odors in the post-strike environment. This was done because we wanted to establish that the elevation of RTF was caused by striking rather than by odors deposited by prey during or after presentation to the rattlesnakes. After this issue was settled (see Chiszar & Scudder, 1980, for a review), we systematically manipulated the chemical contents of the post-strike environment by placing petri dishes containing various materials into snake cages just after strikes were delivered. With one exception (Scudder et al., under review) these manipulations never altered the magnitude of SICS. Even the exception turned out to be a very small effect. Accordingly, we concluded that the *magnitude* of SICS was a modal action pattern (Barlow, 1977) which was released by striking and which was not altered by chemical properties of the post-strike environment.

Most important, none of these experiments involved the presentation of a rodent trail to rattlesnakes. Therefore, we could not say that snakes were better at following trails after S than NS experiences. It is entirely possible that snakes would follow trails equally well after either type of mouse presentation.

However, Dullemeijer (1961) claimed that C. ruber was best at trailing rodents after striking than at any other time. Since this is an important point and since Dullemeijer presented no quantitative data to support this claim, we designed a study to compare rattlesnake trailing ability after S and NS encounters with mice (Golan et al., 1982). The results were quite clear. Six specimens of C. viridis confined 75% of their tongue flicks to the immediate vicinity of mouse trails if they had struck mice prior to being exposed to the trails; these same snakes made only chance encounters with the trails if they had not struck. In fact, the snakes moved steadily along the trails after S presentations, and they rarely moved at all after NS presentations.

Golan et al. also observed their snakes in post-strike environments containing no mouse trails. Interestingly, S presentations were followed by equally high RTFs whether or not a mouse trail was present. This adds further support to the assertion that the magnitude of SICS is a modal action pattern which is released by striking. Only the direction of the tongue flicking is under taxic control of chemical cues in the post-strike environment. A rattlesnake which has struck a mouse will emit a high RTF; if a trail happens to be available, the snake will encounter and follow it through these taxic effects.

These data of Golan et al. leave no doubt that trailing is facilitated by SICS. It is now interesting to consider the situation of a rattlesnake that has struck a rodent from ambush. Such a snake will always face a choice between two trails: the one left by the mouse as it wandered into the snake's striking range (i.e., before the mouse was envenomated), and the one left by the mouse after it was envenomated. Following the first trail will entail a waste of time and energy, whereas following the second trail will quickly lead to a meal (assuming no kleptoparasitism). Hence, it seems reasonable to hypothesize that natural selection should have favored snakes with the abilities (1) to discriminate between E vs NE trails, and (2) to select the former.

Baumann (1927, 1928, 1929) found that Vipera aspis discriminated between trails made by E vs NE mice, and Lisa Golan attempted to obtain similar data for rattlesnakes. Specimens of several species (C. v. viridis, C. horridus atricaudatus, C. e. enyo, C. d. terrificus) were placed into pens (180 x 66 x 81 cm) identical to those described by Golan et al. (1982). The snakes were confined in one end by a partition containing a guillotine door.

When a snake had been fasted for one week it was exposed to either an S or an NS presentation of a mouse. Then the guillotine door was opened and the snake was exposed to two heavy lines (each paralleled by 2 fine lines, 2 cm from the heavy ones). The heavy lines formed a V, each leading to a rock, behind which was either an E or an NE mouse. In two experimental conditions (one S and one NS), chemical trails were made as follows: Along one branch of the V was rubbed an E mouse (ventral surface of mouse in contact with the line), while an NE mouse (same sex litter mate of the E mouse) was similarly rubbed along the other branch. Trails were made and mice were placed behind rocks prior to opening the guillotine door. In two more conditions (one S and one NS), no chemical trails were applied to the V.

The experimental design was a 2x2 combination of S vs NS with presence vs absence of the two chemical trails during the post-presentation period. Sessions lasted until snakes located and began to ingest a mouse or for a maximum of 20 min (if a snake did not locate a mouse during this time, a score of 1200 sec was assigned as latency to find the mouse). Dependent variables included various latency measures and the number of sec that the snakes' heads were positioned between the respective pairs of fine lines. Fig. 4a shows that snakes exhibited trailing behavior only after S presentations (see also Table 1, measures 3, 4, 5, & 9). Only rarely did a snake show any interest in trails after NS presentations. Fig. 4b shows that after S presentations the snakes preferred the E trail, whereas after NS presentations the snakes exhibited no preference. After NS presentations the E and NE trails were contacted only by chance. Each wing of the V occupied about 9% of the floor surface, and after NS presentations, the trails were each contacted about 9% (or less) of the time that snakes were in the apparatus. (See Table 1, measures 3, 6, & 10). Table 1 (measures 7, 8) also presents data replicating SICS.

A total of 16 snakes located (and grasped) mice within 20 min; all but 2 of these were in S conditions ($X^2 = 8.62$, df = 1, $P < .05$ for S vs NS effect). Thirteen snakes (81.2%) selected E trails and located E mice ($X^2 = 6.25$, df = 1, $P < .05$ for E vs NE effect).

This experiment once again revealed that rattlesnakes are excellent at following rodent trails, that trails are followed best after snakes have struck mice, and that trails made with E mice are preferentially followed. Our next step was to replicate this study with slight modifications in procedure. Karen Estep, Richard Feiler, and Kathryn Stimac gathered much of the data. The major change was that E vs NE trails formed a Y pattern rather than a V. Hence, instead of bifurcating immediately, the two trails coursed along side-by-side for about 30 cm prior to bifurcating. This arrangement insured that snakes would sample both

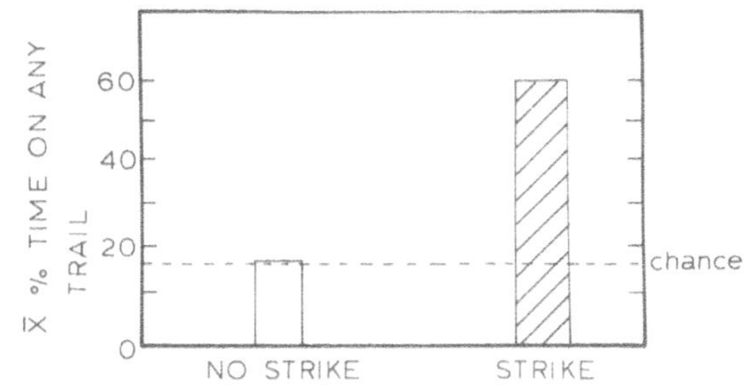

Figure 4a. When rodent trails were available, rattlesnakes followed them after predatory strikes were delivered but not after no-strike presentations of a rodent. See Table 1 for additional data.

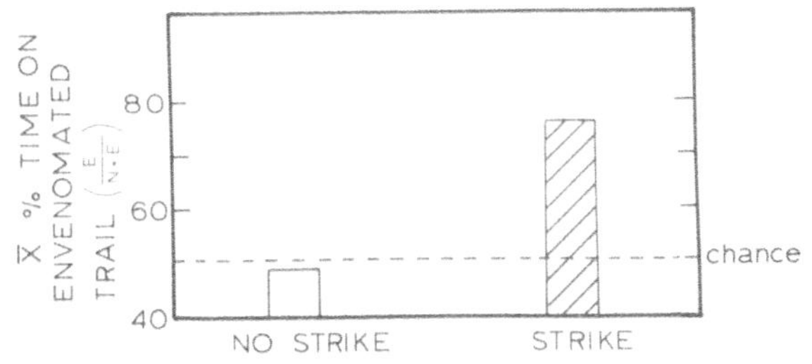

Figure 4b. When two rodent trails (one made by a nonenvenomated mouse and one made by an envenomated one) were available, rattlesnakes clearly preferred the envenomated trail--but only after a predatory strike was delivered. After no-strike presentations, no trailing occurred and no preference for envenomated trails was seen. See Table 1 for additional data.

Table 1. Summary of 10 dependent variables recorded in a simultaneous discrimination experiment in which 30 rattlesnakes were exposed to conditions corresponding to a 2x2 orthogonal combination of Strike vs No-Strike with Presence vs Absence of two trails (E and NE) applied to a V pattern drawn on the floor of the apparatus. The base of the V (10 cm) is called the "Joint" (J) trail, whereas the wings (each 150 cm) are called "Final" trails (F). One wing (randomly selected) contained an E trail (beginning at the base of J) while the other contained an NE trail in appropriate conditions of the experiment. E and NE mice were always placed behind rocks at the end of respective F lines (whether or not chemical trails were applied to them). One-half of the V occupied 9% of the floor area.

	Experimental Conditions			
	No Strike		Strike	
Measure	No chemical trails applied to V (n = 5)	E and NE trails applied to V (n = 6)	No chemical trails applied to V (n = 6)	E and NE trails applied to V (n = 13)
Temporal Measures (sec)				
(1) Mean time to contact J trail	966.0^{a}	787.6^{a}	342.0^{b}	217.0^{b}
and (2) locate mouse	1200.0^{a}	960.0^{a}	1044.0^{a}	514.7^{b}
(3) Mean time on F trails (total includes J trail)	E 0.0^{a}	27.8^{b}	4.0^{a}	51.7^{b}
	NE 0.0^{a}	0.8^{a}	0.2^{a}	25.9^{b}
	total 16.2^{a}	32.2^{a}	76.0^{a}	177.5^{b}
and (4) mean % time on J & F	1.3^{a}	10.1^{a}	7.3^{a}	*37.2^{b}
(5) Mean % "available time" on J & F (ie., from first contact with J to locating mouse)	1.3^{a}	18.6^{b}	21.5^{b}	*59.1^{c}
and (6) Mean % F time on E wing	0.0^{a}	48.0^{b}	30.0^{b}	*71.7^{c}
Tongue Flick Measures				
(7) Mean RTF per min	4.8^{a}	11.0^{a}	32.3^{b}	41.7^{b}
and (8) Mean RTF during "available time"	4.1^{a}	14.6^{a}	35.7^{b}	57.7^{b}
(9) Mean % TFs directed to trail	J 4.7^{a}	20.5^{a}	15.6^{a}	*44.1^{b}
	F 0.0^{a}	22.5^{b}	2.9^{a}	*73.9^{c}
and (10) Mean % TFs on E wing	0.0^{a}	50.0^{b}	16.6^{a}	*71.8^{c}

Data were analyzed by ANOVA (unequal n; Weiner, 1971) followed by non-orthogonal contrasts. Numbers (within a row) with the same superscript do not differ significantly at the .05 level. Percentages marked with * are significantly higher than chance by t-test comparing tabled value with 9%.

trails before reaching the choice point. All other details were comparable to those of the previous study and Golan et al. (1982). Again, rattlesnakes followed rodent trails very well after striking mice (see Table 2), but the present discrimination task proved to be harder for rattlesnakes than the previous one. Although there was numerical evidence that E trails were followed more effectively than NE trails (see Table 2, esp. measures 3-7), the differences were not statistically reliable. In fact, exactly 50% of the snakes selected E trails.

Because the two simultaneous-discrimination (Sim-D) studies produced mixed results, we tried an alternative methodology. Rattlesnakes were again exposed to E and NE trails, but this time we used a successive-discrimination (Suc-D) paradigm (see Chiszar et al., 1980). Each of 10 snakes was tested twice (ITI = 1 wk). During one test the snake was allowed to strike a mouse, after which it was exposed to an E trail. During the other test the snake was again allowed to strike a mouse, but it was then exposed to an NE trail. Dependent variables and general procedures were similar to those described in Golan et al. (1982). Five snakes received the E trail on week one and the NE trail on week two; the remaining snakes received the reverse order. If E trails are more effective stimuli than NE trails, then snakes should perform better when exposed to the former in a Suc-D study. Better performance on E trails in a Sim-D study could occur not only because E trails are more effective stimuli than NE trails, but also because (1) a contrast effect dependent on simultaneous presentation of the two kinds of stimulus-trails may enhance the difference between them, or (2) snakes selecting the E trails are better trailers than those selecting NE trails. In the latter case, poor performance on NE trails could very well derive from the animals selecting such trails rather than from the trails themselves. The only way to assess this possibility is to compare E and NE trails in a Suc-D study where <u>all</u> animals are "forced" to follow both trails. If snakes perform better on E than NE trails in this situation, the difference must derive from stimulus properties of the trails. If snakes perform equally well on the two trails, then the E vs NE differences found in the first Sim-D study must have been based upon chance or upon (momentary or constitutional) differences in trailing abilities between snakes selecting E and NE trails, respectively.

Results of the Suc-D study are summarized in Table 3. It is clear that snakes made contact with and followed E and NE trails (see especially the mean % available time on trail and mean % TFs on trail, measures 5, 11). Although several measures indicated numerically better performance on E than NE trails, none of the E vs NE differences were statistically reliable. Accordingly we conclude that E trails are not more effective stimuli than the NE trails used in this study. In the first Sim-D experiment some

Table 2. Summary of 10 dependent variables in a simultaneous discrimination experiment. Only one condition of 4 reported in Table 1 was re-run: Strike-E and NE trails available in the post-strike environment. Subjects were 14 rattlesnakes. Trails were made in a Y pattern (see text); see Table 1 for definitions of various terms and symbols. Data are shown separately for snakes that selected NE and E trails (and mice). E and NE trails each occupied 11.3% of floor area.

Measure	Snakes selecting NE trail (n = 7)	Snakes selecting E trail (n = 7)	Absolute values of ts (df = 12) comparing snakes that followed E and NE trails. Probabilities are for 2-tailed tests.
Temporal Measures (sec)			
(1) Mean time to contact J trail and	146.3	338.8	1.20, P > .05
(2) locate mouse	528.0	759.2	1.02, P > .05
(3) Mean time on F trails,	34.0	61.5	2.08, .10 > P > .05
mean time on J & F and	106.0	300.5	1.31, P > .05
(4) Mean % time on J & F	30.8*	38.7*	0.61, P > .05
(5) Mean % available time on J & F	55.31*	60.4*	0.36, P > .05
(6) Mean % available time on F	45.1*	76.2*	1.78, .10 > P > .05
Tongue Flick Measures			
(7) Mean RTF per min (baseline)	0.3[a]	0.6[b]	0.54, P > .05
(8) Mean RTF per min (post-strike)	42.8	33.6	1.01, P > .05
(9) Mean % TFs on J trail	60.3*	51.0*	0.63, P > .05
(10) Mean % TFs on F trail	68.7*	76.3*	0.44, P > .05

* significantly greater than chance by t test comparing tabled value to 11.3%

[a] 0.3 vs 42.8, t = 6.88, P < .01
[b] 0.6 vs 33.6, t = 4.99, P < .01

These comparisons indicate that RTF increased after the strike.

Table 3. Summary of 11 dependent variables recorded in the successive discrimination experiment in which 10 rattlesnakes were exposed to E and NE trails on successive weeks. Each trail was about 180 cm long and occupied about 13% of the total area of the floor of the apparatus. See Golan et al. (1982) for methodological details.

Measure	Trail: NE	Trail: E	Absolute values of ts (df = 9) comparing E and NE performances. Probabilities are for 2-tailed tests.
Temporal Measures (sec)			
1. Mean time to contact trail	612.5	477.6	0.72, $P > .05$
2. Mean time to locate mouse	782.3	778.5	0.02, $P > .05$
3. Mean time on trail	80.8	149.7	1.46, $P > .05$
4. Mean % time on trail	13.8	19.8	0.76, $P > .05$
5. Mean % available time on trail	47.0[a]	52.0[b]	0.33, $P > .05$

[a] comparing 47.0% to chance (13%) gives $t = 3.97$, $P < .01$
[b] comparing 52.0% to chance (13%) gives $t = 4.63$, $P < .01$
These comparisons indicate that snakes followed trails once the trails were contacted.

Measure	Trail: NE	Trail: E	Absolute values of ts (df = 9)
Tongue Flick Measures			
6. Mean RTF per min (baseline)	3.1[c]	1.4[d]	1.02, $P > .05$
7. Mean RTF (strike to contacting trail)	29.1	33.6	0.92, $P > .05$
8. Mean RTF per min (post contact)	66.4	52.7	2.05, $P > .05$
9. Mean RTF per min while on trail	65.3	54.6	1.43, $P > .05$
10. Mean RTF per min while off trail	66.6	58.4	1.26, $P > .05$
11. Mean % TFs on trail	46.7[e]	50.3[f]	0.25, $P > .05$

[c] 3.1 vs 29.1; $t = 5.39$, $P < .01$
3.1 vs 66.4; $t = 17.04$, $P < .01$
[d] 1.4 vs 33.6; $t = 5.80$, $P < .01$
1.4 vs 52.7; $t = 7.52$, $P < .01$
These comparisons indicate that RTF increased after the strike.

[e] comparing 46.7 to chance (13%) gives $t = 4.56$, $P < .01$
[f] comparing 50.3 to chance (13%0 gives $t = 4.31$, $P < .01$
These comparisons indicate that snakes followed trails once the trails were contacted.

snakes clearly preferred E trails, and this probably indicates that some snakes can discriminate the two trails. However, we cannot claim this ability to be general across our specimens. That is, we cannot claim the trails to have differential stimulus properties which are generally perceptible to rattlesnakes.

While we continue to believe that venom may contain components which can enhance perceptibility of E trails and E mice, it also is quite clear that rattlesnakes are excellent at utilizing trails made with NE (but dead) mice. This latter fact may derive from our methods of sacrificing NE mice (we have used nitrogen and CO_2 asphyxiation as well as cervical dislocation). Perhaps these methods cause the animals to release alarm pheromones or other trauma-induced substances, and perhaps these chemicals are the cues utilized by rattlesnakes during the post-strike (trailing) period. It may be that painful aspects of envenomation can exacerbate these chemical changes, but some components of them may have occured in our NE mice. This would account for the keen ability of our rattlesnakes to follow NE trails as well as for the ability of these and other viperids to distinguish between live and recently killed mice (Baumann, 1927, 1928, 1929).

These ideas do not make the E vs NE experiments any less interesting, rather they transfer the locus of theoretical attention from effects of venom <u>per se</u> to more general chemical changes induced in mice by trauma. Whether rattlesnakes have evolved special "trail marking" substances as part of their venom or whether they have evolved the ability to detect trauma-related cues, these snakes are clearly able to locate dead rodent prey by following their chemical trails. Interestingly, Bruce Means and O. Greg Brock of the Tall Timbers Research Station (Tallahassee, Fla.) have found that eastern diamondback rattlesnakes (<u>C. adamanteus</u>) are also excellent at trailing recently envenomated rodent prey, and these researchers recognize the same interpretative difficulties as have just been described (personal communication).

One way to attack this problem is to make trails with urine from non-stressed and stressed (but not envenomated) mice. If rattlesnakes follow the latter but not the former, then it could be concluded that the snakes are attending to alarm-pheromone type substances. Of course, this conclusion could be strengthened if urine trails from envenomated and nonenvenomated (but stressed) mice are followed equally well. Experiments along these lines are currently in progress in this laboratory.

COMPARATIVE STUDIES OF SICS

We would like to conclude this chapter with a brief discussion of the phylogenetic generality of SICS. Table 4 summarizes

Table 4. Species known (or strongly suspected) to exhibit the Strike-Release Strategy, SICS, and/or Trailing Behavior (at least under some conditions)

Species	Reference
Family Elapidae:	
Oxyuranus microlepidotus, *O. scutellatus*	Shine & Covacevich, 1982
Naja mossambica pallida and *N. n. kaouthia*	Radcliffe et al., 1982; Chiszar et al., under review
Family Viperidae: (subfamily Viperinae)	
Vipera ammodytes, *V. latifii*, *V. raddei*, *V. xanthina*	Chiszar et al., 1982
V. aspis	Baumann, 1927, 1928, 1929; Naulleau, 1964, 1965, 1966, 1967
V. berus	Wiedemann, 1932
Bitis arietans, *B. gabonica*, *Eristocophis macmahoni*	Chiszar et al., 1982
Echis carinatus	Finstrom & Wintin (personal communication)
(subfamily Crotalinae)	
Agkistrodon bilineatus, *A. piscivorus*	Chiszar et al., 1979, & Chiszar et al., 1982
Bothrops schlegeli and *Calloselasma rhodostoma*	Finstrom & Wintin (personal communication)
Crotalus atrox	Gillingham & Clark, 1981
C. adamanteus	Means & Brock (personal communication) and Brock, 1980
C. durissus, *C. enyo*, *C. horridus*, *C. lepidus*, *C. molossus*, *C. pricei*, *C. ruber*, *C. scutulatus*, *C. triseriatus*, *C. viridis*, and *Sistrurus catenatus*, *S. miliarus*	Chiszar et al., 1982, 1978, 1980, 1981; Dullemeijer, 1961; Scudder, 1982; present paper

the taxa in which SICS has been experimentally verified. Clearly, the phenomenon is widespread among the Viperidae (perhaps indicating a common origin), and several reports (Radcliffe et al., 1982; Shine & Covacevich, 1982) strongly suggest that SICS also occurs in the Elapidae. These latter reports are of great interest because it is suspected that viperids arose from some primitive elapid (Cope, 1896; Bogert, 1943; Dowling, 1959, Johnson, 1956; Marx & Rabb, 1965). Accordingly, if contemporary rodent-feeding elapids exhibit SICS, they may have derived this behavior from the same ancestors as did the viperids. That is, SICS may be homologous in the Viperidae and the Elapidae.

To begin an exploration of this issue we interviewed several reptile curators and keepers at major American zoological parks, and it was learned that cobras typically hold on to rodent prey after striking. However, it is common practice to offer snakes dead prey that represent relatively small meals. It is not surprising that such items are held. Accordingly, we acquired a sample of neonatal red spitting cobras (*Naja mossambica pallida*; Stimac et al., 1982) through a research loan from Mr. John Behler and the New York Zoological Society, and we began by determining (when the snakes were 4 mo old) the probability of holding vs releasing rodent prey as a function of prey size (i.e., age-wt). Very small rodents were always struck and held whereas larger ones were typically released (Radcliffe et al., 1982). This study was replicated with a more extensive range of prey sizes when the snakes were 12 mo old (Chiszar et al., under review), and the results were much the same. Furthermore, after large rodents were released, RTF increased as the snakes began searching for the prey. Next, we executed a series of SICS studies much like those described in the introduction. Results were (1) that NS presentations produced larger elevations in RTF than we have ever seen in rattlesnakes, (2) that part of this effect was attributable to disturbance that attended mouse presentations and part of the effect was clearly prey directed (i.e., visual cues arising from mice activate chemosensory searching), and, (3) that S presentations produced significantly higher RTFs than NS presentations. We were also fortunate to have access to a sample of adult monacled cobras (*Naja naja kaouthia*; provided by Tracy Miller), and SICS was clearly demonstrated in these animals using our standard procedures.

Experiments similar to those described in the section on trailing are currently being conducted with our *N. m. pallida*, and results so far indicate that these cobras follow rodent trails. Interestingly the *N. m. pallida* exhibit a mixture of visual and chemical orientations during their pursuit of envenomated prey. The animals typically begin by flicking their tongues at the trail and proceeding along it for several cm, but then they frequently "stand up" (i.e., raise the anterior 1/3 of the body perpendicular to the substrate without flaring the hood) and scan the environment as if looking for the prey. Rattlesnakes never show the latter behavior. Most often the cobras settle back to the substrate and relocate the chemical trail, but the "standing" posture occurs several more times before the prey is finally located. It seems clear that chemical cues are utilized but that visual ones would be preferred if they were available.

More comparative data are needed, particularly on relatively primitive elapids. Nonetheless, we offer the hypotheses that SICS and trailing behavior will turn out to be common among these animals, although these effects will probably not be as robust as they are in viperids. If this is true, and if SICS in these families can be homologized, then two conclusions will follow. First, since viperids probably arose from elapids 20-70 million years ago (Marx & Rabb, 1965), SICS must have evolved prior to this time. Second, SICS may have been a part of the original adaptive suite of characteristics which enabled the earliest venomous snakes to specialize on rodent prey.

Our guess is that even the most primitive rodent feeders were reluctant to strike and hold their prey, but they probably had especially toxic venom (as do all elapids; Brown, 1973; Minton, 1974), and they may have held their prey for a few seconds in order to deliver a strong dose of venom. The prey would not get far after being released, and the snake would not need keen trailing abilities to locate the carcass. It can be imagined that heightened sensitivity to chemical cues (groundborne or airborne) and/or good vision would be sufficient. Thus, SICS but not trailing behavior might have characterized such snakes, particularly if rodents only formed part of an otherwise eclectic diet (as is the case for many elapids). Increased specialization on rodents would mean increased risk of tissue damage from rodent teeth and claws during the initial envenomation response, and thus could have led to several kinds of natural selection: (1) larger, stronger heads with heavy scales to armor the snakes against effects of rodent struggling, (2) quicker acting venom, and (3) instantaneous injection and release of the rodent. All three specializations exist, but the latter is probably the most creative since it requires not only SICS but also effective trailing skills because the prey is likely to wander relatively far from the site of attack. Consequently, snakes which began to exhibit rapid venom injection followed by immediate release of prey (many viperids and at least some elapids) would be under strong selective pressure to follow trails. Perhaps these snakes have elaborated SICS into a more refined ability by allowing the modal action pattern to be brought under the control of taxic cues arising from chemicals deposited by wounded rodents. Perhaps some of these advanced snakes have also added components to their venom which enhance the taxic control of venom trails.

Clearly, these speculations go considerably beyond the existing data base. Additional comparative data on SICS and trailing are much needed, particularly in proteroglyphs and opistoglyphs. The role of vision in the post-strike behavior of cobras must be carefully analyzed by factorial experiments which manipulate both visual and chemical information (and/or the visual and chemical sensory channels). Finally, the functional stimuli arising from rodent trails must be determined and quantified before much progress can be made on the question of whether or not envenomation makes a special contribution to the perceptibility of such trails.

ACKNOWLEDGEMENTS

The research here reported was made possible by four grants from the M. M. Schmidt Foundation. Mice were donated by H. P. Alpern, R. Clark, and the Institute for Behavior Genetics (U. Colo., Boulder). We thank John L. Behler and the New York Zoological Society for the specimens of N. m. pallida, and we thank each of the following persons for their friendship and collaboration: Karen Estep, Richard Feiler, Mathew Finstrom, Lisa Golan, Barbara O'Connell, Thomas Poole, Hobart Smith, Kathryn Stimac, BJ Willcox, and Kenneth Winton. Reprint requests should be sent to D. Chiszar, Department of Psychology, University of Colorado, Boulder, Colorado, 80309.

REFERENCES

Barlow, G. W., 1977, Modal action patterns, in: "How Animals Communicate," J. Sebeok, ed., Indiana University Press, Bloomington, Indiana, 98-134.

Baumann, F., 1927, Experimente über den Geruchssinn der Viper, Rev. Suisse Zool., 34:173-184.

Baumann, F., 1928, Über den Nahrungserwerb der Viper, Rev. Suisse Zool., 35:233-239.

Baumann, F., 1929, Experimente über den Geruchssinn und den Beuteerwerb der Viper (Vipera aspis L.), Zeitschr. vergl. Physiol., 10:36-119.

Bogert, C. M., 1943, Dentitional phenomena in cobras and other elapids with notes on adaptive modifications of fangs, Bull. Amer. Mus. Nat. Hist., 81:285-360.

Brock, O. G., 1980, Predatory behavior of eastern diamondback rattlesnakes (Crotalus adamanteus): Field enclosure and Y-maze laboratory studies emphasizing prey trailing behavior. Ph.D. dissertation, Florida State University.

Brown, J. H., 1973, "Toxicology and Pharmacology of Venoms from Poisonous Snakes," C. C. Thomas Publ., Springfield, Illinois.

Burghardt, G. M., 1970, Chemical perception in reptiles, in: "Communication by Chemical Signals," J. W. Johnston, Jr., D. G. Moulton, and A. Turk, eds., Appleton-Century-Crofts, New York, 241-308.

Chiszar, D., Duvall, D., Scudder, K., and Radcliffe, C. W., 1980, Simultaneous and successive discriminations between envenomated and nonenvenomated mice by rattlesnakes (Crotalus durissus and C. viridis), Behav. Neural Biol., 29:518-521.

Chiszar, D. and Radcliffe, C. W., 1977, Absence of prey-chemical preferences in newborn rattlesnakes (Crotalus cerastes, C. enyo, and C. viridis), Behav. Biol., 21:146-150.

Chiszar, D., Radcliffe, C. W., O'Connell, B., and Smith, H. M., 1980, Strike-induced chemosensory searching in rattlesnakes (Crotalus enyo) as a function of disturbance prior to presentation of prey. Trans. Kan. Acad., Sci., 83:230-234.

Chiszar, D., Radcliffe, C. W., O'Connell, B., and Smith, H. M., 1981, Strike-induced chemosensory searching in rattlesnakes (Crotalus viridis) as a function of disturbance prior to presentation of rodent prey, Psychol. Rec., 31:57-62.

Chiszar, D., Radcliffe, C. W., O'Connell, B., and Smith, H. M., 1982, Analysis of the behavioral sequence emitted by rattlesnakes during feeding episodes II. Duration of strike-induced chemosensory searching in rattlesnakes (Crotalus viridis, C. enyo), Behav. Neural Biol., in press.

Chiszar, D., Radcliffe, C. W., and Scudder, K. M., 1977, Analysis of the behavioral sequence emitted by rattlesnakes during feeding episodes I. Striking and chemosensory searching. Behav. Biol., 21:418-425.

Chiszar, D., Radcliffe, C. W., and Smith, H. M., 1978, Chemosensory searching for wounded prey by rattlesnakes is released by striking: A replication report, Herpetol. Rev., 9:54-56.

Chiszar, D., Radcliffe, C. W., Smith, H. M., and Bashinski, H., 1981, Effect of prolonged food deprivation in response to prey odors by rattlesnakes, Herpetologica, 37:237-243.

Chiszar, D. and Scudder, K. M., 1980, Chemosensory searching by rattlesnakes during predatory episodes, in: "Chemical Signals: Vertebrates and Aquatic Invertebrates," D. Müller-Schwarze and R. M. Silverstein, eds., Plenum Press, New York, 125-139.

Chiszar, D., Scudder, K. M., and Knight, L., 1976, Rate of tongue flicking by garter snakes (Thamnophis radix haydeni) and rattlesnakes (Crotalus v. viridis, Sistrurus catenatus tergeminus, and S. c. edwardsi) during prolonged exposure to food odors, Behav. Biol., 18:273-283.

Chiszar, D., Simonsen, L., Radcliffe, C. W., and Smith, H. M., 1979, Rate of tongue flicking by cottonmouths (Agkistrodon piscivorus) during prolonged exposure to various food odors, and strike-induced chemosensory searching by the cantil (Agkistrodon bilineatus). Trans. Kan. Acad. Sci., 82:49-54.

Chiszar, D., Stimac, K., Poole, T., Miller, T., Radcliffe, C. W., and Smith, H. M., under review, Strike-induced chemosensory searching in cobras (Naja naja kaouthia, N. mossambica pallida).
Chiszar, D., Taylor, S. V., Radcliffe, C. W., Smith, H. M., and O'Connell, B., 1981, Effects of chemical and visual stimuli upon chemosensory searching by garter snakes and rattlesnakes, J. Herpetol., 15:415-424.
Cope, E. D., 1896, The classification of the Ophidia, Trans. Amer. Phil. Soc., 18:186-219.
Cowles, R. B. and Phelan, R. L., 1958, Olfaction in rattlesnakes, Copeia, 1958:77-83.
Dowling, H. G., 1959, Classification of the Serpentes: A critical review, Copeia, 1959:38-52.
Dullemeijer, P., 1961, Some remarks on the feeding behavior of rattlesnakes, Kon. Ned. Acad. Wetenschap. Series C, 64:383-396.
Duvall, D., Chiszar, D., Trupiano, J., and Radcliffe, C. W., 1978, Preference for envenomated rodent prey by rattlesnakes, Bull. Psychon. Sci., 11:7-8.
Duvall, D., Scudder, K. M., and Chiszar, D., 1980, Rattlesnake predatory behavior: Mediation of prey discrimination, and release of swallowing by odors associated with envenomated mice, Anim. Behav., 28:674-683.
Estep, K., Poole, T., Radcliffe, C. W., O'Connell, B., and Chiszar, D., 1981, Distance traveled by mice after envenomation by a rattlesnake (C. viridis), Bull. Psychon. Soc., 18:108-110.
Gans, C., 1966, The biting behavior of solenoglyph snakes -- Its bearing on the pattern of envenomation, Proc. Internat. Symp. Venom. Anim., Sao Paulo, Brazil, Instituto Butantan.
Golan, L., Radcliffe, C. W., Miller, T., O'Connell, B., and Chiszar, D., 1982, Prey trailing by the prairie rattlesnake (Crotalus v. viridis), J. Herpetol., in press.
Gillingham, J. C. and Baker, R. R., 1981, Evidence for scavenging behavior in the western diamondback rattlesnake (Crotalus atrox), Zeit. Tierpsychol.
Gillingham, J. C. and Clark, D. L., 1981, An analysis of prey searching behavior in the western diamondback rattlesnake, Crotalus atrox, Behav. Neural Biol., 32:235-240.
Johnson, R. C., 1956, The origin and evolution of the venomous snakes, Evolution, 10:56-65.
Klauber, L. M., 1956, "Rattlesnakes -- Their Habits, Life Histories, and Influence on Mankind," Univ. Calif. Press, Berkeley, Cal.
Kubie, J. L., 1977, The role of the vomeronasal organ in garter snake prey trailing and courtship, unpublished Ph.D. dissertation, School of Graduate Studies, Downstate Medical Center, New York.

Marx, H. and Rabb, G. B., 1965, Relationships and zoogeography of the viperine snakes (family Viperidae), Fieldiana Zool., 44:161-206.

Meredith, M., 1980, The vomeronasal organ and accessory olfactory systems in the hamster, in: "Chemical Signals: Vertebrates and Aquatic Invertebrates," D. Müller-Schwarze and R. M. Silverstein, eds., Plenum Press, New York, 303-326.

Meredith, M. and Burghardt, G. M., 1978, Electrophysiological studies of the tongue and accessory olfactory bulb in garter snakes, Physiol. Behav., 21:1001-1008.

Minton, S. A., 1974, "Venom Diseases," C. C. Thomas Publ., Springfield, Ill.

Naulleau, G., 1964, Premières observations sur le comportement de chasse et de capture chez les vipères et les couleuvres, La Terre et la Vie, 1:54-76.

Naulleau, G., 1965, La biologie et le comportement predateur de Vipera aspis au laboratoire et dans la nature, (thèse), Bull. Biol. France Belgique, 99:395-524.

Naulleau, G., 1966, La biologie et le comportement predateur de Vipera aspis au laboratoire et dans la nature, thèse, Paris, P. Fanlac.

Naulleau, G., 1967, Le comportement de predation chez Vipera aspis, Rev. Comp. Animal, 2:41-96.

O'Connell, B., Chiszar, D., and Smith, H. M., 1981, Effect of poststrike disturbance on strike-induced chemosensory searching in the prairie rattlesnake (Crotalus v. viridis), Behav. Neural Biol., 32:343-349.

O'Connell, B., Chiszar, D, and Smith, H. M., 1982, Single vs. multiple predatory strikes by prairie rattlesnakes (Crotalus viridis), Bull. Md. Herp. Soc., in press.

Radcliffe, C. W., Chiszar, D., and O'Connell, B., 1980, Effects of prey size on poststrike behavior in rattlesnakes (Crotalus durissus, C. enyo, and C. viridis), Bull. Psychon. Soc., 16:449-450.

Radcliffe, C. W., Stimac, K., Smith, H. M., and Chiszar, D., 1982, Effects of prey size on poststrike behavior of juvenile red spitting cobras (Naja mossambica pallida), Trans. Kan. Acad. Sci., in press.

Scudder, K. M., 1982, Mechanisms mediating the sequential aspects of predatory episodes in Crotalid snakes, Ph.D. dissertation, University of Colorado, Boulder.

Scudder, K. M., Chiszar, D., and Smith, H. M., The effect of environmental odors on strike-induced chemosensory searching by rattlesnakes, under review.

Scudder, K. M., Poole, T., O'Connell, B., and Chiszar, D., in preparation, Ontogeny of strike-induced chemosensory searching and trailing behavior in neonatal rattlesnakes (Crotalus horridus and C. viridis).

Shine, R. and Covacevich, J., 1982, Ecology of the highly venomous snakes: The Australian genus *Oxyuranus* (Elapidae), *J. Herpetol.*, in press.

Stimac, K., Radcliffe, C. W., and Chiszar, D., 1982, Prey recognition learning by red spitting cobras, *Naja mossambica pallida*, *Bull. Psychon. Soc.*, in press.

Weiner, B. J., 1971, "Statistical Principles -- Experimental Design," McGraw-Hill, New York.

Wiedemann, E., 1932, Zur Biologie der Nahrungsaufnahme der Kreuzotter, *Vipera berus L.*, *Zool. Anz.*, 97:278-286.

FOSSIL AND COMPARATIVE EVIDENCE FOR POSSIBLE CHEMICAL SIGNALING IN THE MAMMAL-LIKE REPTILES

David Duvall, Michael B. King and Brent M. Graves

Department of Zoology and Physiology
University of Wyoming
Laramie, Wyoming 82071

INTRODUCTION

The mammal-like reptiles, or the Synapsida, hold an important position in the study of tetrapod vertebrate evolution for several reasons. First, the pelycosaurs and their therapsid derivatives, the two groups that comprise the synapsids, probably evolved from the cotylosaurs (the "stem reptiles") (Carroll, 1969), and gave rise much later to the mammals (Fig. 1). Indeed, the mammal-like reptiles were the dominant terrestrial tetrapods throughout the Permian and the Triassic, the period from roughly 300 to 190 million years ago (Romer, 1966). Given this position in phylogeny, these extinct members of the Class Reptilia have always held the interest of those interested in the origins and derivation of mammals (e.g., see Simpson, 1959; Crompton and Jenkins, 1979).

There is much more of interest in studying the synapsids than simply their position in phylogeny, and this is probably why most students of tetrapod evolution refer to them as "mammal-like". For example, "mammal-like" endothermy--if not homeothermy--probably arose in the synapsids (see Bennett and Ruben, 1982). It also is likely that hairs of some sort arose in this group (Watson, 1931; Brink, 1957; Van Valen, 1960). Evidence of a soft, possibly secretory epidermis, not covered by horny scales also has been uncovered for certain therapsid taxa (Chudinov, 1968). A secondary palate homologous with that of mammals also appeared in therapsids (Romer, 1959, 1966). Relevant to points made later in this paper, it also has been observed that some young therapsids probably possessed "milk teeth" (Brink, 1957) and snout foramina necessary for dextrous, muscular and, perhaps, sensitive lips (Hotton, Pers. Comm.), both of which suggest that some sort of "suckling", and even maternal care may have characterized patterns of therapsid or even early mammalian parental investment (Brink, 1957; Duvall,

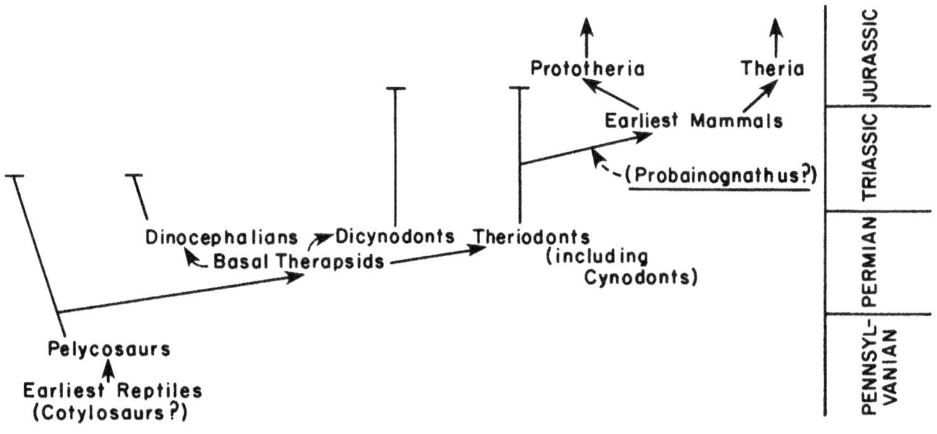

Fig. 1. A dendrogram of tentative relations of basal reptiles, synapsids, and mammals. Redrawn and modified from J.A. Ruben, 1981, Paleobiol., 7:413-417.

1982a; Guillette, 1982; Pond, 1977).

Our goal here is to draw attention to another likely aspect of the paleobiology of synapsids, and therapsids in particular. We address the possibility that these animals probably produced, detected, and responded to chemical signals. In developing this inferential case, we draw upon a parsimonious application of the comparative method, appropriate morphological fossil evidence, a consideration of biology of extant tetrapods, a rapidly increasing body of knowledge about mammal-like reptiles (see Roth et al., 1982), and current ethological evidence about chemical signaling, particularly in extant squamate reptiles and mammals. (For an expanded consideration of issues raised here, see Duvall, 1982a).

CHEMICAL SIGNAL RECEPTION

The importance of vertebrate olfaction in the evolution of the vertebrate brain in particular is well known (Romer, 1959; Jerison, 1973). What has become more clear in recent years, however, is the importance of the vomeronasal organ (VNO; or the Jacobson's organ; Jacobson, 1811; cf. Parsons, 1971) in the life histories of a wide range of tetrapod vertebrates (e.g., see Burghardt, 1980; Shorey, 1976; Stoddart, 1980a). All tetrapods examined in regard to this character thus far possess paired VNOs or embryologues either as adults or embryos, respectively (Parsons, 1959a,b, 1967, 1970, 1971). In the latter case, these chemosensory structures are present as small evaginations from the developing nasal chamber (Parsons, 1959a,b). Only testudines differ (slightly) in VNO embryology, indicative of their early divergence from tetrapod stock (Parsons, 1967). The VNO embryologues eventually disappear prior to the completion of development in crocodilians, birds, many primates (including Homo sapiens), cetaceans, and in some bats (Allison, 1953; Negus, 1958; Tucker and Smith, 1976). Apparently, all other tetrapods possess VNOs as adults, and these are present in the palatal region as ventral portions of the paired nasal chamber mucosae (amphibians, some reptiles), blind pouches which connect directly with the oral chamber (some reptiles), blind pouches which connect directly with the nasal cavities (some mammals), or blind, tube-like pouches which connect with paired nasopalatine canals which pass through paired incisive foramina in the secondary palate, and connect nasal and oral cavities (some mammals) (Meredith, 1980; Negus, 1958; Parsons, 1971).

In certain mammals, such as equids, the nasopalatine canal does not connect nasal and oral cavities, and the VNOs gain access to chemical signals via the nasopalatine canal connection with the nasal cavities. Nevertheless, in the vast majority of extant mammals, complete nasopalatine canals exist, and these provide oral and nasal routes of chemical signal access to the VNOs (Estes, 1972; Meredith, 1980; Negus, 1958; Poduschka, 1977).

The occurrence of VNOs at some point in the life-histories of members of every tetrapod class suggests that this is a phylogenetically old system. As Bertmar (1981) has noted, the tetrapod vomeronasal organ was likely derived in early amphibians or terrestrial tetrapods from something like the folds or sinuses of the fish olfactory organ (cf. Broman, 1920; Duvall, 1982a). Many investigators also have noted and considered several strikingly conservative and similar features of tetrapod VNOs. In addition to similar embryology and gross anatomical placement of the VNO in the palatal region, are the findings that (1) where Bowman's glands and cilia are almost always found in nasal sensory mucosa, they never occur in the VNO sensory mucosa, (2) the point noted in (1) is the case in all three tetrapod classes that possess both olfactory systems as adults (that is, in amphibians, reptiles, and mammals), (3) the nasal sensory mucosa sends afferents to the main olfactory bulb and the VNO sensory mucosa projects first to the accessory olfactory bulb, again, in members of all of the three tetrapod classes noted in (2), and (4) projections into deeper brain centers from the two regions of the olfactory bulb retain significant autonomy or separation, with those associated with VNO sensory mucosa forming more direct connections with areas of the brain associated with sexual function (at least, in those reptiles and mammals examined in this regard; see Burghardt, 1980; Halpern, 1976; Meredith, 1980; Wang and Halpern, 1980). Such observations have led several investigators interested in the evolution of tetrapod vertebrate olfaction, and the vomeronasal system in particular, to hypothesize that the latter is a homologous character, based upon virtually any application of principles or criteria verifying phylogenetic continuity (i.e., ultimate derivation from a common ancestor that possessed the character; see Atz, 1970; Boyden, 1947; Greene & Burghardt, 1978; Simpson, 1961).

Consistent with this comparative interpretation of the derivation of the tetrapod VNO, are findings of many investigators interested in how this system functions in behavior, also suggestive of phylogenetic continuity (see Duvall, 1982a). Based upon their studies of rattlesnakes, Cowles and Phelan (1958) hypothesized that the nasal and VNO olfactory systems of these squamates served somewhat different chemosensory and ethological functions. It was suggested that nasal olfaction was more of a "distance sensing" system, most responsive to the presence of chemical cues. The tongue-VNO system, conversely, was suggested to be less sensitive to the mere presence of chemical cue information, but more capable of discriminating differences between types of chemical signals (i.e., more "qualitative"). Once the nasal system detected the presence of a chemical cue, it also was hypothesized to switch-on the more discriminating tongue-VNO system. Although tested only indirectly by these investigators, in two separate studies Duvall (1981, 1982a) has shown that these hypotheses do seem to hold for a distant squamate relative of rattlesnakes, the western fence lizard, *Sceloporus occidentalis biseriatus*. Pheromones also were investigated with the tongue-VNO system more frequently than control chemical cues. This functional relationship between nasal and VNO olfaction likely will be found to be common to other squamates as well (see Duvall, 1981; Simon, In press). Johns (1980) also has described evidence that argues strongly for a similar interpretation of functional and ethological aspects of nasal and VNO olfaction in some mammals (cf. Johns et al., 1978; Burghardt, 1980; Duvall, 1982a). Therefore, at least for some squamates and mammals, nasal olfaction may be *more* of a distance sensing, or orienting, chemosensory adaption than is the VNO assemblage, the latter being more discriminating of types or qualities of chemical signals. The studies described above on certain mammals and reptiles (and others, yet to be considered; see below), indicate that for certain conspecific chemical signals, or pheromones, the VNO may play a *relatively* greater functional, physiological, and ethological role in signal recipients than the nasal olfactory system. In no way does this argument confound or contradict the established importance of the VNO in detecting chemical signals in the context of feeding (e.g., see Kubie and Halpern, 1979; Halpern and Frumin, 1979). Again, all we are suggesting -- and this is not a new idea -- is that some closely and distantly related tetrapods rely on the VNO assemblage to a greater *relative* extent than nasal olfaction, when responding to certain conspecific chemical signals, or pheromones. Hence, to the extent that synapsids possessed a VNO, they may have done likewise.

Perhaps even more indicative of a special and "old" role for the VNO in the reception of some pheromones (or even non-food allelochemicals; see Weldon, 1980; Weldon and Burghardt, 1979), is evidence that this system is functionally atuned to the reception of reproductive pheromones in some very distantly-related tetrapods. As far as we know, this possibility was first noted by W. Whitten (personal communication; Johns, 1980. cf.; Poduschka and Firbas, 1968). For example, evidence has been obtained that insectivores such as hedgehogs (Poduschka, 1977), mice and rabbits (Whitten, Pers.Comm.), rats (Johns et al., 1978, some ungulates (Estes, 1972), hamsters (Meredith, 1980; Powers and Winans, 1975), guinea pigs (Wysocki et al., 1980), spiny lizards (Duvall, 1981, 1982b; cf. Simon, In press), and garter snakes (Kubie et al., 1978; Heller and Halpern, Under Review), rely on the VNO to a greater relative extent than nasal olfaction for the reception and eventual response to reproductive and certain other social priming and releasing (or signaling) pheromones. Again, this is not so surprising given the direct connection of the VNO with deep brain centers important in sexual function (Burghardt, 1980; Halpern, 1976, 1980; Meredith, 1980).

Nevertheless, one wonders just how general is this phenomenon? Will it be found that other reptiles such as testudines and the tuatara employ their VNO as some sqamates do? What about other squamates? What about members of the three orders of amphibians, all of which possess the VNO? What about other mammals, such as those that do and do not display flehmen actions (Estes, 1972; Poduschka, 1977)? And of course, what about groups ancestral to extant tetrapods, such as the mammal-like reptiles? If they possessed the VNO, and we will argue that they probably did (see below), then they may have employed it just as their derivatives the mammals and several, more distantly related extant reptiles do today. Just as the past is a key to the present, the present is a path to the past.

Evidence for a Mammal-like Reptile Vomeronasal Organ

Functional anatomical, embryological, and even ethological continuity in the tetrapod vomeronasal organ suggest that we are observing an old and homologous chemosensory system in action when we study the biology of the VNO. By definition, this implies that mammal-like reptiles possessed VNOs since mammals do. Nevertheless, what fossil and morphological evidence can be mustered in support of this evolutionary hypothesis? Actually, a good deal is available.

Perhaps one of the most important morphological characters in this regard, is the presence of incisive foramina in the maxillary/premaxillary region of the secondary palate of mammals. This trait is often detectable even in fossil therapsid skulls with very little of the rock matrix removed (see Duvall, 1982a). This almost always indicates the presence of complete nasopalatine canals in mammals, and an oral route to the paired VNOs (Allison, 1953; Eisenberg and Kleiman, 1972; Estes, 1972; Meredith, 1980; Negus, 1958).

Fig. 2. A palatal view of the advanced cynodont Probelesodon. Incisive foramina indicated by arrow. From Duvall (1982a), courtesy of The Smithsonian Institution Press.

What about the mammal-like reptiles? Not much can be hypothesized about the pelycosaurs based upon fossil morphology, since they had not evolved a secondary palate. However, as soon as the secondary palate begins to appear in fossil remains of the therapsids, an accepted homology with the same structure in mammals, so do maxillary/premaxillary incisive foramina (see Broom, 1937; Duvall, 1982a). We have observed difinitive and steep-walled incisive foramina in intact skulls and secondary palate fragments of certain anomodont and several theriodont (e.g., see Fig. 2) therapsids. Indeed, in the important cynodont theriodont taxon, the line that eventually gave rise to the Mesozoic mammals (Crompton and Jenkins, 1979; but cf. Simpson, 1959), incisive foramina in the secondary palate are the rule. As might be expected, skulls and palate fragments of the advanced cynodont Probainognathus jenseni (and the likely and earliest mammalian decendants of this therapsid; Crompton and Jenkins, 1979) possess incisive foramina (Duvall, 1982a).

Of interest in this regard, as well, are published descriptions of acid-cleaned therapsid skulls (that is, those with virtually all of the rock matrix removed from areas such as the nasal cavity). This technique allowed King (1981) and Cluver (1971), studying the dicynodonts Dicynodon trigonocephalus and Lystrosaurus, respectively, to uncover ossified grooves on either side of a semi-ossified nasal

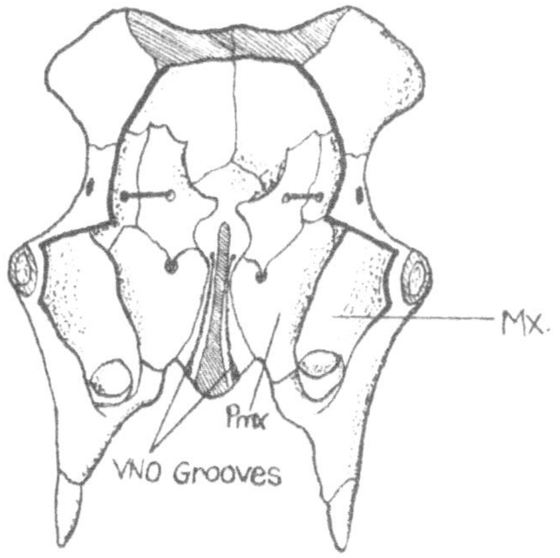

Fig. 3. An anterior and dorsal, and cut-away view of the nasal cavity of the dicynodont Lystrosaurus. Cluver (1971) and King (1981) have interpreted the grooves indicated as the likely loci of the paired VNOs in life. Mx, maxilla; Pmx, premaxilla. Redrawn and modified from Cluver (1971).

septum, strongly suggestive of a functional VNO in life (Fig. 3). Chemical signals from the environment would have had to take a nasal cavity route through an incomplete nasopalatine canal to the VNO mucosa, however. This would have been the case because there are no incisive foramina in the highly derived dicynodont palate and snout, since these animals possessed a robust, shearing-type snout and jaw assemblage, for eating hard or very fibrous plant materials (Hotton, 1982). Kemp (1979) also has observed grooves in the nasal cavity of an acid-cleaned specimen of Prosynosuchus delaharpeae, an upper permian and basal cynodont without a complete secondary palate, and, hence, no definitive incisive foramina. All of this morphological evidence supports strongly the hypothesis that the mammal-like reptiles possessed VNOs. If this form of olfaction was employed by therapsids as it is by many extant mammals and reptiles, it also may have played a role in social signaling and not just in feeding behavior.

CHEMICAL SIGNAL PRODUCTION

It now seems axiomatic that a wide structural array of molecules have evolved signal function in animals (e.g., Shorey, 1976; Stoddart, 1980a; Weldon, 1982). Apparently, as long as reliably produced exudates can handle some interplay of thermodynamic and ecological requirements, chemical signal function can be derived. Implicit here is the evolutionary assumption that those molecules that derive signal function originally possessed some other, perhaps, regulatory or metabolic function (Quay, 1977; Shorey, 1976; Weldon, 1982). From our inspection of the literature, it seems that many accept this derivational pathway for pheromone or allelochemical function (see Weldon, 1982). This implies as well that the production of chemical signals at "early stages" in a particular taxon's existence is or was (relatively) energetically inexpensive, since the signal exudate was already present and functioning in some other capacity. The wide range of known and suspected structural configurations for vertebrate and even invertebrate chemical signals would seem to be a result of this likelihood. Nevertheless, structural variability of vertebrate chemical signals may decrease as this channel of signaling becomes more highly derived within taxa (see Wilson, 1975).

In the case of terrestrial tetrapods such as "adult" amphibians, reptiles, and mammals, for example, it is known that some very general exudates (i.e., those derived from metabolic or regulatory events, and still very close to the original preadapted function for the molecule(s)) possess signal function (e.g., see Auffenberg, 1978; Block et al., 1981; Garstka and Crews, 1981; Stevens and Gerzog-Thomas, 1977). For example, we have no evidence, at least not yet, that any of the specialized glands of the spiny lizard cloaca and proctodeum contribute to the mixture of cloacal exudates that contribute to efficacious, chemical signal affected, social behavior (Duvall, 1979, 1981, 1982a,b; cf. Simon, In press).

Extremely specialized and highly derived chemical signals and tissues associated with their production and release also are known in terrestrial tetrapods. For example, highly derived exocrines (*sensu lato*) and associated products with established signal functions are well known in amphibians (Madison, 1977), reptiles (Ross and Crews, 1977), and especially mammals (Stoddart, 1980a,b).

Therefore, it would seem that chemical signaling represents a channel of communication that can be--and obviously has been--tapped frequently in organism phylogeny. So, what about the possibility of mammal-like reptile chemical signal production? First, general exudates such as saliva, sweat, feces, nitrogenous wastes and the like, probably were produced. Such products could have possessed signal function as they were, or at least have provided a pathway for the derivation of more specific structures or mixtures. Quite simply, such chemicals could have possessed signal function for the synapsids. Second, for at least one anomodont therapsid (specifically, the dinocephalid *Estemmenosuchus*; Chudinov, 1968; cf. Duvall 1982a) skin impressions (relics?) have been unearthed in the Soviet Union that indicate the possible presence of a once soft, moist, and non-horny or scaley epidermis. Again, any secretions expressed could have possessed signal function, whatever other function they may have served (see Brink, 1957; Duvall, 1982a; Van Valen, 1960; cf. Weldon, 1982).

Of particular interest, however, are additional fossil finds strongly suggestive of the presence of some highly specialized exo-

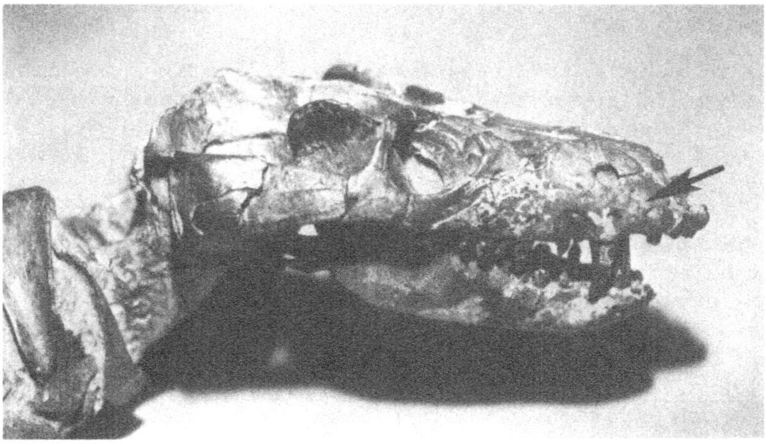

Fig. 4. Lateral view of the skull of the advanced cynodont Thrinaxodon. The maxillary pits indicated by arrow are foramina that may have supplied blood vessel and nerve supply to the roots of snout hairs or vibrissae. From Duvall (1982a), courtesy of The Smithsonian Institution Press.

crines. In the advanced cynodonts, Cynognathus and Diademodon definitive grooves and pits in the bones of the face and snout have been observed (see Brink, 1957). These probably housed specialized exocrines of some sort (Brink, 1957; Grine et al., 1979).
Clearly defined but small foramina in the anterior snout region of Thrinaxodon are believed to have housed vibrissae or hairs of some sort and probably blood and nerve supply to the "hair roots" (Watson, 1931; Brink, 1957; Duvall, 1982a; see Fig. 4; cf. Van Valen, 1960). If these "hairs" were at all like those of extant mammals, then a variety of skin glands could have produced an exudate that bathed or affected the hair, the surrounding skin, or both. In addition to any regulatory, predatory, or even defensive function such face and snout exudates may have served (see Brink, 1957; Van Valen, 1960), surely chemical signal function also may have been involved (Duvall, 1982a). Perhaps some of the advanced therapsids such as Thrinaxodon used presumptive snout exocrine glands and their exudates in some form of active marking, possibly by visibly rubbing their faces on significant objects in their environments. Many extant mammals and squamate reptiles employ some specialized exocrine structures in this way (Berry,

1974; Eisenberg and Kleiman, 1972; Stoddart, 1980a,b; Simon, In press; cf. Duvall, 1982a).

SOME LIMITS FOR BEHAVIORAL INFERENCE

The question of what kinds of ethological actions may have characterized therapsid life, and how chemical signaling might have figured in, is at once interesting and frustrating. The latter is the case, because we will never know anything with certainty in this regard. Nevertheless, it does seem reasonable to construct some very general inferential limits of complexity for social behavior, at least, within which possible pheromonally- or allelochemically-mediated synapsid actions might have fallen (see Duvall, 1982a). For example, simple aggregation or avoidance events mediated by chemical signals could be viewed as a "lower or lesser", social or group (see Wilson, 1975) level of behavior. Non-breeding skinks (lizards) exhibit simple aggregation (Eumeces fasciatus) and avoidance (Scincella lateralis) actions (Duvall et al., 1980), in addition to responsiveness to sexual and aggression releasing pheromones during the breeding season (see Fitch, 1954). Many snakes likewise respond to conspecific chemical signals with aggregation and avoidance (King et al., Under Review; Porter and Czaplicki, 1974). In fact, one might be hard pressed to isolate many macrosmotic organism taxa that do not employ chemical signaling systems for the group functions of simple avoidance, aggregation or dispersal, in addition to any other class of behavioral events mediated by such social cues. Given this, it seems possible that the mammal-like reptiles may have employed chemical signals to mediate such group-level behavioral actions as well, just as many of their extant reptilian and mammalian cousins do today.

More highly derived forms of chemical signaling also may have been characteristic of the mammal-like reptiles, such as active marking. We may view such possibilities as our inferential "upper limit." The morphological evidence suggestive of snout exocrines in some cynodonts may be viewed as evidence in support of this possibility. Perhaps signaling or releasing pheromones were employed in sexual and agonistic encounters, just as they commonly are in several extant amphibians, reptiles, and mammals (Madison, 1977; Simon, In press; Stoddart, 1980a). Priming pheromones are known to be important in mammalian reproductive biology, and probably are important in some extant reptiles also (e.g., see Fitch, 1954; Duvall et al., 1982), although the latter deserve more work in this regard. Furthermore, a good deal of work has appeared, indicating that pheromones are critical in the parent-offspring nexus in mammalian parental care (Blass and Teicher, 1980; Leon, 1979; Singh and Hofer, 1978). Such cues also appear to be important in the maternal actions of many lizards (Duvall et al., 1979; Evans, 1959; Noble 1933). That pheromones may have played some role in the likely patterns of parental investment in some of the advanced cynodonts is taken up next.

AGGREGATION PHEROMONES AND THE DERIVATION OF VENTRUM NUTRITIONAL GLANDS

Although the hypothesis that mammalian mammary glands were derived from skin glands of some sort is not new (Darwin, 1859), it continues to be accepted (e.g., see Long, 1969; Romer and Parsons, 1977). Furthermore, the debate about whether or not mammary glands were derived from sweat or sebaceous glands is an old one as well (see Long, 1969, for a discussion). Most investigators favor the former hypothesis (cf. Long, 1969). The basic argument revolves around similarities in sweat and mammary gland morphology and embryology.

We, too, accept this "morphological hypothesis," but we feel that it is at least possible that certain behavioral and chemosensory events on the part of cynodont pups and their mothers (or, in some of the earliest mammals) may have played a role in the derivation of ventrum nutritional (milk secreting?) glands from the first, and probably quite simple, sweat-type skin glands (Duvall, 1982a; cf. Graves et al., Under Review, for a detailed consideration). That such an event may have occurred in the advanced cynodonts and not in the earliest mammals, however, seems a solid possibility (e.g., see Brink, 1957; Olson, 1959; Van Valen, 1960).

We offer the hypothesis (cf. Duvall, 1982a; Graves et al, Under review) that the earliest mammal-like reptile skin exudates probably served a skin maintenance function (e.g., Chudinov, 1968), that such secretions exuded from the region of female's ventrums (among cynodonts?) next derived chemical signal function (probably through an "aggregation" or ventrum location action) and that these exudates only then began to take on nutritive value or function. In other words, we propose that a pheromonally-mediated, mother-young aggregation event was intermediate between, and perhaps partly responsible for, the derivation of ventrum, nutrient producing glands, from more simple skin glands. Interestingly, oxytocin, the neurohypophysial octapeptide that controls milk ejection, also seems to control exudation of a "nipple location" pheromone in rats (Hofer et al., 1976; Singh and Hofer, 1978; but cf. Blass and Teicher, 1980). Conversely, it may be argued that nutrient function arose first, and that only then chemical cues associated with the ventrum secondarily derived signal function. The latter might occur through the positively reinforcing, energetic and nutritive value of an exudate like milk, for example. Either of the above two hypotheses seem possible, but we do not favor the latter because we feel that it is not the most parsimonious interpretation of available evidence (Graves et al., Under Review).

There exist several reasons for this opinion. First, the evolutionary change in traits typically can be tied in with previously existing adaptations, or preadaptations. In most derivational schemes, the most gradual sequence of hypothesized preadaptation and resulting adaption will be the most parsimonious--especially when hard data are scanty. In our opinion, the evolutionary jump from a simple sweat-type, skin gland to a mammary gland is a very big one. Indeed, too big, without certain qualifications. Why, for example, would neonates have licked the ventrum in the first place? Did certain sweat glands simply start dripping milk? Did a mammal-like brood patch just appear from sweat glands? We think not. Rather, young animals may first have licked their mother's ventrum so that they simply might find or identify it or her. Perhaps lingual and mouthing actions facilitated chemical signal reception via an oral route to the vomeronasal organ. Next, a coevolution of sorts, involving the mother's ventrum on the one hand and the neonates lingual mouthing actions of the "aggregation factor" on the other may have ensued. This "coevolution" might have resulted in a gradual increase in the complexity of the ventrum skin glands, such that some fraction of the total mixture secreted eventually derived some nutritive value for young. At the same time, repeated licking actions by young of the mother's ventrum might have led to the derivation of a more muscular, flexible and sensitive oral opening in the young (such as lips), something the fossil record tells us that some cynodonts probably possessed. Milk teeth originally may have been derived in this context as well. Eventually, a sweat-type gland, modified enough to release an aggregation factor may have become complex enough to exude nutrients. Indeed, some have suggested that the advanced cynodonts or the earliest mammals may have exuded such materials onto a "brood patch" of sorts (Hopson, 1973; Graves et al., Under Review; Guillette, 1982) which only then may have given rise to more discrete secretory structures such as mammary glands. At any rate, at some point in time in this coevolutionary derivation mother cynodonts or early mammals began to produce something like milk, and neonates probably began to show mammal-typical ventrum licking or suckling. In sum, we suggest that some sort of aggregation factor, derived from a mother cynodont's (or an early mammal's) ventrum, may have facilitated the derivation of these "mammalian" characters.

CONCLUSIONS

Our purpose has been to draw attention to several lines of evidence strongly suggestive of possible chemical signaling in the extinct mammal-like reptiles, especially the therapsids. Comparative analysis of extant tetrapod morphology and behavior, fossil evidence, and parsimonious inference indicate that it is possible that synapsids possessed structures for conspecific chemical signal reception (the VNO), likely chemical cue production, and a possible range of chemical-cue mediated social actions not unlike those of certain extant tetrapods. As such, this paper represents evolutionary inference, based upon a broad paleobiological approach that can be strengthened or weakened as new fossil evidence and comparative analyses (behavioral and morphological) come to light. Attempts to include within this approach the study of "paleoethology" have proven fruitful, and seem to have increased in popularity in recent years (e.g., Bakker, 1980; Duvall,

1982a; Galton, 1970; Greene and Burghardt, 1978; Hopson, 1975; MacLean, 1978; Molnar, 1977). Colbert (1958) may have most accurately stated the importance of this overall approach in a germinal paper on this problem, when he noted: "Of course we can only infer the behavior patterns of extinct animals, partly through a knowledge of their morphology and partly with analogy with their living relatives. Such inference is useful, however, and truly necessary if we are to gain a well-rounded knowledge of the correlation between morphology and behavior in modern animals. The past is a key to the present, in the study not only of the physical evolution of life but of mental development and behavior as well (p. 46)."

REFERENCES

Allison, A.C., 1953, The morphology of the olfactory system in the vertebrates, Biol. Rev., 28:195-244.

Atz, J.W., 1970, The application of the idea of homology to behavior, in: "Development and Evolution of Behavior," L.R. Aronson, E. Tobach, D.S. Lehrman, and J.S. Rosenblatt, eds., W.H. Freeman, San Francisco.

Auffenberg, W., 1978, Social and feeding behavior of Varanus komodoensis, in: "Behavior and Neurology of Lizards," N. Greenberg, and P.D. MacLean, eds., NIMH, Poolesville, Maryland.

Bakker, R.T., 1980, Dinosaur heresy--dinosaur renaissance: Why we need endothermic archosaurs for a comprehensive theory of bioenergetic evolution, in: "A Cold Look at the Warm-blooded Dinosaurs," D.K. Thomas and E.C. Olson, eds., Westview Press, Boulder.

Bennett, A.F., and Ruben, J., 1982, Ambient temperature and evolution of basal metabolic rate during the reptilian-mammalian transition, in: "The Paleobiology of the Mammal-like Reptiles," J.J. Roth, E.C. Roth, N.H. Hotton, and P.D. MacLean, eds., The Smithsonian Institution Press, Washington, D.C., In Press.

Berry, K.H., 1974, The ecology and social behavior of the chuckwalla, Sauromalus obesus obesus Baird, Univ. California Publ. Zool., 101:1-60.

Bertmar, G., 1981, Evolution of vomeronasal organs in vertebrates, Evolution, 35:359-366.

Blass, E.M., and Teicher, M.H., 1980, Suckling, Science, 210:15-22.

Block, M.L., Volpe, L.C., and Hayes, M.J., 1981, Saliva as a chemical cue in the development of social behavior, Science, 211:1062-1064.

Boyden, A., 1947, Homology and analogy, a critical review of the meanings and implications of these concepts in biology, Amer. Midl. Nat., 37:648-669.

Brink, A.S., 1957, Speculations on some advanced mammalian characteristics in the higher mammal-like reptiles, Palaeontol. Afr. 4:77-96.

Broman, I., 1920, Das Organon vomero-nasale Jacobsoni, ein Wassergeruchsorgan!, Anat. Hefte., 58:137-191.

Broom, R., 1937, On the palate, occiput and hind foot of Bauria cynops Broom, Amer. Mus. Nov., No. 946:1-6.

Burghardt, G.M., 1980, Behavioral and stimulus correlates of vomeronasal functioning in reptiles: Feeding, grouping, sex, and tongue use, in: "Chemical Signals," D. Müller-Schwarze, and R.M. Silverstein, eds., Plenum, New York.

Carroll, R.L., 1969, Origin of reptiles, in: "Biology of the Reptilia, Vol. 1," C. Gans, A.d'A. Bellairs, and T.S. Parsons, eds., Academic Press, New York.

Chudinov, P.K., 1968, Structure of the integuments of theriomorphs. Doklady Acad. Nauk SSSR, 179:207-210. (Translation by Amer. Geol. Inst.)

Colbert, E.H., 1958, Morphology and behavior, in: "Behavior and Evolution," A. Roe, and G.G. Simpson, eds., Yale University Press, New Haven, Connecticut.

Cowles, R.B., and Phelan, R.L., 1958, Olfaction in rattlesnakes, Copeia, 1958:77-83.

Cluver, M.A., 1971, The cranial morphology of the dicynodont genus Lystrosaurus, Ann. S. Afr. Mus., 56:155-274.

Crompton, A.W., and Jenkins, F.A., 1979, Origin of mammals, in: "Mesozoic Mammals," J.A. Lillegraven, Z. Kielan-Jaworowska, and W.A. Clemans, eds., University of California Press, Berkeley.

Darwin, C., 1859, "The Origin of Species by Means of Natural Selection, or "The Preservation of Races in the Struggle for Life," Reprinted by Doubleday and Co., Inc., Garden City, New York, 1960.

Duvall, D., 1979, Western fence lizard (Sceloporus occidentalis) chemical signals. I. Conspecific discriminations and release of a species-typical visual display, J. Exp. Zool., 210:321-326.

Duvall, D., 1981, Western fence lizard (Sceloporus occidentalis) chemical signals. II. A replication with naturally breeding adults and a test of the Cowles and Phelan hypothesis of rattlesnake olfaction, J. Exp. Zool., 218:351-361.

Duvall, D., 1982a, A new question of pheromones: Aspects of possible chemical signaling and reception in the mammal-like reptiles, in: "The Paleobiology of the Mammal-like Reptiles," J.J. Roth, E.C. Roth, N.H. Hotton, and P.D. MacLean, eds., The Smithsonian Institution Press, Washington, D.C., In Press.

Duvall, D., 1982b, Western fence lizard (Sceloporus occidentalis) chemical signals. III. An experimental ethogram of conspecific body licking, J. Exp. Zool., In press.

Duvall, D., Guillette, L.J., Jr., and Jones, R.E., 1982, Environmental control of reptilian reproductive cycles, in: "Biology of the Reptilia, Vol. 12," C. Gans, and F.H. Pough, eds., Academic Press, London, In Press.

Duvall, D., Herskowitz, R.L., and Trupiano-Duvall, J., 1980, Responses of five-lined skinks (Eumeces fasciatus) and ground skinks (Scincella lateralis) to conspecific and interspecific chemical cues, J. Herpetol., 14:121-127.

Duvall, D., Trupiano, J., and Smith, H.M., 1979, An observation of maternal behavior in the Mexican desert spiny lizard, Sceloporus rufidorsum, Trans., Kansas Acad, Sci., 82:60-62.

Eisenberg, J.F., and Kleiman, D.G., 1972, Olfactory communication in mammals, in: "Annual Review of Ecology and Systematics, Vol. 3," R.F. Johnston, P.W. Frank, and C.D. Michener, eds., Annual Reviews, Inc., Palo Alto, California.

Estes, R.D., 1972, The role of the vomeronasal organ in mammalian reproduction, Mammalia, 36:315-341.

Evans, L.T., 1959, A motion picture study of maternal behavior of the lizard, Eumeces obsoletus Baird and Girard, Copeia, 1959: 103-110.

Fitch, H.S., 1954, Life history and ecology of the five-lined skink, Eumeces fasciatus, Univ. Kansas Publ. Mus. Nat. Hist., 8:1-156.

Galton, P.M., 1970, Pachycephalosaurids--dinosaurian battering rams, Discovery, 6:23-32.

Garstka, W.R., and Crews, D., 1981, Female sex pheromone in the skin and circulation of a garter snake, Science, 214:681-683.

Graves, B., Galvin, R., and Duvall, D., Under Review, Chemical signals as the immediate preadaption for neonate utilization of skin gland nutritive secretions, Evolution.

Greene, H.W., and Burghardt, G.M., 1978, Behavior and phylogeny: Constriction in ancient and modern snakes, Science, 200:74-77.

Grine, F.E., Mitchell, D., Gow, C.E., Kitching, J.W., and Turner, B.R., 1979, Evidence for salt glands in the Triassic reptile Diademodon (Therapsida: Cynodontia), Palaeont. Afr., 22:35-39.

Guillette, L.J.,Jr., 1982, The evolution of viviparity and placentation in reptiles: An hypothesis for the evolution of viviparity in mammal-like reptiles, in: "The Paleobiology of the Mammal-like Reptiles," J.J. Roth, E.C. Roth, N.H. Hotton, and P.D. MacLean, eds., The Smithsonian Institution Press, Washington, D.C., In Press.

Halpern, M., 1976, The efferent connections of the olfactory bulb and accessory olfactory bulb in the snakes, Thamnophis sirtalis and Thamnophis radix, J. Morphol., 150:553-578.

Halpern, M., 1980, The telencephalon of snakes, in: "Comparative Neurology of the Telencephalon," S.O.E. Ebbesson, ed., Plenum, New York.

Halpern, M., and Frumin, N., 1979, Roles of the vomeronasal and olfactory systems in prey attack and feeding in adult garter snakes, Physiol. Behav., 22:1183-1189.

Heller, S.B., and Halpern, M., Under Review, Laboratory observations of aggregative behavior of garter snakes, Thamnophis sirtalis: The roles of visual, olfactory and vomeronasal senses, J. Comp. Physiol. Psychol.

Hofer, M.A., Shair, H., and Singh, P.J., 1976, Evidence that maternal ventral skin substances promote suckling in infant rats, Physiol. Behav., 17:131-136.

Hopson, J.A., 1973, Endothermy, small size, and the origin of mammalian reproduction, Amer. Nat., 107:446-452.

Hopson, J.A., 1975, The evolution of cranial display structures in hadrosaurian dinosaurs, Paleobiology, 1:21-43.

Hotton, N.H., 1982, Dicynodonts and other primary consumers, in: "The Paleobiology of the Mammal-like Reptiles," J.J. Roth, E.C. Roth, N.H. Hotton, and P.D. MacLean, eds., The Smithsonian Institution Press, Washington, D.C., In Press.

Jacobson, L., 1811, Description anatomique d'un organ observé dans les mammifères, Ann. du Mus. d'Hist. Nat. Paris., 18:412-424.

Jerison, H.J., 1973, "Evolution of the Brain and Intelligence," Academic Press, New York.

Johns, M.A., 1980, The role of the vomeronasal system in mammalian reproductive physiology, in: "Chemical Signals," D. Müller-Schwarze, and R.M. Silverstein, eds., Plenum, New York.

Johns, M.A., Feder, H.H., Komisaruk, B.R., and Mayer, A.D., 1978, Urine-induced reflex ovulation in anovulatory rats may be a vomeronasal effect, Nature, 272:446-448.

Kemp, T.S., 1979, The primitive cynodont Procynosuchus: Functional anatomy of the skull and relationships, Phil. Trans, Royal Soc London, B. Biol. Sci., 285:73-122.

King, G.M., 1981, The functional anatomy of a Permian dicynodont, Phil. Trans. Royal Soc. London, B. Biol. Sci., 291:243-322.

King, M.B., McCarron, D., Duvall, D., Baxter, G., and Gern, W., Under Review, Avoidance of conspecific but not interspecific chemical signals by prairie rattlesnakes, Crotalus viridis viridis, J. Herpetol.

Kubie, J.L. and Halpern, M., 1979, Chemical senses involved in garter snake prey trailing, J. Comp. Physiol. Psychol., 93:648-667.

Kubie, J.L., Vagvolgyi, A., and Halpern, M., 1978, Roles of the vomeronasal and olfactory systems in courtship behavior of male garter snakes, J. Comp. Physiol. Psychol., 92:627-641.

Leon, M, 1979, Mother-young reunions, in: "Progress in Psychobiology and Physiological Psychology, Vol. 8," J.M. Sprague, and A.N. Silverstein, eds., Academic Press, New York.

Lillegraven, J.A., 1979, Reproduction in Mesozoic mammals, in: "Mesozoic Mammals," J.A. Lillegraven, Z. Kielan-Jaworowska, and W.A. Clemens, eds., The University of California Press, Berkeley.

Long, C.A., 1969, The origin and evolution of mammary glands, Bioscience, 19:519-523.

MacLean, P.D., 1978, Why brain research on lizards?, in: "Behavior and Neurology of Lizards," N. Greenberg, and P.D. MacLean, eds., NIMH, Poolesville, Maryland.

Madison, D.M., 1977, Chemical communication in amphibians and reptiles, in: "Chemical Signals," D. Müller-Schwarze, and M.M. Mozell, eds., Plenum, New York.

Meredith, M., 1980, The vomeronasal organ and accessory olfactory system in the hamster, in: "Chemical Signals," D. Müller-Schwarze and R.M. Silverstein, eds., Plenum, New York.

Molnar, R.E., 1977, Analogies in the evolution of combat and display structures in ornithopods and ungulates, Evol. Theory, 3:165-190.

Negus, V., 1958, "The Comparative Anatomy and Physiology of the Nose and Paranasal Sinuses," Livingstone Ltd., Edinburgh and London.

Noble, G.K., and Mason, E.R., 1933, Experiments on the brooding habits of the lizards Eumeces and Ophisaurus, Amer. Mus. Novitates, 619: 1-29.

Olson, E.C. 1959, The evolution of mammalian characteristics, Evolution, 13:344-353.

Parsons, T.S., 1959a, Nasal anatomy and the phylogeny of reptiles, Evolution, 13:175-187.

Parsons, T.S., 1959b, Studies on the comparative embryology of the reptilian nose, Bull. Mus. Comp. Zool., Harvard Univ., 120:101-277.

Parsons, T.S., 1967, Evolution of the nasal structure in the lower tetrapods, Amer. Zool., 7:397-413.

Parsons, T.S., 1970, The nose and Jacobson's organ, in: "Biology of the Reptilia, Vol. 2, Morphology B," C. Gans and T.S. Parsons, eds., Academic Press, London.

Parsons, T.S., 1971, Anatomy of nasal structures from a comparative viewpoint, in: "Handbook of Sensory Physiology, Vol. IV, Chemical Senses, Part 2, Olfaction," L.M. Beidler, eds., Springer-Verlag, New York.

Poduschka, W., 1977, Insectivore communication, in: "How Animals Communicate," T.A. Seboek, ed., Indiana University Press, Bloomington.

Poduschka, W., and Firbas, W., 1968, Das Selbstbespeicheln des Igels, Erinaceus europaeus Linnè, 1758, steht in Verbindung zur Funktion des Jacobsonschen Organes, Z. Säugetierk., 33:160-172.

Pond, C.M., 1977, The significance of lactation in the evolution of mammals, Evolution, 31:177-199.

Porter, R.H., and Czaplicki, J.A., 1974, Response of water snakes (*Natrix r. rhombifera*) and garter snakes (*Thamnophis sirtalis*) to chemical cues, Anim. Learn. Behav., 2:129-132.

Powers, J.B., and Winans, S.S., 1975, Vomeronasal organ: Critical role in mediating sexual behavior in the male hamster, Science,187:961-963.

Romer, A.S., 1959, "The Vertebrate Story," The University of Chicago Press, Chicago.

Romer, A.S., 1966, "Vertebrate Paleontology," 3rd Ed., The University of Chicago Press, Chicago.

Romer,A.S., and Parsons, T.S., 1977, "The Vertebrate Body," 5th Ed., Saunders College Publishing, Philadelphia.

Ross, P., Jr., and Crews, D., 1977, Influence of the seminal plug on mating behaviour in the garter snake, Nature, 267:344-345.

Roth, J.J., Roth, E.C., Hotton, N.H., and MacLean, P.D., eds., 1982, "The Paleobiology of the Mammal-like Reptiles," The Smithsonian Institution Press, Washington, D.C., In press.

Shorey, H.H., 1976, "Animal Communication by Pheromones," Academic Press, New York.

Simon, C.A., In press, Lizard Chemoreception: A review, Amer. Zool.

Simpson, G.G., 1959, Mesozoic mammals and the polyphyletic origin of mammals, Evolution, 13:405-414.

Simpson, G.G., 1961, "Principles of Animal Taxonomy," Columbia University Press, New York.

Singh, P.J., and Hofer, M.A., 1978, Oxytocin reinstates maternal olfactory cues for nipple orientation and attachment in rat pups, Physiol. Behav., 20:385-389.

Stevens, D., and Gerzog-Thomas, D.A., 1977, Fright reactions in rats to conspecific tissue, Physiol. Behav., 18:47-51.

Stoddart, D.M. 1980a, "The Ecology of Vertebrate Olfaction," Chapman and Hall, New York.

Stoddart, D.M. 1980b, Aspects of the evolutionary biology of mammalian olfaction, in: "Olfaction in Mammals," Zool. Soc. London, Symp., No. 45, D.M. Stoddart, ed., Academic Press, London.

Tucker, D., and Smith, J.C., 1976, Vertebrate olfaction, in: "Evolution of Brain and Behavior in Vertebrates," R.B. Masterson, M.E. Bitterman, C.B.G. Campbell, and N. Hotton, eds., Lawrence Erlbaum Assoc., Hillsdale, N.J.

Van Valen, L., 1960, Therapsids as mammals, Evolution, 14:304-313.

Wang, R.T., and Halpern, M., 1980, Light and electron microscopic observations on the normal structure of the vomeronasal organ of garter snakes, J. Morphol., 164:47-67.

Watson, D.M.S., 1931, On the skeleton of a Bauriamorph reptile, Proc., Zool. Soc. London., Part 3:1163.

Weldon, P.J., 1980, In defense of "kairomone" as a class of chemical releasing stimuli, J. Chem. Ecol., 6:719-725.

Weldon, P.J., 1982, The evolution of alarm pheromones, in: "Chemical Signals," D. Müller-Schwarze, and R.M. Silverstein, eds., Plenum, New York. In press.

Weldon, P.J. and Burghardt, G.M., 1979, The ophiophage defensive response in crotaline snakes: Extension to new taxa, J. Chem. Ecol., 5:141-151.

Wilson, E.O., 1975, "Sociobiology," Belknap/Harvard University Press, Cambridge.

Wysocki, C.L., Wellington, J.L., and Beauchamp, G.K., 1980, Access of urinary non-volatiles to the mammalian vomeronasal organ, Science, 207:781-783.

Acknowledgments. We thank N.H. Hotton, A.W. Crompton, and T. Kemp and G. King, of the Smithsonian Institution, The Museum of Comparative Zoology (Harvard University), and the Oxford University Natural History Museum,respectively, for the use of specimens in their care. We thank Mary David for typing this manuscript.

Note added in proof. For a thorough and recent consideration of bat olfaction see K.P. Bhatnagar, 1980, The chiropteran vomeronasal organ: Its relevance to the phylogeny of bats, in: "Proceedings of the Fifth International Bat Research Conference," D.E. Wilson, and A.L. Gardner, eds., Texas Tech Press, Lubbock.

SNAKE TONGUE FLICKING BEHAVIOR: CLUES TO VOMERONASAL SYSTEM FUNCTIONS

Mimi Halpern and John L. Kubie

Departments of Anatomy and Cell Biology and Physiology
Downstate Medical Center
Brooklyn, N.Y. 11203

For the past ten years we have been studying the function of the garter snake vomeronasal system. The snake vomeronasal organ is remarkably large and contains more sensory neurons than the snake's main olfactory apparatus. It appears to detect sexual scents, conspecific scents, and prey odors. These odors are critically important in mating, aggregation, prey trailing, prey attack and prey ingestion (Burghardt & Pruitt, 1975; Halpern & Frumin, 1979; Heller & Halpern, 1983; Kubie & Halpern, 1979; Kubie, Vagvolgyi & Halpern, 1978; Wilde, 1938). The snake vomeronasal organ is situated in a rather odd position for an external chemoreceptor: the walnut-shaped organs sit paramidline above the roof of the mouth and the tiny openings are not directed towards the outside world (or the nasal cavity, as in many vertebrates) but downward towards the roof of the mouth.

It has long been thought that the unusual features of the snake tongue are specialized to deliver odorants to the vomeronasal organ (Kahmann, 1932; McDowell, 1972). Two of these specializations are the tongue's shape, which is long, slender and bifid; and its behavior, which is to flick regularly, especially when the snake is moving or following a trail. The snake tongue travels in a canal in the roof of the mouth known as the *fenestra vomeronasalis* into which the vomeronasal ducts open (McDowell, 1972). Recent evidence suggests that pairs of pads on the floor of the mouth may push the tongue tips and/or odorants up towards the openings of the vomeronasal organs (Clark, 1981; Gillingham & Clark, 1981). Kahmann (1932) initially demonstrated that when he permitted a snake to tongue flick carbon black, and killed the snake shortly thereafter, concentrations of carbon were found in the vomeronasal organs.

More recently we attempted a Kahmann-type experiment, but instead of using carbon black we used a radioactive non-volatile amino acid, ^{3}H-proline (Halpern & Kubie, 1980). The proline was detected by autoradiography. In normal animals we found high concentrations of proline in the vomeronasal organs. When the tongue tips were cut off, lower levels of proline accumulated in the vomeronasal organs as long as the tongue stub or lips actually touched the source of proline. Suturing closed the vomeronasal ducts eliminated proline delivery. We took these data as strong support of the idea that tongue flicking is a behavioral mechanism for delivering odorants, including non-volatile odorants, to the vomeronasal organs.

This paper addresses itself to the behavioral process of tongue flicking. Tongue flicking appears to represent a sense-seeking behavior. The prime purpose of the tongue flick is not to manipulate the outside world or to propel the animal through the outside world--as we conceive is the purpose of most motor acts--but to collect information from the outside world. In this sense tongue flicking is something like sniffing, ear movements and eye movements, all of which are primarily sense-seeking behaviors. As such, the snake's strategy of tongue flicking ought to tell us something about the function of the vomeronasal organ.

We describe below three studies that represent extensions of our previous work with prey odors and courtship behavior in garter snakes. In each case the role of the tongue is the primary focus of interest. In the first study we analyze the components of tongue flicking during trailing and in an open field apparatus to determine the relative contribution of different components of the tongue flick to our previously reported observation that tongue flick rate increases as a function of prey extract concentration. The second study concerns the role of the tongue in transporting pheromone molecules to the vomeronasal organ and the effect of interfering with this process on male courtship behavior. The final study deals with the issue of the efficacy of airborne odorants in initiating increased tongue flicking. These studies are not a comprehensive treatise on tongue flicking; they do, however, demonstrate the power and importance of analyzing tongue-flick behavior in studying the garter snake vomeronasal system.

FILM ANALYSIS OF TONGUE FLICKING DURING TRAILING AND OPEN FIELD TESTING

Garter snakes will accurately follow earthworm-extract trails in a 2 or 4 choice maze to receive earthworm-bit rewards (Kubie & Halpern, 1975; 1978). When tested with earthworm extracts of varying concentration the snakes most accurately follow the higher extract concentrations. Tongue-flick rates are highest when snakes follow high concentration extract trails, and lowest when the trail is dilute or removed from direct lingual contact (Kubie & Halpern, 1978). Individual tongue flicks of garter snakes vary considerably in duration and degree of tongue extension. The distance of individual tongue excursions can vary dramatically, as can the time the tongue is out of the snake's mouth and the pattern of tip movements while the tongue is extruded.

The increased tongue-flick rates of trailing could be accomplished by either decreasing the time the tongue is extended or decreasing the between tongue flick interval i.e., the time the tongue remains in the mouth between successive flicks. The following study investigated these alternatives and, in addition, correlated the degree of tongue flick extension with trail concentration and testing conditions (open field vs. maze).

Subjects

The subjects were 9 adult garter snakes of the species *Thamnophis sirtalis sirtalis* and *Thamnophis radix*. Seven of these animals were from a group used in a previous experiment (Kubie & Halpern, 1978) and the reader is referred to that publication for details of subject selection, housing and care.

Apparatus

The apparatus was a four-choice star maze (see Kubie & Halpern, 1978) with wood sides, a Plexiglas floor, and a removable Plexiglas roof.

Earthworm extract, prepared by the method of Wilde (1938) and Burghardt (1966) at a concentration of 6 g earthworm per 20 ml of water (1x), was stored frozen in capped vials. An extract of 18 g per 20 ml of water (3x) was also prepared. On test days, one vial of 1x extract and one vial of 3x extract were thawed. The 1x extract was serially diluted 1:2 with distilled water five times to make extract concentrations of 1/3, 1/9, 1/27, 1/81, and 1/243 the original.

PROCEDURE

The snakes had been previously trained to follow an earthworm-extract trail down a randomly selected arm of the four-choice maze to receive a worm-bit reward. They were additionally trained with seven different extract concentrations during each of 20 test sessions as reported in Kubie and Halpern (1978).

Each snake's trailing was filmed on the day following the last training session. Prior to photography the maze was placed on the floor and the snake given two "warm-up" 1x trials. The maze was used in a one-choice (error-free) configuration for photography with the open alley alternating between the two far alleys. All filmed trials were run with the Plexiglas roof of the maze removed. The maze was illuminated with a tungsten lamp, and the snake was photographed with the camera (Beaulieu 4008 ZMII super 8) set at 36 frames per second. The camera was hand held and controlled so that the width of the alley filled the frame. The snake was fed an earthworm bit in the goal box on each trial, and the maze was washed with water between trials. For seven snakes the first photographed trials used either 1/9th extract trails or 1x trails. Enough trials (2 to 3) were run with one trail concentration to expose a cartridge of Kodachrome film. The second set of filmed trials was run with distilled water trails. Enough trials were run to expose a second cartridge of film. A third set of filmed trials were run with 1x (1/9th) earthworm extract trails until a third cartridge of film was exposed. Thirty minutes after completion of the filmed trailing, the snake was placed in an open arena (57 x 57 cm) and photographed whenever it moved until a fourth cartridge of film was exposed. At the end of each film session the camera was calibrated by filming a running stop-watch for five seconds and by timing 1000 frames on the camera's frame counter.

For the two additional snakes training and filming were as described above except that four concentrations (3x, 1/3x, 1/27x and 1/243x) of earthworm extract were used instead of the original two.

Film Analysis

The film was analyzed on a super 8 film editor or projector. Each cartridge of film was considered a unit of analysis. For each tongue-flick the following data were recorded:

1. Tongue-flick initiation interval (TFII) -- the number of frames between the initiation of one tongue flick and the initiation of the next (tongue flick duration plus the between tongue-flick interval).

2. Tongue-flick duration-the number of frames the tongue was visible.

3. Between tongue-flick interval -- the number of frames between the tongue's disappearance on one tongue flick and re-appearance on the next.

4. Tongue-flick excursion type --

I - only tongue tips visible
II - black body of tongue visible; red part of the tongue, which starts about 2-3 mm behind the bifurcation, not visible
III - less than 3 mm of the red part of the tongue visible
IV - more than 3 mm of the red part of the tongue visible

The frequency distribution for each of these measures was tabulated for each roll of film and graphed as a percentage distribution.

Tongue-flick initiation intervals were shorter on extract trials than on distilled water or open field trials (Figure 1). Similarly, tongue-flick durations and between tongue-flick intervals were shorter on extract trials than on distilled water or open field trials (Figures 2 and 3). The most consistent finding was shortened tongue-flick durations on high extract concentration trials (Table 1 and Figure 2) with a decrease in duration of almost 50 percent occurring between distilled water trials and 1x or 3x trials. For most animals (e.g. Figure 3), but not all animals (e.g. Table 1), between tongue flick intervals also shortened as a function of extract concentration with a 25 percent decrease in between tongue flick intervals typically occurring between distilled water trials and 1x or 3x trials.

The majority of tongue flicks exhibited on 3x, 1x and 1/3x trials were of the shorter excursion types (type I and type II). On distilled water trials there were many more of the longer, type III and type IV, excursions and in the open field 85 to 90% of the tongue-flicks were of the III and IV type (Figure 4 and Table 1).

Figures 1 to 6 illustrate these results for one snake that was extensively trained prior to complete bilateral olfactory nerve transection and tested as well following surgery. (For details of surgery and histological verification of lesions refer to Kubie & Halpern, 1979). Postoperatively this animal continued to exhibit short-duration, short-excursion tongue flicks when following concentrated odor trails (Figures 5 and 6).

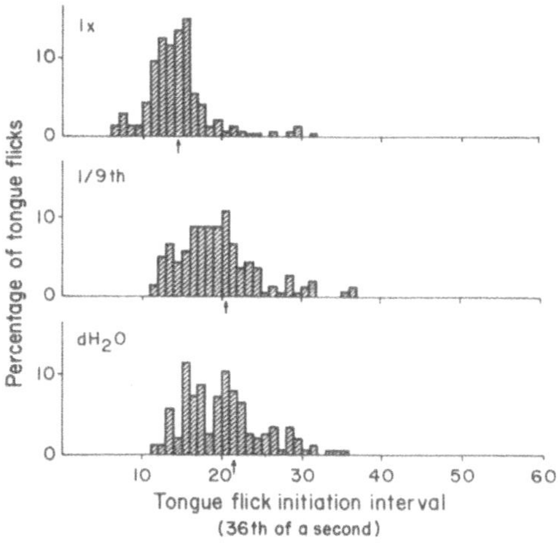

Fig. 1. Tongue-flick initiation intervals for female X1 plotted as a percentage of total tongue-flick initiation intervals on each of three types of trials: 1x, 1/9x and dH_2O. Arrows indicate mean intervals.

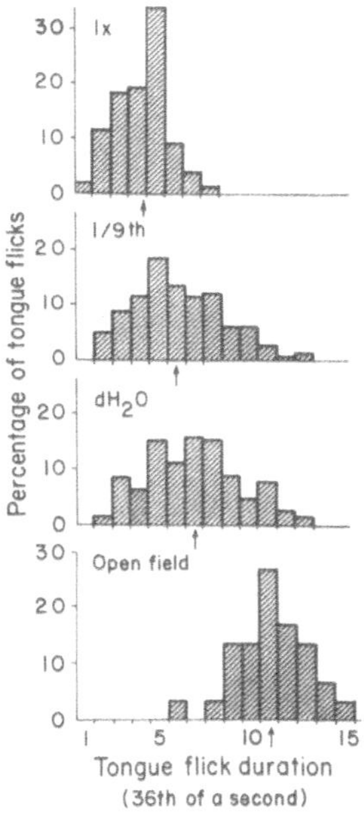

Fig. 2. Tongue-flick durations for female XL plotted as a percentage of total tongue flick durations on each of four trial types 1x, 1/9, dH_2O and open field. Arrows indicate mean durations.

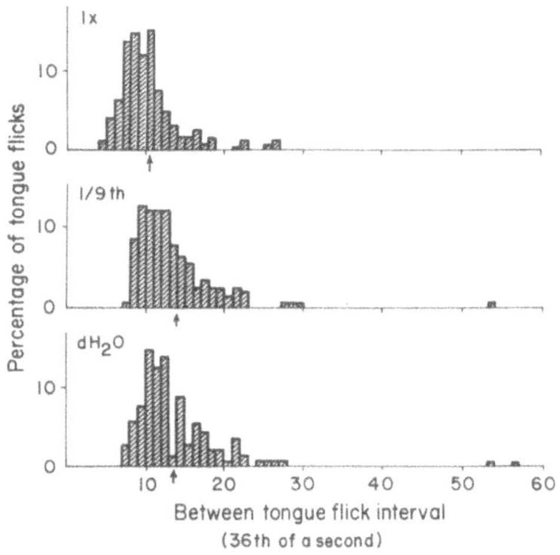

Fig. 3. Between tongue-flick intervals for female XL plotted as a percentage of total between tongue-flick intervals on each of three types of trials: 1x, 1/9x and dH_2O.

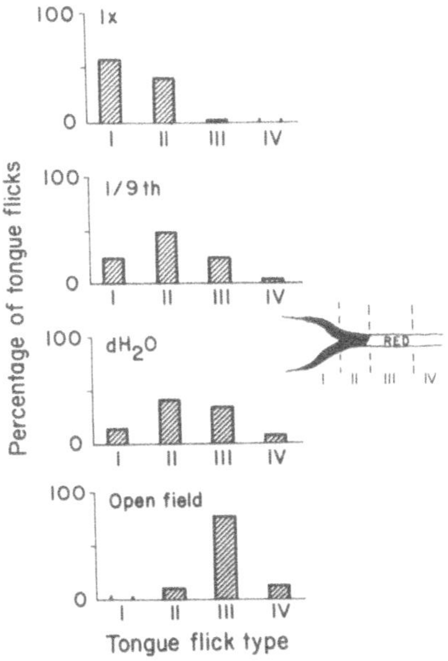

Fig. 4. Percentage of tongue flicks falling into each of four tongue-flick categories for female XL plotted as a percentage of total tongue flicks on each of four types of trials.

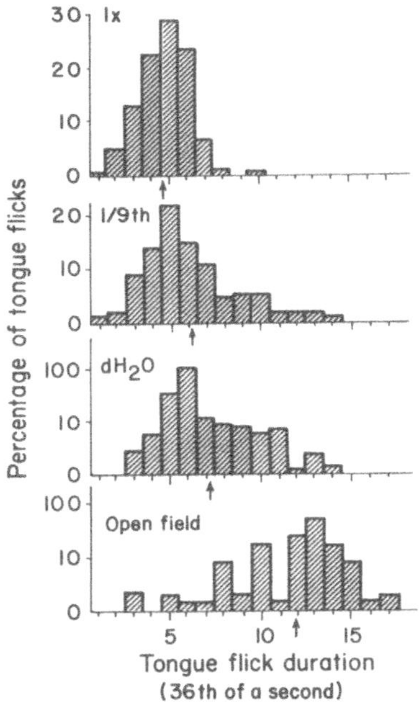

Fig. 5. Tongue-flick durations for female XL following complete bilateral olfactory nerve transection.

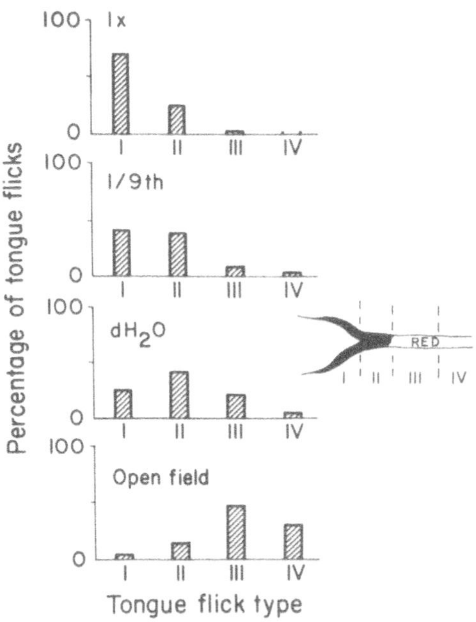

Fig. 6. Percentage of tongue flick falling into each of four tongue-flick categories for female XL following complete bilateral olfactory nerve transection.

Table 1. Film analysis of trailing trials (3x, 1/3x, 1/243x and dH_2O)and open field (OF) testing of snake S2. Duration of tongue flicks (in frames; 1/36th of a second) and interval between tongue flicks (in frames) are indicated as are the percentage of tongue flicks classified I - IV (see text). The 1/27x trials were eliminated because film data were unreliable.

	Duration		Between Interval		Type (%)			
Trail Type	Mean	S.D.	Mean	S.D.	I	II	III	IV
3x	4.4	1.70	21.2	9.35	62.3	29.0	6.5	2.2
1/3x	6.3	2.37	22.2	9.28	20.8	57.1	20.8	1.3
1/243x	7.6	1.72	22.6	11.53	2.0	34.0	62.0	2.0
dH_2O	8.1	2.78	20.4	10.50	4.1	20.3	40.5	35.1
OF	11.2	2.80	34.0	9.90	4.2	12.5	35.4	47.9

S.D. = Standard deviation

Discussion

These results suggest that the positive relation between tongue-flick rate and earthworm-extract concentration previously reported by us (Kubie & Halpern, 1978) is the result of both a decrease in tongue-flick duration and between tongue-flick interval. However, the major change that occurs as earthworm concentration is decreased is an increase in tongue-flick duration. Longer duration tongue-flicks are characterized by longer extension of the tongue. These longer duration, long extension tongue-flicks were most commonly observed in the open field trials. Keiser (1975) has hypothesized that long-excursion tongue-flicks serve to accumulate detectable amounts of odorant on the tongue in a stimulus-poor environment. The exploratory type III and IV tongue-flicks would likely have three disadvantages in trailing high concentration trails; first, they sample too large an air and ground space and would not give precise localization of an odor source; second, they would be too likely to pick up amounts of odorants which would saturate the discriminative range of the system; and third, they would occur less frequently than shorter tongue-flicks and would, therefore, provide fewer units of spatial information within a given time interval. Short-excursion, short-duration tongue flicks ought to aid a snake in trail following while long-excursion, long-duration tongue-flicks ought to aid a snake during non-specific exploratory sampling.

PERIPHERAL BLOCK OF TONGUE-DELIVERED SEXUAL ODORS TO THE VOMERONASAL ORGAN

The dorsal skin of female garter snakes contains a pheromone-like substance to which males respond during the breeding season (Aleksiuk & Gregory, 1974; Crews, 1976; Garstka & Crews, 1981; Kubie, Cohen & Halpern, 1978; Kubie, Vagvolgyi & Halpern, 1978; Noble, 1937). A male, encountering an estrus female, tongue flicks the dorsal surface of the female, moving rapidly up and down her back, tongue flicking continuously and pressing his chin along her dorsal surface. Tongue-flick delivery of pheromonal substances to the vomeronasal organ of male garter snakes appears to initiate the chin pressing behavior characteristic of male courtship (Blanchard & Blanchard, 1941; Crews, 1976; Kubie et al., 1978; Noble, 1937). We have never observed any active response on the part of the female to the male's behavior. As the chin-pressing behavior of the male progresses he straightens the female's body, aligning his body alongside hers. After a time, the male moves his cloacal region in apparent search for the cloacal region of the female. In our studies trials are terminated at this point.

Noble (1937) suggested that a special dermal sense organ under the chin of the snake was stimulated during chin pressing and that this sense organ had to be functional for courtship to occur. However, during chin pressing tongue flicks do occur although they are very difficult to observe without close-up, high speed cinematography. When observed on film they are very short excursion, rapid tongue flicks, similar to those seen during trailing of high concentration prey extracts. We had discovered that injection of small amounts of xylocaine near the course of the hypoglossal nerve caused anesthesia to the chin and lower jaw and, for a shorter interval, blocked tongue flicking. The present study used this technique to determine if the tongue flick delivery of odorants to the vomeronasal organ is essential to the elicitation and maintenance of chin pressing. We also reinvestigated the role of somatosensory stimulation which occurs during chin pressing on continued courtship displays.

Subjects and Hormone Treatment

Eleven male Thamnophis radix, weighing 40 to 93 grams and nine females Thamnophis radix weighing 40 to 110 grams at the beginning of the experiment were obtained from Chicago Zoological Supply Co. The males were preselected by screening them for courtship displays. Only those animals courting females were utilized in this study.

Males were lightly anesthetized with Brevital sodium (10 mg/kg body weight) subcutaneously (Wang et al., 1977) and a 5 mg ($\pm$.3mg) pellet of crystalline testosterone propionate (Sigma Chemical Co.) was implanted subcutaneously in each male. Intact females were injected daily for 5 consecutive days with 40 ug per 75 gms of body weight of estradiol benzoate (Sigma Chemical Co.) dissolved in sesame oil. The injection sequence was followed by nine days without injections.

Housing

During the course of the experiments the snakes were kept in ten gallon glass aquaria in groups of 2 to 5. Males and females were segregated. The experimental room was maintained on a 14 hour light, 10 hour dark cycle with temperatures at 31°C during the light part of the cycle and 23°C during the dark part of the cycle. No attempts were made to control humidity. Water was available continuously and snakes were fed once a week.

Testing Procedures

Each male was tested daily with each female. We placed the female in the test tank and permitted her to adapt to the tank for one minute. The male was then gently placed in the test tank with the female, observed for five minutes or until the male exhibited cloacal searching movements. Males who failed to court any female were returned to their home aquaria. Males who courted females were subjected to the following series of tests: 1) repeat of a normal trial - no intervention; 2) just prior to the trial subcutaneous injections of .01cc xylocaine was delivered lateral to the chin shields bilaterally; 3) normal trial; 4) just prior to the trial .01cc xylocaine was injected subcutaneously at midbody levels; 5) tape was affixed to the chin region of the animal. Alternate animals received xylocaine injections to the body before xylocaine injections to the chin to control for order effects. Normal trials always followed the trials with xylocaine injections to the chin. Following this series of trials we either sutured closed the snakes' vomeronasal ducts or placed sham sutures into the soft tissues of the roof of the mouth (Kubie & Halpern, 1979). For duct suture the snake was given a local injection of xylocaine paravertebrally in the cranial region. The snake was secured on its back with tape and its mouth propped open with a stick. Four independent sutures were placed with 9-0 Ethilon^{R} tying together the cartilaginous ridge of the lateral side of the fenestra vomeronasalis of one side of the snake's mouth with the corresponding ridge on the other side. In the sham suture procedure we placed two sutures into each of the lateral ridges leaving the two ridges independent. Following testing, we removed the sutures from those animals with duct sutures and placed sham sutures while those animals with sham sutures had the sham sutures removed and duct sutures placed.

During test trials with courting males we used a six channel event recorder to note 1) normal tongue flicking 2) "sloppy" tongue flicks - (poorly coordinated long excursion tongue flicks which result from xylocaine-induced block of the hypoglossal nerve) 3) the position of the male's head on the back of the female 4) chin pressing 5) full courtship short of intromission.

Results

All eleven male snakes subjected to the control procedures, i.e. xylocaine along the body surface or sham duct suture behaved as they did on normal (non-intervention) trials. Placing autoclave tape on the snake's chin to remove somatic sensation from the chin did not interfere with courtship (Figure 7).

Only two procedures disrupted male courtship: xylocaine injected subcutaneously along the chin region and suturing closed the vomeronasal ducts. Immediately after injection of xylocaine into the chin a male placed into the testing arena will tongue flick a female he encounters and begin chin pressing. Within 30 seconds (usually 10 seconds) however, the tongue flicking will become sloppy i.e. each flick is long in duration and excursion, poorly coordinated and the tongue touches the substrate along a large portion of its length. Eventually tongue flicking ceases altogether (Figure 7). The appearance of sloppy tongue flicks coincides with a cessation of chin pressing and movement of the male snake off the female's back. The male will continue to move around the arena chancing on the female, but is unresponsive to her. If the experimenter places the male on the female the male snake will move away or just rest there making no attempt to explore the female's surface. After a variable period (four to ten minutes) sloppy tongue flicking resumes followed rapidly within 15 to 20 seconds by normal tongue flicking. Chin pressing follows almost as soon as the male encounters the female and courtship ensues. The time course of the entire courtship display after chin injection is very variable (Figure 7) as if may be under relatively normal conditions. Contrast, for example, the tracings in Figures 7 and 8.

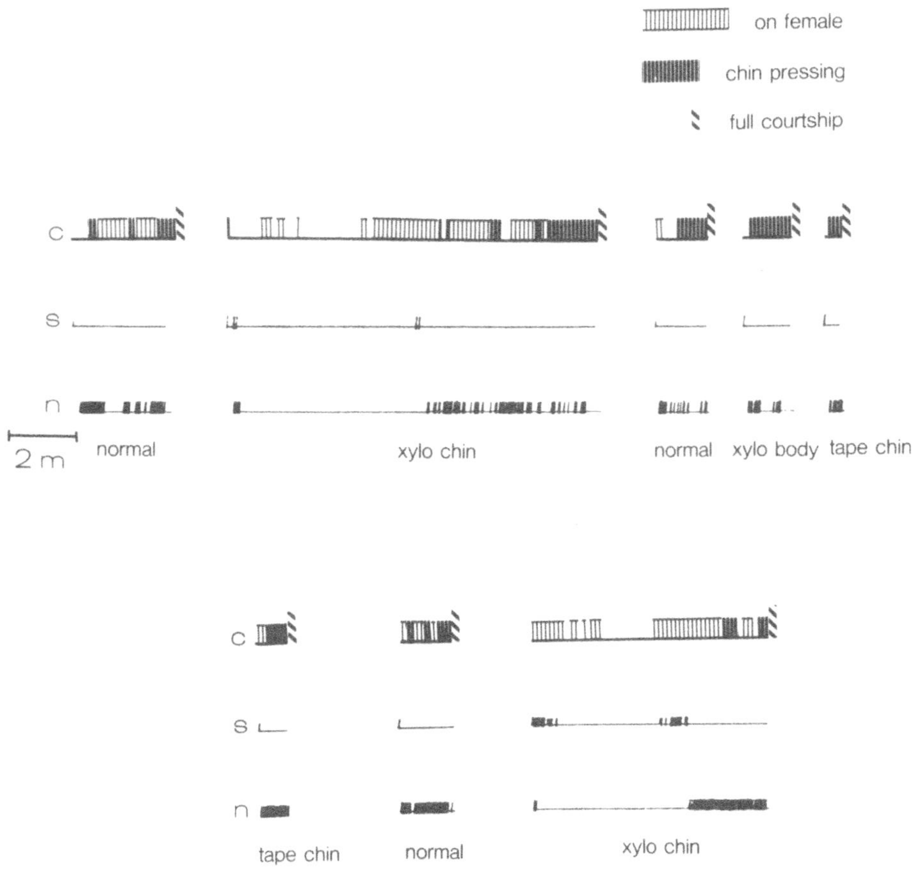

Fig. 7. Event recorder tracings of tongue flicks and courtship behavior during normal, control and experimental trials of testosterone-treated males with estradiol-treated females. Top rows illustrate normal (n) tongue flicks, sloppy (s) tongue flicks and position (on female) or courtship behavior (chin pressing or full courtship) of male on normal trials, with xylocaine injected into the chin (xylo chin) or into the body (xylo body) or tape placed on chin (tape chin). Bottom set of traces from another snake during three trials as marked. Time indicator represents two minutes.

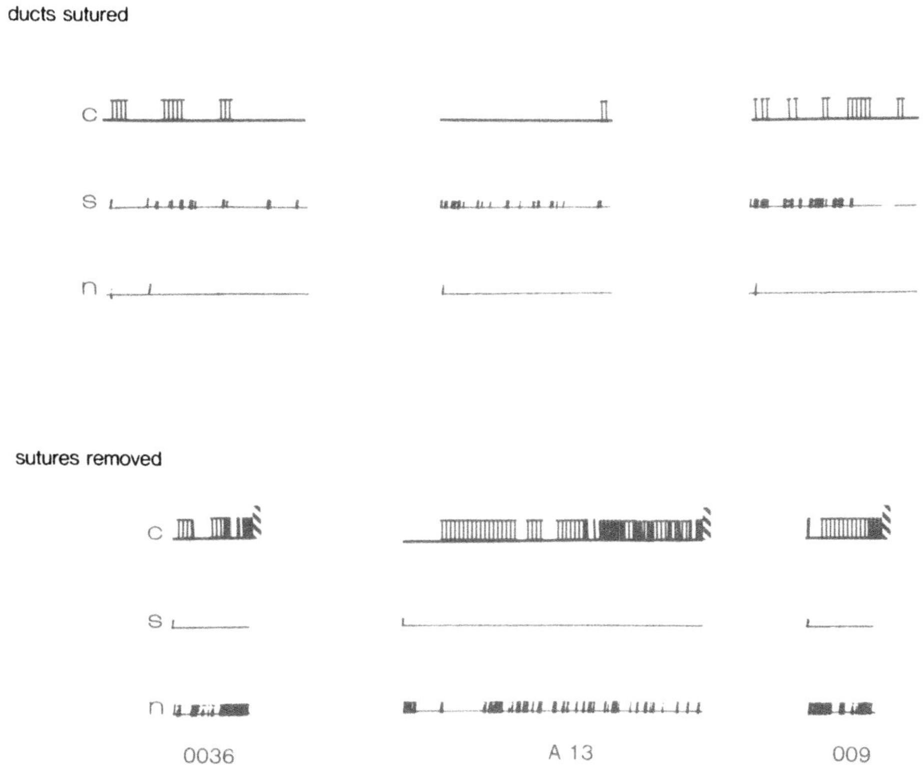

Fig. 8. Event recorder tracings of courtship behavior during trials when the vomeronasal ducts of three males (0036, A13 and 009) were sutured closed (above) and following removal of duct sutures (below). Abbreviations and symbols as in Fig. 7.

After each trial prior to which xylocaine had been injected into the chin, we tested the snake for sensitivity to pain in the chin and lip region by squeezing the skin of the chin, lower lip and upper lip with a fine forceps. In all cases, at the end of the trial the lower lip and chin were still anesthetized but squeezing the upper lip led to immediate head withdrawal. In all these cases normal tongue flicking and full courtship was observed despite the absence of somatic sensitivity in the chin and lower lip region.

None of the snakes with their vomeronasal ducts sutured closed chin-pressed or courted a female (Figure 8). The tongue flick patterns of these snakes were not entirely normal; occasionally the fine tips got caught in the sutures and the tongue protrusion was not as smooth as normally observed. However, some snakes did not show any obvious deficit in tongue flicking yet they did not court.

Discussion

This study demonstrates the critical role of the tongue in delivering sex-related pheromonal stimuli to the vomeronasal organs of garter snakes. This finding supports our previous report (Kubie et al.,1978) that male snakes with vomeronasal nerve lesions do not court estrus females. Since vomeronasal deafferentation deprives the accessory olfactory bulb of a possibly important source of tonic influence, the results of that experiment did not specifically address the issue of the role of purely sensory events in male courtship behavior. The present study utilizes two different, reversible procedures for preventing odorant access to the vomeronasal organ. With both of these procedures, i.e. suturing the vomeronasal ducts closed or inhibiting tongue flicking by xylocaine injection, male snakes failed to court although control procedures had no effect. Thus it is clear that the tongue plays a critical role in delivering pheromonal substances to the vomeronasal organ.

Initially we thought it possible that chin pressing was a behavioral mechanism for delivering odorants to the vomeronasal organ without tongue flicking. We had, in fact, demonstrated that odorants can gain access to the vomeronasal organs even when the tongue is surgically removed (Halpern & Kubie, 1980). However, using close range, high speed cinematography it is clear that short excursion, very rapid tongue flicks are a constant feature during chin pressing. Furthermore, a snake who is in the process of chin pressing immediately leaves the back of the female and "searches" rapidly from side to side as soon as tongue flicking is blocked by xylocaine. This suggests that chin pressing is an intense tongue-flick delivery process similar to prey trailing delivery of odorants via rapid, short excursion tongue flicks.

Two observations in the present study argue strongly against Noble's (1937) assertion that tactile stimuli from the chin are critical to male courtship displays. Tape placed on the chin did not affect male courtship. Similarly, although xylocaine's anesthetic effects on the skin surface outlasted its paralyzing effect on the tongue musculature, courtship resumed as soon as coordinated tongue movements were evidenced. Thus the sensory information from the chin could not have been critical for manifestation of this behavior.

TONGUE FLICK CHANGES IN RESPONSE TO AIR-BORNE PREY ODORS

As indicated above, the tongue flick behavior of snakes is capable of delivering substances to the vomeronasal organ (Halpern & Kubie, 1980; Kahmann, 1932) but does not deliver odorants to the olfactory epithelium. At present the best evidence available strongly supports the idea that the tongue must be in direct contact with the stimulus source to deliver molecules from that substance to the vomeronasal organ (Halpern & Kubie, 1980; Kubie & Halpern, 1978; Sheffield, Law & Burghardt, 1968). Air-borne odorants would presumably stimulate the olfactory, but not the vomeronasal system since direct contact with the stimulus source would not be possible. Cowles and Phelan (1958) postulated that when air-borne odors are perceived by snakes searching movements are initiated which are accompanied by increased tongue flicking, leading to proximal detection of the odor source by the vemeronasal system. Some evidence suggests that earthworm extract contains volatiles to which garter snakes respond with increased tongue flicking and attack (Burghardt, 1977; 1979).

The present study (Halpern, Heller and Vagvolgyi, 1981) was designed to provide a quantitative assessment of the effect of airborne prey odors on tongue flick behavior of garter snakes.

Subjects

Seventeen adult garter snakes of several species of the genus *Thamnophis* were utilized in this experiment. Most of the animals ate both goldfish and earthworms prior to and during the experiment.

Procedure

The test apparatus consisted of a 22.5 x 10.6 x 11.3 cm Plexiglas box with an intake port and outflow port located on opposite walls. The Plexiglas box was contained within a larger, opaque cardboard box with a small viewing hole and 2 tube access holes (Figure 9).

Pressurized air was passed in series through activated charcoal, loosely packed fiber glass and a glass fiber filter (Millipore AP 150250) at a rate of 10 liters per minute. This air was then directed by use of 2-way valves to the intake ports of bottles containing distilled H_2O, distilled H_2O with an earthworm or distilled H_2O with a goldfish. For some animals earthworm extract or goldfish extract prepared by the Wilde (1938) or Burghardt (1966) method was used instead of the prey itself. The outflow tube of the stimulus bottles led to the intake port of the test chamber. A constant vacuum (set to match the air flow rate) pulled air from the test chamber through the outflow port of the chamber (Figure 9).

The snakes were tested four to six days, one session per day. Each session began with a ten minute acclimation period in the testing box followed by one presentation of each stimulus for one minute. Stimulus presentations were separated by five minute intertrial intervals. The order of stimulus presentations was randomly assigned with the constraint that each stimulus appeared in each ordinal position an equal (or near equal) number of times.

Tongue flicks (the number of tongue excursion from the mouth) were visually monitored and tallied on a hand counter for the minute preceding, the minute during and the minute following stimulus presentation.

Results

The results are graphically presented in Figure 10. Tongue flick rates differed during presentation of the three stimuli for each snake. Tongue flick rates during presentation of either prey odor were almost twice as high as during control trials. These differences were statistically significant ($F(2,32)=6.50$, $p<.05$). Furthermore, tongue flick rates after stimulus presentation were significantly elevated over prestimulus levels on earthworm ($t=1.70$; 30df; $p<.05$) and fish ($t=2.02$; 30df; $p<.05$) trials, but not on distilled water trials. There were no significant differences in tongue flick rate before stimulus presentation in the 3 stimulus conditions nor were the tongue flick rates before, during and after dH_2O significantly different from each other.

Discussion

These results clearly demonstrate that garter snakes are capable of detecting air-borne odorants from prey even when the snakes cannot make physical contact with the odorant source. These findings support the idea of Cowles & Phelan (1958) that air-borne odorants, probably acting through the olfactory system, activate the animal, initiating several behavioral changes including increased tongue flicking and searching movements. It is likely that when the tongue comes into direct contact with the source of

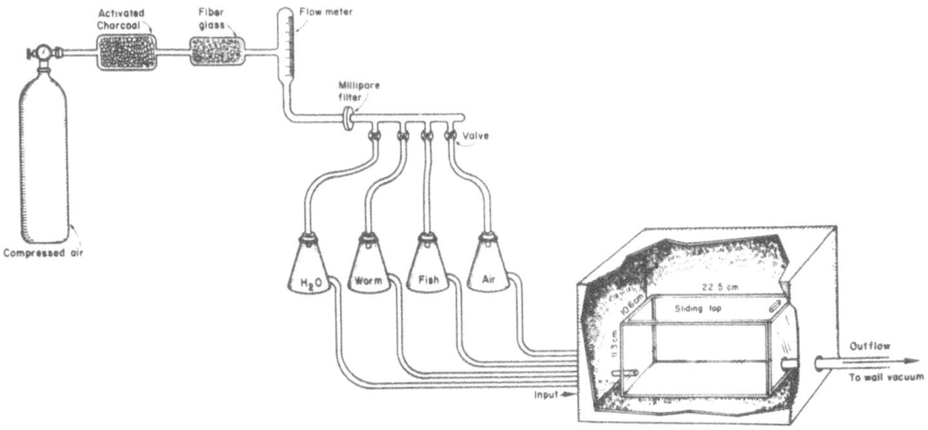

Fig. 9. Apparatus used in delivering air-borne odorants to snakes.

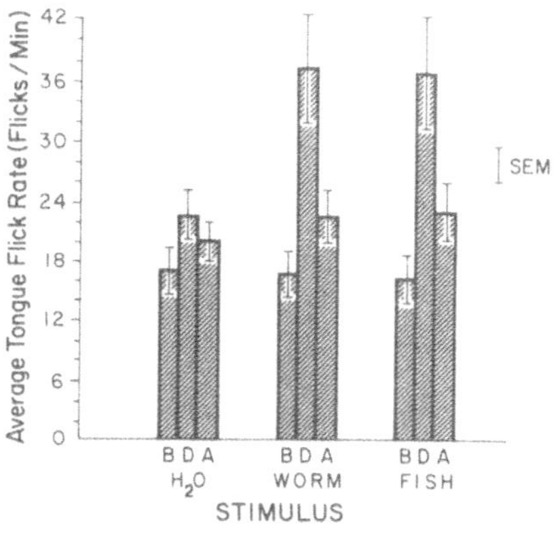

Fig. 10. Average tongue-flick rate before (B), during (D) and after (A) stimulus delivery.

the odorant the vomeronasal system is stimulated resulting in a series of responses that culminate in prey attack. Tongue flick rates during air-borne odorant delivery were never as high as those found during trailing.

Previous studies from this laboratory (Kubie & Halpern, 1978) revealed that garter snakes have great difficulty following an earthworm extract trail when direct lingual contact with the trail is prevented by placing the trail under a perforated "false" floor. One might expect that if snakes are capable of responding to air-borne odorants arising from prey, as demonstrated above, that they should be capable of using the same sensory information in following trails removed from direct contact with the tongue. The two tasks however, are different. The trailing task requires a simultaneous discriminated approach response to the earthworm extract trail and inhibition of an approach response to the water trail whereas the odorized air experiment presents stimuli successively and requires that the animal simply increase the frequency of a naturally occurring behavior. In a natural environment these two models of response may occur successively in response to changing environmental cues. Thus an immobile snake rarely emits tongue flicks until some stimuli, e.g. olfactory or visual, elicit searching or exploratory movements. These movements, typically accompanied by tongue flicking, lead the animal into the discriminative mode of responding, i.e. approaching increasingly concentrated prey odors and avoiding areas with less concentrated prey odors.

It is likely that the initial, nondiscriminative, increased tongue-flick rate and movement is signalled by stimuli activating the olfactory system whereas the vomeronasal system is critically involved in discriminative responding to proximal cues emanating from prey.

GENERAL DISCUSSION

The studies we have described are aimed at analyzing the functional relationship between the snake's tongue and its vomeronasal system. In the first study we analyzed tongue-flick responses to potent vomeronasal stimuli -- earthworm extract trails. In the second study we analyzed the disruptive effects of a tongue-flick block on male mating behavior. In the third study we analyzed tongue flick responses to air-borne prey odors (detected by the olfactory and/or the vomeronasal systems). The relationship between the tongue and the vomeronasal system is two-directional: in one direction, tongue-flick delivery of odorants to the vomeronasal system seems critical for functional vomeronasal activation, as in mating; in the other direction, intense odor stimulation increases the tongue-flick rate and affects the pattern of tongue flicking.

The functional relationship between the snake tongue and vomeronasal system is of interest on three levels. The first is the ethological level. How do snakes use this system? Clearly, both tongue flicking and the vomeronasal apparatus are highly developed in snakes. Our studies and others suggest that snakes use this system in a variety of important behaviors ranging from prey trailing to mating. In addition, different groups of snakes have developed different uses or styles of use for the system. For instance, the tongue flicking of vine snakes is exceedingly slow and may be adapted for both a slow approach to prey and maximal accumulation of odorants on the tongue (Keiser, 1975). The question of why snakes of all vertebrates find the tongue-vomeronasal system of such value is largely a mystery (although see Gove, 1979, for interesting discussion).

The second level of interest is the comparative level. In what ways are the snake tongue and vomeronasal system similar to the tongue and vomeronasal system of other vertebrates? At this early stage it is our impression that the snake vomeronasal system is more unusual in the extent of its development than in the uses to which it is put. Perhaps the same is true of snake tongue flicking. It has been suggested that a variety of mammals use their tongue to deliver odorants to the vomeronasal organs (Epple, 1974). If most or all adequate stimuli are non-volatile odorants, as appears to be the case in several mammals (Beauchamp et al., 1979), then tongue flicking or licking might be effective odorant delivery processes. If lingual delivery of odorants to the vomeronasal apparatus is common in mammals, then the relationship of the vomeronasal system to the sense of taste becomes more fascinating. At present, little is known of interrelationships between these two sensory systems.

Finally, the third level of interest is the neurological level. Much of the recent interest in the vomeronasal system stemmed from findings of segregated central pathways of the olfactory and vomeronasal system for amphibians (Herrick, 1921; Scalia, 1972; Scalia et al., 1968), reptiles (Halpern, 1976; Heimer, 1969; Kowell, 1974; Ulinski & Peterson, 1981) and mammals (e.g., Broadwell, 1975; Davis et al., 1978; Scalia & Winans, 1975; Skeen & Hall, 1977; Winans & Scalia, 1970). The findings of Winans and Scalia (1970) originally led to the formulation of the dual olfactory hypothesis, i.e. that two non-overlapping chemosensitive systems exist from the periphery through the telencephalon and into the diencephalon. The behavioral corollary of this hypothesis, that distinctive functional roles exist for the two systems, is still under active investigation and has found considerable experimental support (Halpern & Frumin, 1980; Heller & Halpern, 1983; Johns et al., 1978; Kaneko et al., 1980; Kubie & Halpern, 1979; Kubie et al., 1978; Marques, 1979; Meredith et al., 1980; Powers & Winans, 1975; Powers et al., 1979; Reynolds & Keverne, 1979; Winans & Powers, 1977). These studies indicate that pheromonal communication is primarily dependent on a functional vomeronasal system.

The close relationship between a snake's vomeronasal system and its tongue suggests that an oligosynaptic central pathway exists linking the two. For instance, the finding that intense vomeronasal stimulation in two situations (mating and trailing) leads to fast, short excursion tongue flicking suggests excitatory influences from the accessory olfactory bulb or nucleus sphericus to parts of the hypoglossal nucleus. The finding that air-borne odorants from prey elicit increased tongue flicking, but not the extremely rapid tongue flicks of high extract trailing, suggests that other, parallel pathways may be involved in this response. Although this reflex model of the relationship between odorants and tongue flicking may be simplistic, we feel it should be explored further.

As mentioned above, tongue flicking can be depicted as a sense-seeking behavior. Tongue flick rate and pattern are greatly affected by vomeronasal stimulation which puts the tongue flick in a feedback configuration. We are on the threshold of being able to apply a systems analysis approach to understanding an apparently simple behavior -- prey trailing. This approach has been particularly fruitful in the study of eye movements. With respect to prey trailing we know some of the key elements: the vomeronasal organ, its central neural targets, the tongue and its motor innervation. We know some simple relationships among the elements: high concentration trails cause tongue flick rates to increase, and cause tongue excursions to become shorter; electrical stimulation of some central neural structures (in the lizard) cause tongue flicking (Distel, 1976). Many questions remain. Can a snake compare odors delivered from its two tongue tips? Is the width of the spread of the tips affected by the trail? What is the role of tactile feedback from the tongue in tongue flicking? How are head movements and locomotor patterns affected by trail information? Currently we are following the trail of some of these questions.

ACKNOWLEDGEMENTS

The research described in this chapter was generously supported by NIH grants NS 11713 and 1 F32 NS 06152. We are grateful to Jeffrey Halpern, Steven Heller, Robert Somma and Alice Vagvolgyi who contributed to the work described and to Rose Kraus for typing the manuscript.

REFERENCES

Aleksiuk, M., and Gregory, P.T., 1974, Regulation of seasonal mating behavior in Thamnophis sirtalis parietalis. Copeia, 1974: 681-688.

Beauchamp, G.K., Wellington, J.L., Wysocki, C.J., Brand, J.G., Kubie, J.L., Smith, III, A.B., 1979, Chemical communication in the guinea pig: urinary components of low volatility and their access to the vomeronasal organ. in: "Chemical Signals Vertebrates and Aquatic Invertebrates," Müller-Schwarze, D. and Silverstein, R.M., eds., Plenum Press, New York.

Blanchard, F.N., and Blanchard, F.C., 1941, Mating of the garter snake, Thamnophis sirtalis sirtalis (Linnaeus). Mich. Acad. Sci., Arts and Letters, Papers 27:215-234.

Broadwell, R.D., 1975, Olfactory relationships of the telencaphalon and diencephalon in the rabbit. I. An autoradiographic study of the efferent connections of the main and accessory olfactory bulbs. J. Comp. Neurol., 163:329-346.

Burghardt, G.M., 1966, Stimulus control of the prey attack response in naive garter snakes. Psychonom. Sci., 4:37-38.

Burghardt, G.M., 1977, The ontogeny, evolution, and stimulus control of feeding in humans and reptiles. in:""The Chemical Senses and Nutrition," Kare, M.R., and Maller, O., eds. Academic Press, New York.

Burghardt, G.M., 1979, Behavioral and stimulus correlates of vomeronasal functioning in reptiles: feeding, grouping, sex, and tongue use. in: "Chemical Signals Vertebrates and Aquatic Invertebrates," Müller-Schwarze, D. and Silverstein, R.M., eds. Plenum Press, New York.

Burghardt, G.M., and Pruitt, G.H., 1975, The role of the tongue and senses in feeding of naive and experienced garter snakes. Physiol. Behav., 14:185-194.

Clark, D.L., 1981, The tongue-vomeronasal transfer mechanism of snakes. Micron., 12:299-300.

Cowles, R.B., and Phelan, R.L., 1958, Olfaction in rattlesnakes. Copeia., 77-83.

Crews, D., 1976, Hormonal control of male courtship behavior and female attractivity in the garter snake (Thamnophis sirtalis sirtalis). Horm. Behav., 7:451-460.

Davis, B.J., Macrides, F., Youngs, W.M., Schneider, S.P., and Rosene, D.L., 1978, Efferents and centrifugal afferents to the main and accessory olfactory bulbs in the hamster. Brain Res. Bull., 3:59-72.

Distel, H., 1976, Behavioral responses to the electrical stimulation of the brain in the green iguana. in: "Behavior and Neurology of Lizards," Greenberg, N., and MacLean, P.D., eds., National Institute of Mental Health, Rockville, Maryland.

Epple, G., 1974, Primate pheromones. in: "Pheromones," Birch, M.C., ed., North-Holland Publishing, Amsterdam.

Garstka, W.R., and Crews, D., 1981, Female sex pheromone in the skin and circulation of a garter snake. Science., 214: 681-683.

Gove, D., 1979, A comparative study of snake and lizard tongue-flicking, with an evolutionary hypothesis. Tierpsychol., 51:58-79.

Halpern, M., 1976, The efferent connections of the olfactory bulb and accessory olfactory bulb in the snakes, Thamnophis sirtalis and Thamnophis radix. J. Morphol., 150:553-578.

Halpern, M., and Frumin, N., 1979, Roles of the vomeronasal and olfactory systems in prey attack and feeding in adult garter snakes. Physiol. Behav., 22:1183-1189.

Halpern, M., Heller, S., and Vagvolgyi, A., 1981, Garter snakes respond to air-borne odorants and increased tongue-flicking. Eastern Psychological Association. 52nd Annual Meeting. New York, N.Y. April 22-25, 1981.

Halpern, M., and Kubie, J.L., 1980, Chemical access to the vomeronasal organs of garter snakes. Physiol. Behav., 24:367-371.

Heimer, L., 1969, The secondary olfactory connections in mammals, reptiles and sharks. Ann. N.Y. Acad. Sci., 167:129-146.

Heller, S.B., and Halpern, M., 1983, Laboratory observations of aggregative behavior of garter snakes, Thamnophis sirtalis. J. Comp. Physiol. Psychol., In Press.

Gillingham, J.C., and Clark, D.L., 1981, Snake tongue flicking transfer mechanisms to Jacobson's organ. Can. J. Zool., 59:1651-1657.

Herrick, C/J., 1921, The connections of the vomeronasal nerve, accessory olfactory bulb and amygdala in amphibia. J. Comp. Neurol., 33:213-280.

Johns, M.A., Feder, H.H., Komisaruk, B.R., and Mayer, A.D., 1978, Urine-induced reflex ovulation in anovulatory rats may be a vomeronasal effect. Nature., 272:446-447.

Kahmann, H., 1932, Sinnesphysiologische Studien an Reptilien: I. Experimentelle Untersuchungen über das Jakobsonische Organ der Eidechsen und Schlangen. Zoologische Jahrbücher, Abt. für Allge. Zool. und Physiol. der Tiere. 51:173-238.

Kaneko, N., Debski, E.A., Wilson, M. C., and Whitten, W.K., 1980, Puberty acceleration in mice. II. Evidence that the vomeronasal organ is a receptor for the primer pheromone in male mouse urine. Biol. Reprod., 22:873-878.

Keiser, E.D., 1975, Observation on tongue extension of vine snakes (genus Oxybelis) with suggested behavioral hypothesis. Herpetologica, 31:131-133.

Kowell, A.P., 1974, The olfactory and accessory olfactory bulbs in constricting snakes: Neuroanatomic and behavioral observations. Unpublished Ph.D. dissertation, University of Pennsylvania, Philadelphia, Pa.

Kubie, J.L., Cohen, J., and Halpern, M., 1978, Shedding of estradiol benzoate treated garter snakes enhances their sexual attractiveness and the attractiveness of untreated penmates. Anim. Behav., 26:562-570.

Kubie, J.L., and Halpern, M., 1975, Laboratory observations of trailing behavior in garter snakes. J. Comp. Physiol. Psychol., 89:667-674.

Kubie, J.L., and Halpern, M., 1978, Garter snake trailing behavior: effects of varying prey extract concentration and mode of prey extract presentation. J. Comp. Physiol. Psychol., 92:362-373.

Kubie, J.L., and Halpern, M., 1979, The chemical senses involved in garter snake prey trailing. J. Comp. Physiol. Psychol., 93:648-667.

Kubie, J.L., Vagvolgyi, A., and Halpern, M., 1978, The roles of the vomeronasal and olfactory systems in the courtship behavior of male garter snakes. J. Comp. Physiol. Psychol. 92: 627-641.

Marques, D.M., 1979, Roles of the main olfactory and vomeronasal systems in the response of the female hamster to young. Behav. Neural Biol., 26:311-329.

McDowell, S.B., 1972, The evolution of the tongue of snakes, and its bearing on snake origins. in: Evolutionary Biology. Vol. 6, Dobzhansky, T., Hecht, M.K., and Steere, W.C., eds., Appleton-Century-Crofts, New York.

Meredith, M., Marques, D.M., O'Connell, R.J., and Stern, F., 1980, Vomeronasal pump: significance for male hamster sexual behavior. Science., 207:1224-1226.

Noble, G.K., 1937, The sense organs involved in the courtship of Storeria, Thamnophis and other snakes. Bull. Amer. Mus. Nat. Hist., 73:673-725.

Powers, J.B., Fields, R.B., and Winans, S.S., 1979, Olfactory and vomeronasal system participation in male hamsters' attraction to female vaginal secretions. Physiol. Behav. 22:77-84.

Powers, J.B., and Winans, S.S., 1975, Vomeronasal organ: Critical role in mediating sexual behavior in the male hamster. Science, 187:961-963.

Reynolds, J., and Keverne, E.B., 1979, The accessory olfactory system and its role in pheromonally mediated suppression of oestrus in grouped mice. J. Reprod. Fertil., 57:31-35.

Scalia, F., 1972, The projection of the accessory olfactory bulb in the frog. Brain Res., 36:409-411.

Scalia, F., Halpern, M., Knapp, H., and Riss, W., 1968, The efferent connections of the olfactory bulb in the frog: a study of degenerating unmyelinated fibers. J. Anat., 103:245-262.

Scalia, F., and Winans, S.S., 1975, The differential projections of the olfactory bulb and accessory olfactory bulb in mammals. J. Comp. Neurol., 161:31-56.

Sheffield, L.P., Law, J.M., and Burghardt, G.M., 1968, On the nature of chemical food sign stimuli for newborn garter snakes. Commun. Behav. Biol. 2:7-12.
Skeen, L.C., and Hall, W.C., 1977, Efferent projections of the main and accessory olfactory bulb of the tree shrew (Tupaia glis). J. Comp. Neurol., 172:1-36.
Ulinski, P.S., and Peterson, E.H., 1981, Patterns of olfactory projections in the desert iguana, Dipsosaurus dorsalis. J. Morph., 168:189-227.
Wang, R.T., Kubie, J.L., and Halpern, M., 1977, Brevital Sodium: An effective anesthetic agent for performing surgery on small reptiles. Copeia, 738-743.
Wilde, W.S., 1938, The role of Jacobson's organ in the feeding reaction of the common garter snake, Thamnophis sirtalis sirtalis (Linn). J. Exp. Zool., 77:445-465.
Winans, S.S., and Powers, J.B., 1977, Olfactory and vomeronasal deafferentation of male hamsters: Histological and behavioral analyses. Brain Res., 126:325-344.
Winans, S.S., and Scalia, F., 1970, Amygdaloid nucleus: New afferent input from the vomeronasal organ. Science, 170: 330-332.

THE ACCESSORY OLFACTORY SYSTEM: ROLE IN MAINTENANCE OF CHEMOINVESTIGATORY BEHAVIOR

Gary K. Beauchamp[1,2], Irwin G. Martin[1], Judith L. Wellington[1] and Charles J. Wysocki[1]

[1]Monell Chemical Senses Center, 3500 Market Street Philadelphia, PA 19104, and [2]Department of Otorhinolaryngology and Human Communication, University of Pennsylvania, Philadelphia, PA 19104

INTRODUCTION

In mammals, the critical role of the vomeronasal organ (VNO) and the accessory olfactory system (AOS) in mediating neuroendocrine responses to chemical signals has now been demonstrated (e.g., Bellringer et al., 1980; Kaneko et al., 1980; Reynolds and Keverne, 1979; Wysocki and Katz, submitted). Behavioral responses are also altered when input to the AOS is disrupted in studies of maternal responsiveness (Fleming et al., 1979; Marques, 1979), male sexual behavior (Powers and Winans, 1975; Winans and Powers, 1977; Meredith et al., 1980) and ultrasonic vocalizations (Wysocki et al., in press). The hypothesis for a central role for the AOS in mammalian reproductive behavior has been substantiated (see Wysocki, 1979).

Less clear are studies on investigatory responses (sniffing/licking frequency, flehmen) elicited by chemical signals. Table 1 presents the results of several selected studies with a variety of species where AOS disruption was attempted and investigatory responses to conspecifics or their odors were studied. In cases where sniffing, licking or flehmen were measured, quantitative decreases occurred in some studies following disruption whereas in other studies, few effects were noted. These studies permit the following conclusions: (1) disruption of the AOS may influence investigatory responses but (2) in the apparent absence of input to the AOS, considerable responsiveness to conspecifics and their odors remains. Perhaps interruption of input to the AOS disrupts critical reproductive events but has little effect on investigatory responses or attractiveness of conspecific odors.

Table 1. Results of Selected Studies on the Effects of Disruption of the Accessory Olfactory System on Investigatory Responses to Chemical Signals.

Species	Method[a]	Behavior	Effect	Reference
Hamster	nc	male invest. of vaginal secretions	depression in 25% no change in 75%	Powers et al., 1979
Hamster	nc	male sniffing female, male licking female genitalia	no effect	Murphy, 1980
Hamster	AOT	male anogenital sniffing of estrous female	depression in all animals	Marques et al., 1982
Goat	blockage[b]	flehmen	no change	Ladewig and Hart, 1980
Cat	blockage	flehmen to various odors	decreases in frequency	Verberne, 1976
Rat	nc	sniffing and licking pups by female	apparent depression of licking	Fleming et al., 1979
Guinea pig	removal	male invest. of urine	depression increasing with time	Beauchamp et al., in press

[a]Methods of disruption include: nc = severing of vomeronasal nerve; AOT = transection of accessory olfactory tracts; blockage = use of obstructing materials to prevent access of odors to the VNO; removal = surgical removal of the VNO.
[b]Possibly partial blockage only.

In this essay, we will discuss our investigations of the roles of the AOS and the main olfactory system in maintenance of the responsiveness of domestic guinea pigs (Cavia porcellus) to conspecific urine odors. Following a brief description of the background for this work, the results of studies on the effects of altering access to the VNO on investigatory responses to urine samples will be described.

BACKROUND

Guinea pigs, like many species, employ odors to communicate (see Beauchamp et al., 1979 and references therein). A distinctive feature of the guinea pig's response to conspecific odors (particularly of the male's response to female urine odors) is a behavior we have called headbobbing. During this behavior, the animal places its nose in direct contact with the urine or other odor source and moves its head up and down rapidly (2-4 cycles/second). In adult males isolated from females, this behavior may continue, without interruption, for up to several minutes following contact with conspecific female urine samples. Often associated with headbobbing are loud vocalizations, scent marking behaviors and, rarely, components of courtship behavior. An extensive series of studies has demonstrated differential responsiveness (as measured by time spent investigating and headbobbing) to urine and secretions from different classes of donors. These studies demonstrated that information about genus, species, sex, age, prior dietary history and individual identity are available in urine samples (see Beauchamp et al., 1979 for references).

Several lines of evidence suggest that the VNO may be involved in mediating responsiveness of guinea pigs to urine odors. First, Planel (1953) reported that nerve cuts interrupting VNO input to the accessory olfactory bulb (AOB) disrupted the ability to locate an estrous female in some but not all of the males studied (however, Matthes, 1932, found no change in headbobbing to female odors following cuts of the vomeronasal nerve in one guinea pig which he studied). Second, responsiveness to urine odors was attenuated when the urine was placed where it could not be contacted, 0.5 cm below a wire screen (Beauchamp, 1973; Beauchamp et al., 1980). Third, information remains in voided, dry urine for weeks or months (Wellington et al., 1981); this agrees with results obtained in other species (e.g. Johnston and Schmidt, 1979; Nyby and Zakeski, 1980). Fourth, chemical studies have implicated non-volatile substances (circa 600 daltons) as mediating a portion of the response to urine odors (Berüter et al., 1973; Beauchamp et al., 1980). Fifth, non-volatile substances in guinea pig urine have access to the vomeronasal organ when the animal is permitted direct contact with the urine (Wysocki et al., 1980; see also Ladewig and Hart, 1980; Schilling, 1980). Our studies also indicated that the route of access for guinea pigs is via the external nares. Following these results, we next investigated changes in response to urine odors following surgical removal of the VNO.

VOMERONASAL ORGAN REMOVAL

In a recent study (Beauchamp et al., in press), adult male guinea pigs were evaluated for responses to female vs. male urine with standard preference tests (Beauchamp, 1973) prior to surgical removal of the VNO. Two groups of animals (equated for their responsiveness to urine) were established (control, $\underline{n}$ = 10; experimental, $\underline{n}$ = 10). The VNO was removed from each experimental male via an approach through the hard palate. This method of eliminating input from the VNO has the advantages that (1) it is done peripherally, not centrally; (2) it spares the septal organ and the primary olfactory system; and (3) it is done under visual control. Control males underwent a sham surgery.

Three to four weeks following surgery, preference tests identical to the presurgery evaluation were conducted. During the succeeding 5 months a variety of types of urine preference tests were conducted including, 5 months following surgery, another male vs. female urine test. Also during this period (3-5 months post-surgery), sexual and social behavior tests were conducted which involved substantial periods of male-female cohabitation. Methods and results of these tests are discussed in detail elsewhere (Beauchamp et al., in press). Finally, approximately 6.3 months after surgery, male vs. female urine tests were again conducted. The three post-surgical male vs. female urine tests employed urine collected from different donors for each test. Several months later, animals were sacrificed, their brains removed for histology and their nasal area decalcified. Three of the animals were found to have incomplete surgery although only small portions ($\leq$ 5%) of the VNO remained in each. Each of these animals had one or a few glomeruli remaining in their accessory olfactory bulbs (AOBs); all other experimental animals exhibited complete degeneration of glomeruli in the AOBs. The conclusions described below are similar regardless of the presence or absence of the data obtained from these animals.

In experimental males, removal of the VNO caused a substantial decrement in response to urine. Surprisingly, this decrement was time dependent as illustrated in Figs. 1 and 2. When tested 3-4 weeks following removal of the VNO, the males' responses to urine were depressed relative to presurgical levels although considerable headbobbing was still evident in all males. Responses at 5 months post-surgery were little altered (Figs. 1 and 2). However, when tested 6.3 months following surgery (after many other urine preference tests not described herein) the investigatory and headbobbing responses elicited by female urine were virtually absent (Fig. 1). Similarily, the response to domestic female urine and wild female urine was attenuated on the second post surgery test relative to the first (Figs. 1 and 2). Repeated testing and/or the passage of time appeared to result in an attenuation of the response to urine odors following removal of the VNO.

Data obtained by other investigators have also implicated a time dependency. Kubie et al. (1979) noted a time dependent deterioration in feeding following VNO deafferentiation in snakes. Additionally, there appears to be a decrement in mating behavior over time and/or repeated

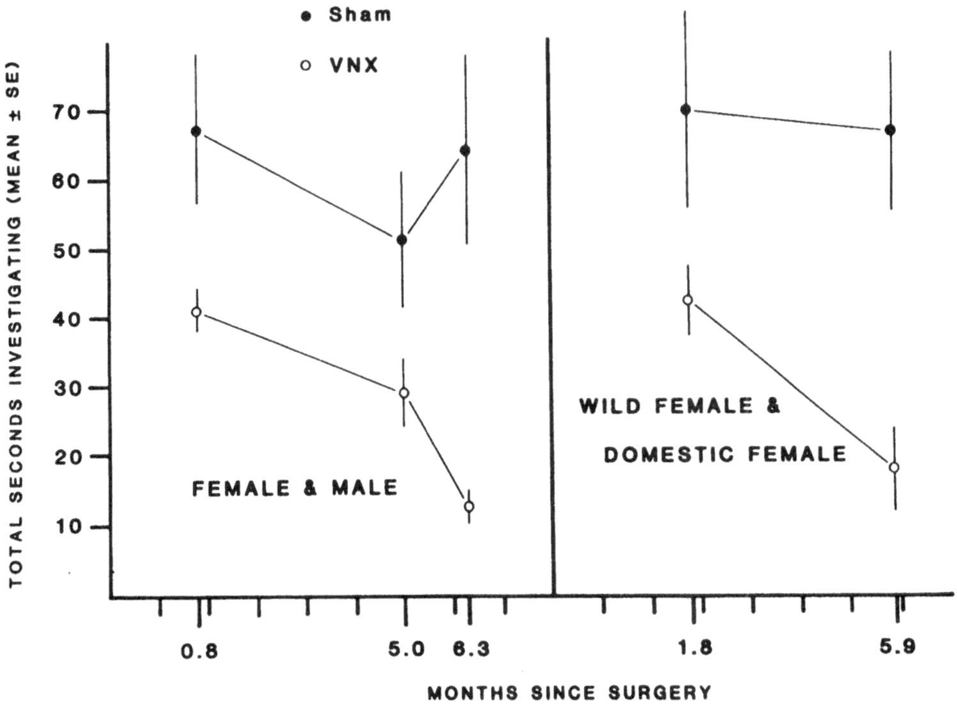

Fig. 1. Mean (± SEM) seconds during 4-min choice tests spent investigating male urine plus female urine (left panel) and wild female (Cavia aperea) urine plus domestic female urine (right panel) by male guinea pigs with (Sham; n = 10) or without (VNX; n = 8) vomeronasal organs. (The number in the experimental group was 8 due to the death of two subjects before the completion of the study.) Note the decline in response to urine odors months after surgery in the VNX group contrasted with an absence of a decline in the Sham group.

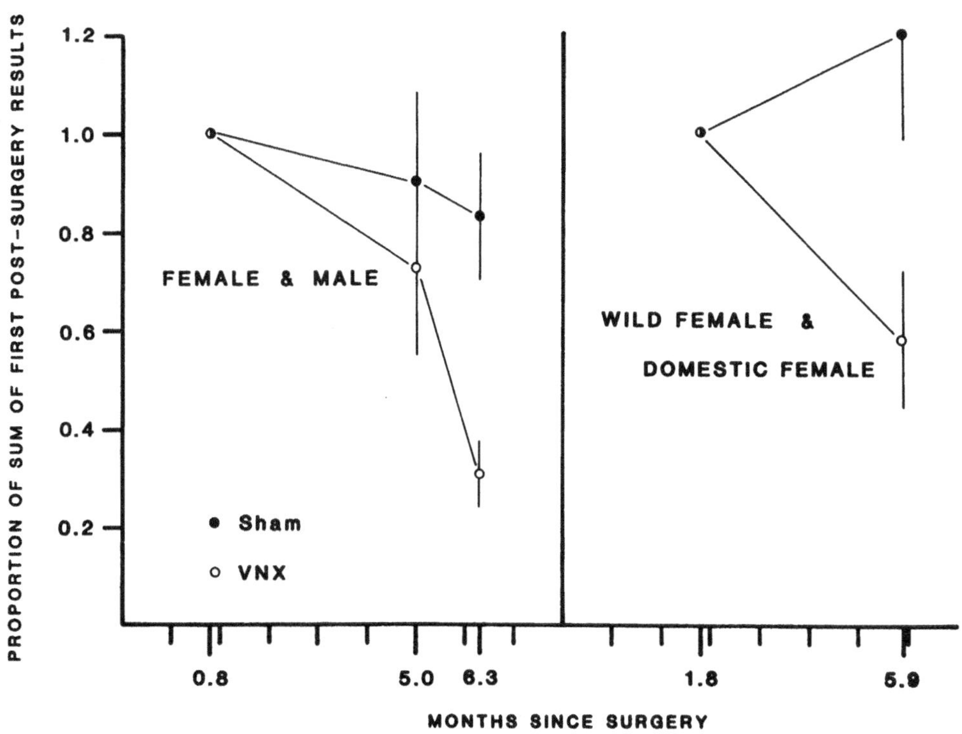

Fig. 2. This figure emphasizes the change in response by animals without vomeronasal organs (VNX) as the result of repeated testing or the passage of time following surgery. The mean number of seconds investigating male urine plus female urine (left panel) or wild female urine plus domestic female urine (right panel) on the first test following surgery was arbitrarily set at 1.0. The mean ratio of responses (± SEM) on the succeeding tests was then computed relative to responses on this first test. The VNX group declined while the Sham surgery group did not.

testing in male hamsters following VNO deafferentiation (Marques et al., 1982 and data from Winans and Powers, 1977, circulated by authors but not appearing in the publication; see also Powers et al., 1979).

The following hypotheses are suggested to explain our results: (1) Consonant with our prior chemical and behavioral studies, components in guinea pig urine detected by the main olfactory system may provide information on sexual identity but these substances are not sufficient to maintain the attractiveness of urine. Immediately following removal of the VNO, these substances were still quite attractive because they had always been perceived in conjunction with the same or other substances detected by the VNO. Following repeated presentation, when only the main olfactory system was intact, the attractiveness of the substances declined - it appeared to extinguish (see Kubie et al., 1979 for a similar hypothesis). This contrasts sharply with the response of intact animals in which no decline with repeated testing and the passage of time was detected when urine donors were varied (see below). Also supporting this extinction hypothesis (see Fig. 1) was the more substantial decline in response observed between 5.0 and 6.3 months after surgery (a period when many urine tests employing female urine were conducted) compared with the period between 0.8 and 5.0 months following surgery (a period when fewer urine tests were conducted). (2) An alternative explanation posits changes in the CNS (e.g., degeneration due to loss of sensory input) which followed this relatively long time course. Such CNS changes may be independent of subsequent exposure to the urine (see Marques et al., 1982, for discussion of similar hypotheses to explain delayed effects of lateral olfactory tract transections on hamster sexual behavior).

We favor the first hypothesis. The greatest change in response to urine odors occurred between the tests at 5.0 months and 6.3 months and it is hard to understand this time course if CNS degeneration or non-specific neurotransmitter changes following VNO removal were involved.

If the VNO is specialized to respond to non-volatile substances (either through its unique structure and location and/or through having specialized receptors) then prevention of access of these non-volatile substances could mimic removal of the VNO. This line of reasoning leads to the prediction that if contact with a chemical signal is prevented, then responsiveness to this signal may extinguish. However, if contact is permitted, responses should not be depressed following repeated presentations. Our prior data (see Beauchamp et al., 1980) support the second prediction; investigatory responsiveness and headbobbing to female conspecific urine odors does not extinguish following repeated presentations over the course of months and years providing different individuals serve as urine donors. The following experiment tested this hypothesis.

PREVENTION OF LARGE-MOLECULE ACCESS

Two groups of adult male guinea pigs were formed such that the responsiveness to female urine samples (seconds investigating or licking or headbobbing) were approximately the same. All animals had had considerable experience with standard testing procedures prior to the initiation of this study. Each animal was then given one 2-min test with urine from an individual female every 3 or 4 days for the succeeding 12 weeks. For the first 10 weeks, testing consisted of placing the urine either on top of a wire screen where it could be contacted (control subjects; $\underline{n} = 6$) or 0.5 cm below the screen where contact was not possible (experimental subjects; $\underline{n} = 7$; see Beauchamp, 1973). Four individual females served as urine donors, one female per test, for the first 12 tests. Thus each subject encountered the urine of each individual female for 2 min once every two weeks (see Table 2 for design). Four new donors were then used for the next 12 tests. During the last 4 weeks of testing, subjects from both groups were tested twice in the contact condition and finally twice in the no contact condition. During testing, the experimenter recorded a standard series of behaviors (Beauchamp, 1973); time spent within 1.0 cm of the sample is the variable which was analyzed.

As we had previously reported (Beauchamp et al., 1980), preventing contact with the urine produced an immediate decrement in investigatory time (Fig. 3, block 0). The prediction that, following repeated testing ("extinction trials"), there would be a decline among those animals lacking contact with urine, but not among those animals contacting the urine, was also supported (Figs. 3 and 4). The proportional decline (Fig. 4) illustrates this most clearly. Relative to the first block of 4 trials (set arbitrarily at 1.0), the experimental (no contact) animals exhibited a substantial decline in response while the control (contact) animals did not (comparison of average proportions in blocks 2,4,6 and 8, Mann-Whitney $U = 0$; $p < 0.001$; see Fig. 4).

When both groups were subsequently given contact (block 9, Fig. 3) their responses did not differ. However, when both groups were prevented from contacting the urine (block 10, Fig. 3) a substantial group difference emerged ($U = 0$; $p < 0.001$). This difference was likely the result of the experimental treatments.

These data support the view that when chemosensory input to the AOS is prevented (first experiment) or limited (second experiment) investigatory responsiveness to conspecific urine odors declines. Furthermore, in both cases this decline appeared to be gradual which is consistent with an extinction hypothesis. In the second experiment, the effect of repeated exposure to urine when the urine could not be contacted was most evident in the last block of trials when both experimental and control animals were offered female urine 0.5 cm below the screen (Fig. 3, block 10). The response of the control group was similar to the response of the experimental group on the first block of 4 trials (Fig. 3, block 0). In contrast, the response of the experimental group on these last trials was not different from their response on the last 3 blocks of four trials during the extinction period (Fig. 3, blocks 4, 6 and 8).

Table 2. Schedule of Female Urine Donors

Test Block[a]	Test Week	Female Urine Donors[b]	
		Monday	Thursday
0	1	A	B
	2	C	D
2	3	A	B
	4	C	D
4	5	A	B
	6	C	D
6	7	E	F
	8	G	H
8	9	E	F
	10	G	H
9	11	E	F
10	12	G	H

[a]Test Block refers to the time in weeks since the first block of 4 trials. See Figures 3 and 4 for further explanation.
[b]Female urine samples were collected from individual females (female A, female B, etc.). Fresh urine was collected for each trial.

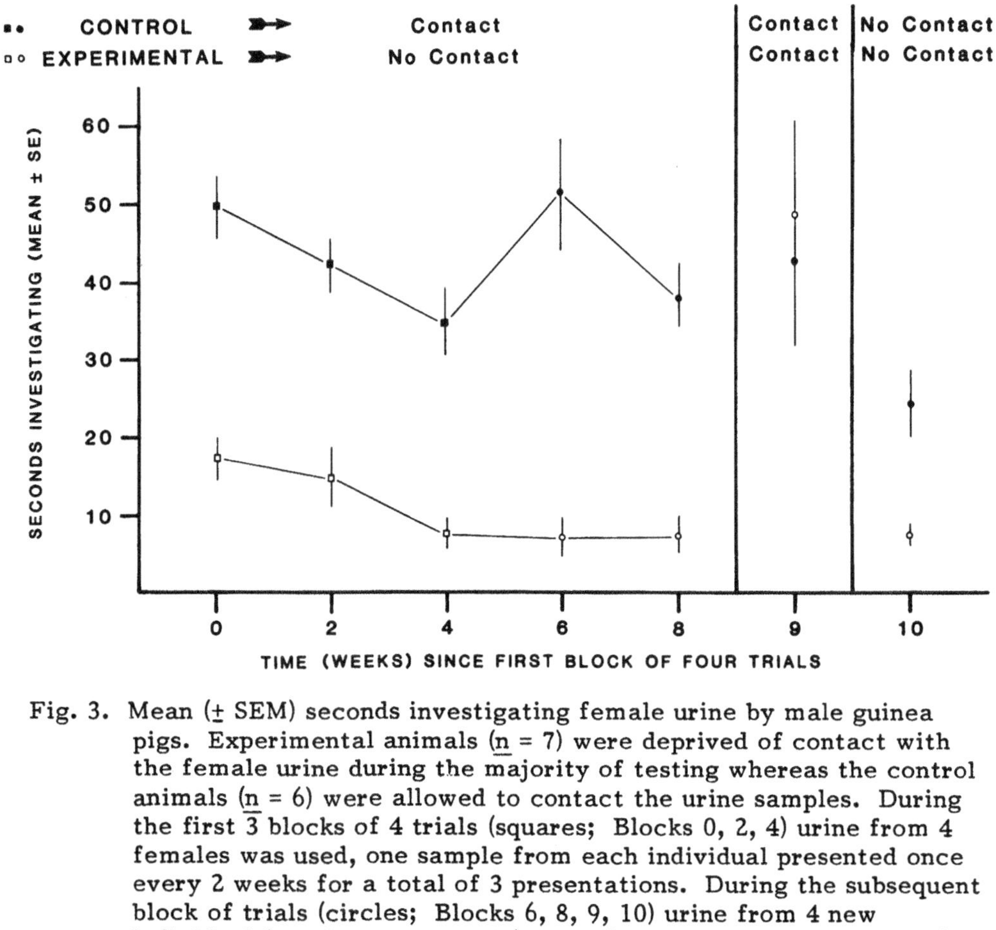

Fig. 3. Mean (± SEM) seconds investigating female urine by male guinea pigs. Experimental animals (n = 7) were deprived of contact with the female urine during the majority of testing whereas the control animals (n = 6) were allowed to contact the urine samples. During the first 3 blocks of 4 trials (squares; Blocks 0, 2, 4) urine from 4 females was used, one sample from each individual presented once every 2 weeks for a total of 3 presentations. During the subsequent block of trials (circles; Blocks 6, 8, 9, 10) urine from 4 new individual females was tested (Table 2). Note that during block 6, when urine from novel females was used, all subjects in the control group exhibited an increased response. During the next block, this response declined in all control subjects. No such changes were evident for the experimental subjects. During block 9 both groups were permitted to contact the urine samples and during block 10 both groups were not permitted contact.

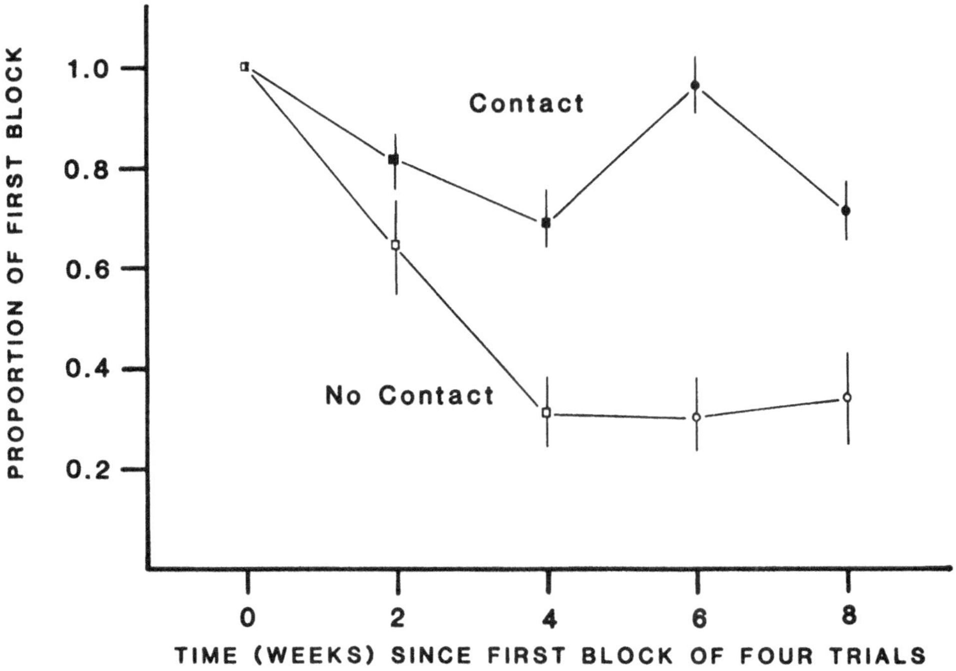

Fig. 4. This figure emphasizes the change in response to female urine samples by animals not permitted contact with the samples. The mean number of seconds investigating female urine on the first block of 4 trials (see Fig. 3) was arbitrarily set at 1.0. The ratio of responses (± SEM) on the succeeding blocks of 4 trials for each animal was computed relative to responses on this first block of 4 trials. The group not permitted to contact the urine (experimental subjects) exhibited a decline while the control group, which was allowed to contact the urine, did not. Squares and circles as per Fig. 3.

One further point is worthy of emphasis. Both experimental and control groups appeared to exhibit a decline in response during the first 3 blocks of 4 trials (blocks 0 to 2 and 2 to 4, Fig. 3). Following the use of a new set of donor females (block 6) every individual in the control group exhibited a increased response; this was not the case for the experimental group. Finally, the last block of 4 trials (8, Fig. 3) found all individuals in the control group declining again relative to the previous 4 trials. These data imply a rather remarkable memory among test males for individual odors. Consider that 0.1 ml of urine from an individual female was presented to each control male for just 2 min once every 2 weeks. Thus, these males were apparently habituating to the odor of individuals they had experienced briefly and many days earlier. These data indicate the extreme sensitivity of these animals to all aspects of conspecific odor and caution one experimentally. The fact that the experimental group exhibited no such alteration in response following use of new female urine donors may also suggest a role for the VNO in the perception of individual odor; further study of this issue is required.

SUMMARY

When male guinea pigs were prevented from contacting female conspecific urine, their interest in investigating these samples declined following repeated testing. This result and the results obtained following removal of the VNO are consistent with the idea that non-volatile molecules, stimulating the VNO, provide sensory input which is rewarding and, further, that the AOS may be a hedonic as well as an analytical sensory system.

ACKNOWLEDGEMENTS

This study was supported by NSF BNS 7906234 to GKB and JLW, NIH 5T32NS07176-02 to IGM, and NIH 5F32NSO6421-02 to CJW. Christine Wojciechowski provided excellent technical assistance.

REFERENCES

Beauchamp, G.K., 1973, Attraction of male guinea pigs to conspecific urine, Physiol. Behav., 10:589.

Beauchamp, G.K., Criss, B.R., and Wellington, J.L., 1979, Chemical communication in *Cavia*: Responses of wild (*C. aperea*), domestic (*C. porcellus*) and F_1 males to urine, Anim. Behav., 27:1066.

Beauchamp, G.K., Martin, I.G., Wysocki, C.J., and Wellington, J.L., Chemoinvestigatory and sexual behavior of male guinea pigs following vomeronasal organ removal, Physiol. Behav., in press.

Beauchamp, G.K., Wellington, J.L., Wysocki, C.J., Brand, J.G., Kubie, J.L., and Smith, A.B., III, 1980, Chemical communication in the guinea pig: Urinary components of low volatility and their access to the vomeronasal organ, in: "Chemical Signals: Vertebrates and Aquatic Invertebrates," D. Müller-Schwarze and R.M. Silverstein, eds., Plenum, New York.

Bellringer, J.F., Pratt, H.P.M., and Keverne, E.B., 1980, Involvement of the vomeronasal organ and prolactin in pheromonal induction of delayed implantation in mice, J. Reprod. Fert., 59:223.

Berüter, J., Beauchamp, G.K., and Muetterties, E.L., 1973, Complexity of chemical communication in mammals: Urinary components mediating sex discrimination by male guinea pigs, Biochem. Biophys. Res. Comm., 53:264.

Fleming, S., Vaccarino, F., Tambosso, L., and Chee, P., 1979, Vomeronasal and olfactory system modulation of maternal behavior in the rat, Science, 203:372.

Johnston, R.E., and Schmidt, T., 1979, Responses of hamsters to scent marks of different ages, Behav. Neural Biol., 26:64.

Kaneko, N., Debski, E.A., Wilson, M.C., and Whitten, W.K., 1980, Puberty acceleration in mice. II. Evidence that the vomeronasal organ is a receptor for the primer pheromone in male mouse urine, Biol. Reprod., 22:873.

Kubie, J.L., and Halpern, M., 1979, Chemical senses involved in garter snake prey trailing, J. Comp. Physiol. Psych., 93:648.

Ladewig, J., and Hart, B.L., 1980, Flehmen and vomeronasal organ function in male goats, Physiol. Behav., 24:1067.

Marques, D.M., 1979, Roles of the main olfactory and vomeronasal systems in the response of the female hamster to young, Behav. Neural Biol., 26:311.

Marques, D.M., O'Connell, R.J., Benimoff, N., and Macrides, F., 1982, Delayed deficits in behavior after transection of the olfactory tracts in hamsters, Physiol. Behav., 28:353.

Matthes, E., 1932, Weitere Geruchsdressuren an Meerschweinchen, Z. vgl. Physiol., 17:464.

Meredith, M., Marques, D.M., O'Connell, R.J., and Stern, F.L., 1980, Vomeronasal pump: Significance for hamster sexual behavior, Science, 207:1224.

Murphy, M.R., 1980, Sexual preference of male hamsters: Importance of preweaning and adult experience, vaginal secretion, and olfactory or vomeronasal sensation, Behav. Neural Biol., 30:323.

Nyby, J., and Zakeski, D., 1980, Elicitation of male mouse ultrasounds: Bladder urine and aged urine from females. Physiol. Behav. 24:737.

Planel, H., 1953, Etudes sur la physiologie de l'organe de Jacobson, Arch. Anat. Histol. Embryol., 36:197.

Powers, J.B., Fields, R.B., and Winans, S.S., 1979, Olfactory and vomeronasal system participation in male hamsters' attraction to female vaginal secretions, Physiol. Behav. 22:77.

Powers, J.B. and Winans, S.S., 1975, Vomeronasal organ: Critical role in mediating sexual behavior of the male hamster, Science, 187:961.

Reynolds, J., and Keverne, E.B., 1979, The accessory olfactory system and its role in the phermonally mediated suppression of oestrus in grouped mice, J. Reprod. Fertil., 57:31.

Schilling, A., 1980, The possible role of urine in territoriality of some nocturnal prosimians, in: Symp. Zool. Soc. Lond., No. 45: "Olfaction in Mammals," D.M. Stoddart, ed., Academic Press, London.

Verberne, G., 1976, Chemocommunication among domestic cats, mediated by the olfactory and vomeronasal senses. II. The relation between the function of Jacobson's organ (vomeronasal organ) and flehmen behavior, Z. Tierpsychol., 42:113.

Wellington, J.L., Beauchamp, G.K., and Smith, A.B., III, 1981, Stability of chemical communicants of gender in guinea pig urine, Behav. Neural Biol., 32:364.

Winans, S.S., and Powers, J.B., 1977, Olfactory and vomeronasal deafferentation of male hamsters: Histological and behavioral analyses, Brain Res., 126:325.

Wysocki, C.J., 1979, Neurobehavioral evidence for the involvement of the vomeronasal system in mammalian reproduction. Neurosci. Biobehav. Rev. 3:301.

Wysocki, C.J., and Katz, Y., The male vomeronasal organ mediates female-induced testosterone surges, submitted.

Wysocki, C.J., Wellington, J.L., and Beauchamp, G.K., 1980, Access of urinary nonvolatiles to the mammalian vomeronasal organ, Science, 207:781.

Wysocki, C.J., Nyby, J., Whitney, G., Beauchamp, G.K., and Katz, Y., The vomeronasal organ: Primary role in mouse chemosensory gender recognition, Physiol. Behav., in press.

FLEHMEN BEHAVIOR AND VOMERONASAL ORGAN FUNCTION

Benjamin L. Hart

Department of Physiological Sciences
School of Veterinary Medicine
University of California
Davis, California

The flehmen or lip-curl behavior displayed by males of most ungulate and felid species is a prominent, but poorly understood, aspect of sociosexual behavior. The behavior is sexually dimorphic in that it is seen much more frequently in males than females during sexual encounters. Also it appears to be under gonadal androgen control since it occurs most frequently during the rutting season and castration reduces the occurrence of flehmen. Males typically display the behavior following olfactory investigation of the anogenital area or freshly voided urine of females. Most theories, and the experimental information available regarding flehmen, have related the behavior to the role of the vomeronasal organ (VNO) in chemosensory detection of estrus.

A point worth emphasizing is that the performance of flehmen may well be a behavioral marker as to when the VNO is being utilized. In a sense, the performance of flehmen might be considered analogous to the behavior of sniffing in the use of the main olfactory system. However, we know that animals may detect chemicals by olfaction without obvious sniffing; the behavior of sniffing is reserved for the most intense olfactory chemoreception. The same may be true of VNO chemoreception; animals could use the VNO without performing flehmen, but flehmen would be performed during the most intense use of the VNO. The other possibility is that the VNO is used only in association with flehmen. In either case more information about the biological aspects of flehmen behavior will contribute to a more complete understanding of VNO function in general.

In this paper I will discuss several anatomical and physiological considerations about the flehmen response. I will then go on to examine what has been learned about the hormonal control of the behavior and finally comment upon the function of flehmen, and its presumed relationship to VNO chemoreception, in sexual behavior.

ANATOMICAL AND PHYSIOLOGICAL CONSIDERATIONS OF FLEHMEN

Flehmen is the typical response of male ungulates after they have sniffed and usually made oral contact with female urine or vaginal secretions. The male elevates his head or at least holds it still while he curls or lifts the upper lip. Some observers have noticed characteristic tongue movements against the hard palate before, during, or after the lip curl grimace. The male

Fig. 1. Characteristic postures and facial grimaces of the flehmen responses in male goats (shown after close investigation of urine), sheep, horses, cats and cattle.

remains motionless during flehmen for as short as a few seconds to as long as minute. As suggested in Figure 1 the behavior is highly stereotyped and can be recognized not only in a variety of ungulates and felids, but in swine (Martys, 1977) certain bats, viverrids, hyenas (Estes, 1972) and marsupials (Gaughwin, 1979). Males may be so engrossed in performing flehmen during a sexual encounter with a receptive female that copulatory attempts may be delayed. Urine and vaginal discharge appear to be the strongest stimuli to evoke flehmen, but other odors, especially unfamiliar ones, can elicit the behavior.

Most of the discussion in the literature on flehmen behavior implies that physical changes in the oral and nasal cavities, brought about by the head elevation and the lip curl, promote the transfer of urine or vaginal secretions from the oral cavity to the VNO. The VNO (also known as Jacobson's organ) is the peripheral sensory organ of the accessory olfactory system and consists of two blind, mucus filled, tubes lying on either side of the base of the nasal septum. In animals examined, including sheep (Kratzing, 1971) and goats (Ladewig and Hart, 1980), the medial wall of the epithelium lining the posterior two-thirds of the organ contains olfactory receptor cells that resemble those of the main olfactory system. The epithelium of the rostral one third resembles non-sensory epithelium of the nasal cavity. In most species which display flehmen, access of materials to the VNO can be either through the nasal or oral cavity. In goats and other ruminants the entrance to the VNO is near the nasal opening of the nasopalatine duct (Figure 2).

One of the intriguing aspects of the study of flehmen, and its relationship to the VNO, is that chemosensory information from the VNO projects to different neural areas of the brain than the main olfactory system. The main olfactory system projects via the main olfactory bulbs to the olfactory tubercles, prepiriform cortex, enthorhinal area and anterior part of the corticomedial amygdala (Devor, 1976; Scalia and Winans, 1975). Vomeronasal nerves, which arise from receptor cells in the VNO epithelium terminate in the accessory olfactory bulbs which then project to the medial nuclei and posteriomedial part of the cortical nuclei of the amygdala. Accessory olfactory tract axons also enter the stria terminalis terminating in the bed nucleus of the stria terminalis (Scalia and Winans, 1975). The projection of the VNO is to central brain areas closely identified with the mediation of male sexual behavior. Lesions of the medial amygdaloid nucleus, stria terminalis or bed nucleus of the stria, for example, disrupt male sexual behavior in rats (Emery and Sachs, 1976; Harris and Sachs, 1975). Furthermore, lesions of the medial preoptic - anterior hypothalamic area, to which the amygdala and bed nuclei project, disrupt male copulatory responses in all species tested to date (Giantonio et al., 1970; Heimer and Larsson, 1966; Hart et al., 1973; Hart, 1974; Slimp et al., 1978).

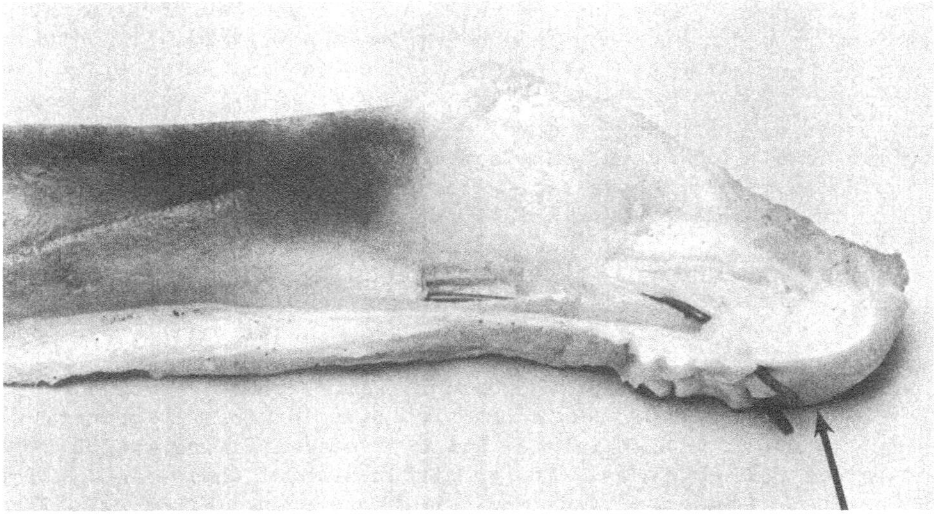

Fig. 2. Dissection of the nasal septum and hard palate of a male goat to demonstrate the relationship between the nasopalatine duct, the ventral part of the nasal cavity and the vomeronasal organ. The arrow points to the dental pad of the hard palate (ruminants have no upper incisors). A dark plastic tubing has been inserted through the nasal palatine duct on each side, into the nasal cavity, and through the slit-like opening of the vomeronasal organ. A window has been cut into the medial side of the nasal septum to show the plastic tube lying in the lumen of the vomeronasal organ.

Although flehmen is not displayed by rodents, canids and most primates, the VNO is found throughout the vertebrate class. Humans, old world primates, and cetaceans reportedly do not have a VNO although the anatomical evidence for this is inconsistent (Wysocki, 1979). As shown by many studies, this organ is used in snakes extensively in chemoreception related to prey capture, aggregation, and reproduction (Burghardt, 1980). The preponderance of neurobehavioral evidence from mammals points to a more restricted involvement of the VNO in reproductive fitness (Wysocki, 1979). Recent experiments in rodents strongly support the notion that the function of the VNO (regardless of whether the animal performs flehmen or not) is for chemosensory analysis of sexually significant materials that are non-volatile or of low volatility. Male hamsters, for example, with intact VNOs but impaired main olfactory systems, are able to detect female hamster vaginal discharges, but only when the material is nuzzled (Powers et al., 1979). Male guinea pigs are able to discriminate between male and female urine, but only if direct contact with the urine is possible (Beauchamp et al., 1980).

In further work on guinea pigs Wysocki et al. (1980), used the fluorescent dye, rhodamine, as a marker substance in urine, and found that the dye reached the VNO only when the subjects were allowed to contact the urine. They proposed that non-volatile materials, like the dye, reached the VNO through the external nares.

Hypotheses about the mechanical effects produced by flehmen range from opening of passages or ducts, to aerodynamic changes and venturi effects. Knappe (1964) suggested that curling of the upper lip during flehmen enlarged the nasal opening of the nasopalatine duct and possibly the entrance to the VNO. He proposed that simultaneous elevation of the head and shrinkage of the erectile tissue surrounding the lumen of the VNO facilitated the influx of materials into the VNO. This theory was later questioned by Dagg and Taub (1970) who dissected the muscles involved in the lip curling. They suggested that the lip curling restricts the external nares so that air is trapped in the nasal cavities for a more thorough investigation by the main olfactory epithelium.

In a paper that drew considerable attention to the flehmen response, Estes (1972) focused on a different mechanism and suggested that in ungulates flehmen facilitated the introduction of materials to the VNO by constricting the external nares so that, with inspiration, air would be pulled through the nasopalatine duct causing mucus to be drawn out of the VNO through a venturi effect. The resulting vacuum in the VNO would cause air to flow into the VNO.

Observations in my laboratory have brought into question the notion that air movement is affected in a major way since we have found the external nares are not occluded during flehmen, and the respiratory rate and pattern are unchanged during flehmen whether the behavior is performed with the animal at rest or after exercise (Ladewig and Hart, 1980). At the present time it is impossible to ascribe any particular role to the facial grimace or the head elevation during flehmen.

In a recent study of tongue movments during flehmen in cattle, Jacobs et al. (1980) noted that bulls stroke their tongues against the hard palate near the openings of the nasopalatine ducts prior to flehmen. The tongue movements would be quite effective in forcing fluids from the rostral part of the oral cavity along grooves in the hard palate and into the nasopalatine ducts (Jacobs et al., 1981).

One mechanism which appears to be clearly involved in VNO function is a pumping mechanism which pulls material into the VNO. Meredith and O'Connell (1979) demonstrated that in hamsters activation of sympathetic nerves causes a contraction of blood vessels in the cavernous tissue surrounding the lumen of the VNO. Since this tissue is anchored to a rigid cartilage capsule, the lumen is dilated and material sucked into the organ. Parasympathetically induced vasodilation presumably results in compression of the lumen and emptying of the VNO. Interruption of the efferent nerves, carried in the nasopalatine nerves, produces deficits in mating in hamsters similar to those produced by cutting vomeronasal nerves (Meredith et al., 1980). The vascular characteristics of the ruminant VNO appear similar to those of the hamster (Ladewig and Hart, 1980), and it is therefore likely that such a pumping mechanism functions simultaneously with the flehmen response.

The evidence pointing to the involvement of flehmen behavior in VNO function is largely indirect but it comes from several lines of observation. First, is the behavioral context in which the behavior is usually displayed. The animal has almost always performed an olfactory investigation immediately prior to the flehmen. VNO chemoreception of fluid borne material is a logical follow-up of olfactory chemoreception, and the neural projections of the accessory olfactory system reinforce the concept that flehmen is performed most frequently in males during encounters with females or their recently voided urine. Flehmen involves lip and tongue movements which would logically affect fluid movement through structures in the oral and nasal cavities.

Finally there is circumstantial evidence to incriminate the flehmen response in VNO function. Estes (1972; personal communication) in his studies of reproductive behavior in East African ungulates has noted that neither the topi nor the hartebeest perform flehmen. Upon dissecting the heads of two topi and one hartebeest Estes found no VNOs.

As a direct approach to determining if there is a movement of fluid from the oral cavity to the VNO during flehmen in goats Ladewig and I (1980) designed an experiment to follow the fluorescent dye, sodium fluorescein, from the oral cavity to the VNO during flehmen. In six male goats flehmen was evoked immediately after the dye was placed in the oral cavity. In the remaining seven subjects (control group) flehmen was not evoked, but the animals were left undisturbed for a period of time corresponding to the flehmen duration in the experimental subjects. Immediately afterwards, each subject was given a lethal dosage of an anesthetic, and when the animals were unconscious, the maxilla was removed by a transverse cut 2 cm rostral to the orbit. The isolated maxilla was placed in liquid nitrogen. Upon freezing the maxilla was cut transversely into sections about 1 cm thick. The sections were examined under ultraviolet light for the presence of the fluorescein dye.

In all 13 subjects, fluorescence was found on the hard palate and extending into the nasopalatine ducts. In four of the experi-

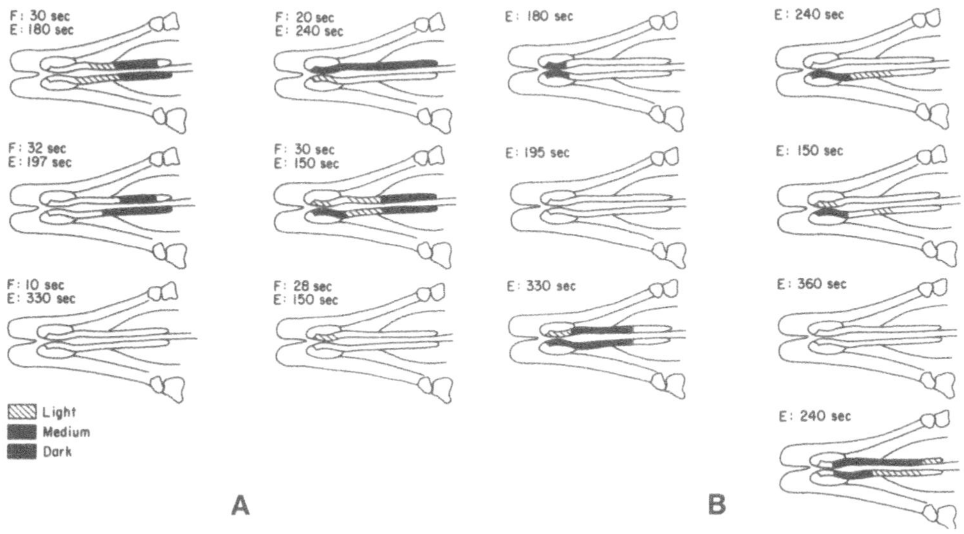

Fig. 3. Diagrammatic illustrations of horizontal sections through the vomeronasal organ of male goats showing extent of fluorescein dye accumulation when the dye was placed in the oral cavity. Subjects depicted in A performed flehmen (F, duration; E, elapsed time from dye application to freezing of maxilla to preserve dye) and control subjects, depicted in B, did not perform flehmen. From Ladewig and Hart, 1980.

mental subjects fluorescein dye was found in the VNOs; the concentration was heaviest in the caudal part of the organs. In the control subjects, no dye was found in the VNOs of two animals, and in the remaining five subjects dye was found in the anterior part of the organ (Figure 3). In the same study we had found that in the VNO the sensory epithelium was present only in the posterior two thirds of the medial wall of the vomeronasal lumen. Thus, if the fluorescein dye represented some biologically significant material, this material would only have reached the sensory epithelium of the VNOs in animals that performed the flehmen behavior.

HORMONAL CONTROL OF FLEHMEN

Since flehmen is an integral part of precopulatory sexual behavior it is not surprising to find that the behavior is both sexually dimorphic and is influenced by changes in blood concentration of androgen. Observations from both the laboratory and field reveal that the behavior is displayed much more frequently by males than females. In studies in my laboratory of sexual encounters in goats we have never seen females display flehmen in response to male urine or body secretions. In groups of females we have seen animals perform flehmen after close investigation of the urine or genital secretions of other females. In fact females can be often induced to perform flehmen by dipping their lips in fresh urine collected from other females (unpublished observation). In field studies on goats Coblentz (1974) frequently observed flehmen performed by males but never by females. In calves Reinhardt et al. (1978) frequently observed flehmen being performed by males but the behavior occurred rarely in females.

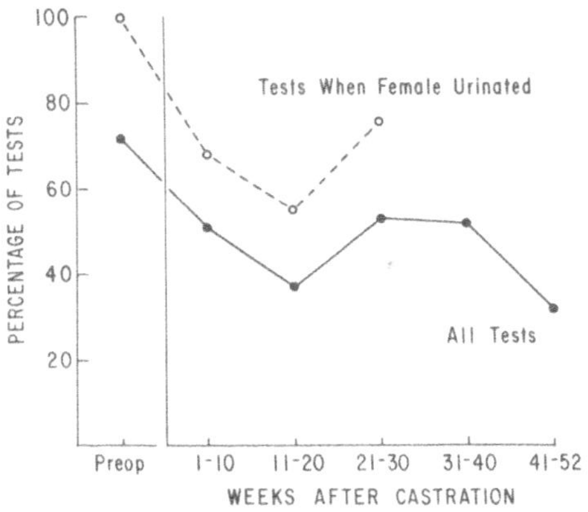

Fig. 4. Effects of castration on frequency of flehmen behavior during mating tests in tropical male goats. Shown are the mean percentages of tests in which at least one flehmen response occurred for all tests (solid line) and for only those tests when the female urinated (broken line). From Hart and Jones, 1975.

The control of flehmen by androgen is evident in several studies. In a study of the effect of castration on sexual behavior of tropical male goats we found that the probability of flehmen was markedly decreased by castration (Hart and Jones, 1975). Figure 4 depicts the effect.

Changes in season, for species in which there are seasonal fluctuations in testosterone, are reported to result in changes in the frequency with which flehmen may be displayed. It is reported that flehmen in black-tailed deer is more frequent during the mating season than at any other time (Müller-Schwarze, 1979).

A clear indication of hormonal influences on the probability of flehmen is evident in a study by Lincoln and Davidson (1977) in

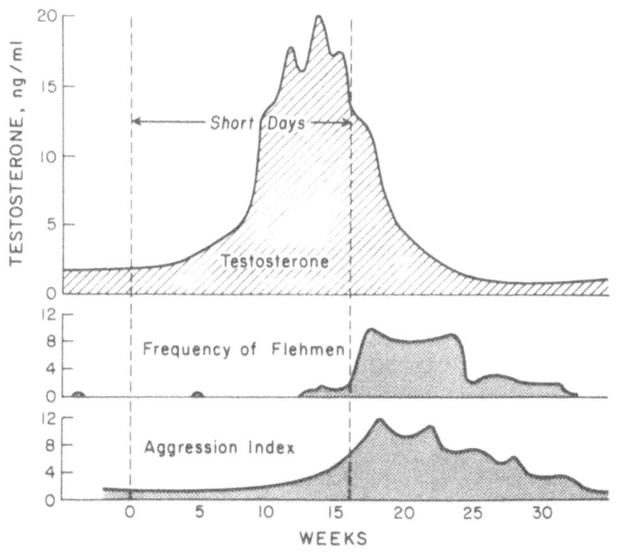

Fig. 5. Illustration of hormonal influence on flehmen behavior in Soay rams housed under artificial lighting and switched from long days to short days. Elevation of blood plasma testosterone preceded the increase in flehmen. Modified from Lincoln and Davidson, 1977.

which males of the ancient breed of Soay sheep were subjected to artificially short days after living under a long-day environment. The change in day length brought on a resurgence of testosterone production followed by increased frequency of flehmen behavior along with measures of increases in aggressive behavior (Figure 5).

Flehmen appears to be controlled by the same hormonal influences as sexual behavior in general. In sheep, as in rodents and many other mammals, estrogen (an aromatized metabolite of testosterone that is believed to stimulate neuronal target cells in the brain) will stimulate male sexual behavior. One report documents that estrogen will also stimulate flehmen behavior in castrated male sheep (Parrott, 1978).

Flehmen behavior is also one of those sex-typical behavioral patterns that is under the developmental influences of perinatal androgen secretion. For example, female sheep that are artificially androgenized by prenatal injections of testosterone into their dams, display an increase in both male sexual behavior and flehmen behavior (Clarke, 1977). Naturally androgenized female cattle, the so-called freemartins, also have a greater tendency to show flehmen behavior than normal females when injected with testosterone in adulthood (Greene et al., 1978).

The hormonal control that is apparent over flehmen behavior, and presumably VNO chemoreception in ungulates, is an interesting contrast to the lack of obvious hormonal control over olfactory chemoreception. Hormonal influences have not been noted in rodents for behaviors such as nuzzling or head bobbing that appear to be related to VNO use. The hormonal control in ungulates discussed here may be mediated at the peripheral level (VNO), or in the brain. Both levels may be affected. At this point no investigator has pointed out any changes in the VNO that result from altering blood hormone concentrations. In fact, no one has reported a sexually dimorphic difference in the gross structure, cytoarchitechure or histochemistry of the VNO, although such differences could exist in some species.

ROLE OF FLEHMEN IN DETECTION OF ESTRUS

Certainly not all the evidence is in, but at this juncture both field studies and laboratory observations strongly support the notion that the function of flehmen behavior is to facilitate the transport of nonvolatile chemosensory materials from the oral cavity to the VNO. Flehmen is much more likely to occur if the mouth or nose has been in direct contact with the stimulus fluid. Coblentz (1974) found in his observation of feral goats on Catalina Island that it is common for a male to stick his muzzle into a stream of female urine just prior to performing flehmen. I have seen the same behavior several times in tropical and temperate-zone male goats in captivity. Müller-Schwarze (1979) found in black-tailed deer that flehmen was six times more likely to occur after the animal contacted the urine than after merely sniffing it. In experimental studies where male goats were presented with urine samples from females, we found that 90 percent of investigations involving oral contact with the urine were followed by flehmen whereas only 17 percent of investigations involving sniffing were followed by flehmen (Ladewig et al., 1980).

It is clear that the main trigger which evokes flehmen is olfactory stimulation with a sexually significant odorant. Placing urine into the oral cavity of goats does not usually evoke flehmen but placing a bit in their nose or misting them in the nose with urine spray will evoke flehmen (Ladewig et al., 1980; unpublished observations). Olfactory bulbectomy virtually eliminates the display of flehmen behavior in goats (Ladewig et al. 1980).

We have also found that surgically blocking the nasopalatine ducts has no effect on the occurrence of flehmen in any test situation (Ladewig et al., 1980). The persistence of the behavior may represent the continuation of a species-typical behavior in the absence of sensory input, just as olfactory bulbectomized animals continue to sniff and perform close investigation. It is also possible that enough material enters the VNO from the nasal cavity to allow for VNO function.

At this point it should be mentioned that in some ungulates, most notably equids, males perform a very distinct flehmen, but there is no connection between the oral cavity and the opening of the VNO in the nasal cavity through a nasopalatine duct. Thus equids represent an exception for everything that has been said about the role of flehmen in transporting materials from the oral cavity to the VNO. Wallach (personal communication) has observed flehmen behavior at close range in stallions that were allowed to investigate urine-soaked tampons. She noted the stallions made nasal contact with the tampons and that there was a copious flow of mucus from the nasal cavity prior to and during olfactory investigation which preceded the flehmen response. She advances the notion that sexually significant chemicals may dissolve in the mucous secretions during close olfactory investigation and that some of the secretions are then transported to the VNO during flehmen.

The possible role of flehmen in the detection of the reproductive state of females is somewhat elusive. For one thing it appears that goats and other ungulates can detect females in estrus or urine from females in estrus by means of regular olfaction alone. In an experiment we conducted, male goats were exposed, in a two-choice test, to urine from female goats in diestrus and female goats in which estrus had been induced by estrogen injections. In almost all tests the subjects approached and sniffed each urine sample.

Interestingly,the diestrous urine elicited both sniffing and close (contact) investigation significantly more frequently than the estrous urine. The higher frequency of olfactory investigation of diestrous urine was due to the subjects going back for a second sniff. Flehmen was elicited significantly more frequently, and for a longer duration, by diestrous urine. In fact, the diestrous urine evoked four times the frequency of flehmen as estrous urine (Figure 6). This experiment revealed that the olfactory discrimination between estrous and diestrous urine had been made prior to flehmen (Ladewig et al., 1980).

The phenomenon of estrous urine being less potent in evoking flehmen than diestrous urine is not unique to our experiment. Müller-Schwarze (1979) found in observations of black-tailed deer that the display of flehmen to urine from females dropped off markedly on the day of estrus. In other tests where estrous and diestrous urine were presented in successive daily trials, the deer

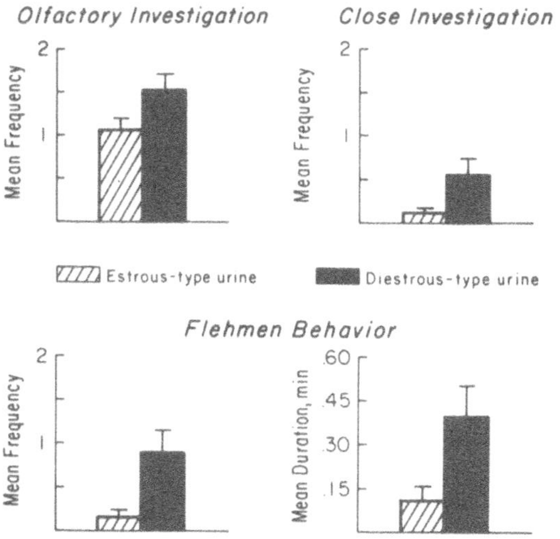

Fig. 6. Olfactory and flehmen responses of male goats to urine from female goats in diestrus or artificially induced estrus. Samples were presented simultaneously in a two-choice test. Frequency is represented as mean (+ SEM) number of responses per test. From Ladewig et al., 1980.

displayed flehmen about equally to both types of urine (Müller-Schwarze, 1979). In wild mountain sheep, Geist (1971) noted that flehmen was evoked more frequently by anestrous ewes than estrous ewes courted by rams. In observations of flehmen behavior in domestic rams (along with other aspects of sexual activity) it was found that flehmen was evoked more frequently by contact with control females not in estrus than contact with females in estrus (Signoret, 1975).

Is the performance of flehmen, and presumed use of the VNO, necessary for a male ungulate to detect estrus? Clearly not. Females in estrus display behavioral signs of their receptive state by visual signals and sexual approaches towards males or at least standing still when a male approaches or mounts. Our two-choice female urine presentation experiment revealed that males discriminated between estrous and diestrous urine by regular olfaction because the flehmen responses were differentially directed (Ladewig et al., 1980). Whatever chemosensory information is gained through flehmen is presumably a refinement or elaboration of olfaction. A logical function of this refined chemosensory analysis is the detection of small quantities of sex attractants during proestrus -- a task that may be too difficult for regular olfaction. Or, as Estes (1972) puts it, "the overriding biological significance of the accessory olfactory organ may be the provision of a relatively foolproof means of ascertaining reproductive status." Estes stresses the obvious importance of the ability of ungulates to detect proestrus since males can then locate and tend or guard such females before they are in standing estrus. In wild ungulates it is common for males to associate with females a few days before they are in estrus. This has been documented, for example, in bison (Lott, 1981) and wild sheep (Geist, 1971).

One could argue, as has Gaughwin (1979), that in some instances, where copulation takes only a second or so, that there has been selection against flehmen at the time a female is in standing estrus. If the male, presumably a dominant one, was guarding a female and distracted himself by performing flehmen, the males nearby that had been attracted to the female might take advantage of his preoccupation during flehmen.

Although it is true that males are unlikely to perform flehmen during initial encounters with estrous females or their urine, they may perform flehmen following copulation. In our study of flehmen in goats (Ladewig et al., 1980), we found that in sexual encounters with estrous females, only about 20 percent of the copulatory responses were preceded by flehmen. Flehmen performed after a copulation was common; 70 percent of all flehmen responses made around estrous females occurred after copulation. Our observation of the occurrence of post-copulatory flehmen is not an isolated finding. Coblentz (1974) noted this in his study of a feral goat population.

The function of post-copulatory flehmen may again reflect a type of refined VNO chemosensory analysis that is beyond the capability of regular olfaction. Through use of the VNO, a male may be able to detect chemical cues as to whether a female is going to be in estrus for several more days (or hours, depending upon species) or if she is near the end of estrus. His behavior, in terms of pursuit of other females or continued guarding of the same female, may be guided accordingly.

CONCLUSION

Flehmen behavior has undoubtedly been noticed by animal handlers since the first days of keeping ungulates in captivity. It is a stereotyped aspect of reproductive behavior that is remarkably similar across species. Assuming the behavior is closely related to VNO function, then it is interesting that for decades it has also been known that the VNO architecture is remarkably alike in species as diverse as goats, cats, hedgehogs and lemurs (Broom, 1915). This point has been made more recently by Estes (1972). Schneider (1930; 1931; 1932a; 1932b) was the first to systemically deal with the comparative aspects of flehmen behavior and to call attention to the similarity of flehmen behavior in various species and the fact that the behavior had the same meaning most of the time. Schneider fully recognized the importance of flehmen in finding a sexual partner and directing the male's attitude towards its partner. Much remains today to be learned about the exact role of flehmen behavior in sexual interactions. In addition to this, we know that the behavior does occur at times other than in sexual encounters. The behavior occurs among animals in all-male groups particularly when an individual investigates the urine voided by another male or himself. In females, flehmen has been observed when they investigate the urine of another female or the birth fluids of a newborn. Thus, if the performance of flehmen is associated with introducing materials into the VNO, then it is undoubtedly true that the VNO sensory epithelium is used to obtain chemosensory information other than that strictly associated with sex attractants.

ACKNOWLEDGEMENT

The preparation of this paper was supported in part by NSF Grant BNS 8103574.

REFERENCES

Beauchamp, G. K., Wellington, J. L., Wysocki, C. J., Brand, J. G., Kubie, J. L., and Smith, A. B. III., 1980, Chemical communication in the guinea pig: urinary components of low volatility and their access to the vomeronasal organ. pp. 327-339. In Müller-Schwarze, D. and Silverstein, R. M. (Eds.) "Chemical Signals Vertebrates and Aquatic Invertebrates," New York, Plenum Press.

Broom, R., 1915, On the organ of Jacobson and its relations in the "Insectivora:" Part I. Tupaia and Gymnura. Proc. Zool. Soc. Lond. 14:157-162.

Burghardt, G. M., 1980, Behavioral and stimulus correlates of vomeronasal functioning in reptiles: feeding, grouping, sex and tongue use. pp. 275-301. In Müller-Schwarze, D. and Silverstein, R. M. (Eds.), "Chemical Signals in Vertebrates and Aquatic Invertebrates," New York, Plenum Press.

Clarke, I. J., 1977, The sexual behaviour of prenatally androgenized ewes observed in the field. J. Reprod. Fert., 49:311-315.

Coblentz, B. E., 1974, Ecology, behavior and range relationships of the feral goat. Unpublished Ph.D. Thesis, University of Michigan, Ann Arbor. 259 pp.

Dagg, A. I., and Taub, A., 1970, Flehmen, Mammalia 34:686-695.

Devor, M., 1976, Fiber trajectories of olfactory bulb efferents in the hamster, J. Comp. Neurol. 166:31-48.

Emery, D. E. and Sachs, B. D., 1976, Copulatory behavior in male rats with lesions in the bed nucleus of the stria terminalis, Physiol. Behav. 17:803-806.

Estes, R. D., 1972, The role of the vomeronasal organ in mammalian reproduction, Mammalia, 36:315-341.

Gaughwin, M. D., 1979, The occurrence of flehmen in a marsupial--the hairy-nosed wombat (Lasiorhinus latifrons), Anim. Behav. 27:1063-1065.

Geist, V., 1971, "Mountain Sheep: A study in Behavior and Evolution", University of Chicago Press, Chicago.

Giantonio, G. W., Lund, N. L., and Gerall, A. A., 1970, Effects of diencephalic and rhinencephalic lesions on the male rat's sexual behavior, J. Comp. Physiol. Psychol. 73:38-46.

Greene, W. A., Mogil, L., and Foote, R. H., 1978, Behavioral characteristics of freemartins administered estradiol, estrone, testosterone and dihydrotestosterone, Horm. Behav. 10:71-84.

Harris, V. S., and Sachs, B. D., 1975, Copulatory behavior in male rats following amygdaloid lesions, Brain. Res. 86:514-518.

Hart, B. L., and Jones, T. O. A. C., 1975, Effects of castration on sexual behavior of tropical male goats, Horm. Behav. 6:247-258.

Hart, B. L., Haugen, C. M., and Peterson, D. M., 1973, Effects of medial preoptic-anterior hypothalamic lesions on mating behavior of male cats, Brain Res. 54:177-191.

Hart, B. L., 1974, Medial preoptic-anterior hypothalamic area and sociosexual behavior of male dogs: A comparative neurophysiological analysis. J. Comp. Physiol. Psychol. 86:328-349.

Heimer, L., and Larsson, K., 1966, Impairment of mating behavior in male rats following lesions in the preoptic-anterior hypothalamic continuum Brain Res. 3:248-263.

Jacobs, V. L., Sis, R. F., Chenoweth, P. J., Klemm, W. R., Sherry, C. J., and Coppock, C. E., 1980, Tongue manipulation of the palate assists estrous detection in the bovine, Theriogenology 13:353-356.

Jacobs, V. L., Sis, R. F., Chenoweth, P. J., Klemm, W. R., and Sherry, C. J., 1981, Structures of the bovine vomeronasal complex and its relationship to the palate: Tongue manipulation Acta. Anat. 110:48-58.
Knappe, H., 1964, Zur Funktion des Jacobsonschen Organs (organon vomeronasale Jacobsoni). Der Zoologische Garten, 28:188-194.
Kratzing, J., 1971, The structure of the vomeronasal organ in the sheep, J. Anat., 108:247-260.
Ladewig, J. and Hart, B. L., 1980, Flehmen and vomeronasal organ function in male goats, Physiol. Behav. 24:1067-1071.
Ladewig, J., Price, E. O., and Hart, B. L., 1980, Flehmen in male goats: role in sexual behavior, Behav. Neural. Biol. 30:312-322.
Lincoln, G. A., and Davidson, W., 1977, The relationship between sexual and aggressive behaviour, and pituitary and testicular activity during the seasonal sexual cycle of rams, and the influence of photoperiod, J. Reprod. Fert., 49:267-276.
Lott, D. F., 1981, Sexual behavior and intersexual strategies in American Bison, Z. Tierpsychol. 56:97-114.
Martys, M., 1977, Das Flehmen der Schweine, Suidae. The flehmen behavior of pigs, Zool. Anz. 199:433-440.
Meredith, M., and O'Connell, R. J., 1979, Efferent control of stimulus access to the hamster vomeronasal organ, J. Physiol. 286:301-316.
Meredith, M., Marques, D. M., O'Connell, R. J., and Stern, F. L., 1980, Vomeronasal pump: significance for male hamster sexual behavior, Science 207:1224-1226.
Müller-Schwarze, D., 1979, Flehmen in the context of mammalian urine communication. pp. 85-96. In Ritter, F. J. (Ed.), "Chemical Ecology: Odour Communication in Animals," Amsterdam, Elsevier.
Parrott, R. F., 1978, Courtship and copulation in prepubertally castrated male sheep (wethers) treated with 17β-estradiol, aromatizable androgens and dihydrotestosterone, Horm. Behav., 11:20-27.
Powers, J. B., Fields, R. B. and Winans, S. S., 1979, Olfactory and vomeronasal system participation in the male hamsters' attraction to female vaginal secretions, Physiol. Behav. 22:77-84.
Reinhardt, V., Mutiso, F. M., and Reinhardt, A., 1978, Social behaviour and social relationships between female and male prepubertal bovine calves (Bos indicus), Appl. Anim. Ethol. 4:43-54.
Scalia, F. and Winans, S. S., 1975, The differential projections of the olfactory bulb and accessory bulb in mammals. J. Comp. Neurol. 161:31-56.
Schneider, K. M., 1930, Das Flehmen. I. Zool. Garten 3:183-198.
Schneider, K. M., 1931, Das Flehmen II. Zool. Garten 4:349-364.
Schneider, K. M., 1932a, Das Flehmen III. Zool. Garten 5:200-226.
Schneider, K. M., 1932b, Das Flehmen Iv. Zool. Garten 5:287-297.
Signoret, J. P., 1975, Influence of the sexual receptivity of a teaser ewe on the mating preference in the ram, Appl Anim. Ethol. 1:229-232.

Slimp, J. C., Hart, B. L., and Goy, R. W., 1978, Heterosexual, autosexual, and social behavior of adult male rhesus monkeys with medial preoptic-anterior hypothalamic lesions, Brain Res. 142:105-122.

Wysocki, C. J., 1979, Neurobehavioral evidence for the involvement of the vomeronasal system in mammalian reproduction. Neuroscience Biobehav, Rev. 3:301-341.

Wysocki, C. J., Wellington, J. L., and Beauchamp, G. K., 1980, Access of urinary nonvolatiles to the mammalian vomeronasal organ, Science 207;781-783.

OLFACTION IN CENTRAL NEURAL AND NEUROENDOCRINE SYSTEMS: INTEGRATIVE REVIEW OF OLFACTORY REPRESENTATIONS AND INTERRELATIONS*

W. B. Quay

Neuroendocrine Laboratory, Department of Anatomy
University of Texas Medical Branch
Galveston, Texas

INTRODUCTION

The ultimate significance of chemical signals resides within the central nervous system. For olfactory signals this has meant classically the rhinencephalon, or "olfactory brain". However, it is increasingly demonstrated that the "olfactory brain" has many other neural and functionally important connections. Therefore the relatively narrow olfactory connotation of "rhinencephalon" is being supplanted by the descriptive designation limbic lobe, or better, limbic system. This lobe or system represents phylogenetically older cortex that borders (limbic) the brainstem, does not have or pass through the 6-layered structural pattern typifying the isocortex, and has major functional relations with a diverse array of relatively primitive physiological and behavioral actions. Many of these actions or responses are considered emotional or emotion-related, from our not unbiased human perspective (Papez, 1937; Pribram and Kruger, 1954; MacLean, 1959, 1960; Bargmann and Schadé, 1963; Adey and Tokizane, 1967; Yutzey et al., 1967; Isaacson, 1974; Wenzel, 1974; Gloor, 1975).

Experimental manipulations of the olfactory organs, their neural connections and the limbic system, are now performed for many different purposes, corresponding to the concerns of different specialty areas in basic and applied science. Some of the result-

*This report is dedicated with gratitude to Dr. Elizabeth C. Crosby, Professor Emeritus of Anatomy, University of Michigan and University of Alabama, on the occasion of her 94th birthday in 1982.

ing work and its proposed interpretation gives the impression that better communication is needed between the specialty areas. An overly narrow footing in areas of the relevant basic sciences can foster assumptions and interpretations that are tenuous at best.

I wish in the present brief review to point out some features in the basic anatomical and physiological relations of the mammalian olfactory connections which appear to be frequently lost sight of, and which are important in the design and interpretation of experiments. Much of this has been inspired by an interest in neuroendocrine mechanisms in mammals, and by a concern for the meaning of neuroendocrine changes following loss or modification of sensory input. A central issue in this review concerns the most frequent surgical manipulation, olfactory bulbectomy. This procedure is employed to produce "anosmic" experimental animals, and animals that are considered useful in particular kinds of studies in such diverse fields as neurophysiology, physiological psychology, neurochemistry, neuropsychopharmacology, neuroendocrinology, reproductive biology and developmental neurobiology.

The essential kinds of information and questions that we wish to consider here are as follows: (1) what are the olfactory organs of mammals and what are their major neural connections? (2) Which olfactory and neural components are involved either directly or indirectly when "olfactory bulbectomy" is performed? (3) What are the direct connections of the main and accessory olfactory bulbs? (4) What are the polysynaptic projections of the olfactory bulbs? (5) Are there functionally significant transneuronal and cybernetic effects of olfactory bulbectomy? (6) What are the major behavioral effects of olfactory bulbectomy? (7) What are the major neuroendocrine effects of olfactory bulbectomy? (8) What are the possible mechanisms of "pineal potentiation" following olfactory bulbectomy? (9) What kinds of improvements in design of experiments are suggested by the information and considerations stemming from these questions?

MAMMALIAN OLFACTORY ORGANS AND THEIR CONNECTIONS

The olfactory system can be conceptualized as consisting of three major components: (1) the peripheral olfactory sensory organs and the nerves or tracts which transmit information from them to the brain; (2) the main and accessory olfactory bulbs, the parts of the brain that receive directly the olfactory fibers from the usually considered peripheral olfactory organs; and (3) the brain regions that directly and/or indirectly (polysynaptically) receive the nerve fiber projections from the main and accessory olfactory bulbs. The brain regions comprising the third of these components give rise to what is often interpreted as an "olfactory response". But this is the result of integration of multiple kinds of input and of neural interactions. Such an "olfactory response" may be detected through measurement of electrophysiological, behavioral, endocrine or other kinds of changes, from the molecular and cellular to the whole animal level.

The specificity of the response in relation to the olfactory input is an important but usually unstated question. There is as well the question of which of the peripheral olfactory organs contribute significantly to the input necessary for the particular response being studied.

Five "olfactory" organs or regions have been noted in mammals (Leonard and Tuite, 1981):

(1) The main olfactory mucosa contains neurosensory olfactory cells that send their unmyelinated axons in fascicles (collectively = the "olfactory nerve") to the main olfactory bulb (MOB), with some evidence of a topographical correspondence of mucosal area of origin and bulbar region to which the fibers project.

(2) Axons of neurosensory cells of the vomeronasal organ (VNO) form fascicles (collectively = vomeronasal nerve) that pass over the medial side of the MOB and terminate in the accessory olfactory bulb (AOB).

(3) The paired septal olfactory organ (= organ of Rudolfo-Masera) lies at either side of the nasal septum near the union of the nasal chambers and the nasopharyngeal canal. Its nerve fibers have been traced to the caudal part of the MOB. From its anatomical position the septal organ has been proposed to continuously monitor respired air and to trigger odor sampling behaviors, such as active siffing, and opening access routes to the VNO.

(4) A "nervus terminalis" (NT), long known (Huber and Guild, 1913) but still poorly understood functionally, innervates both the main olfactory epithelium and the VNO. Its nerve fascicles, along with those from the VNO extend along the medial surface of the MOB. The fibers terminate within the brain beyond the olfactory bulbs. In man diffuse terminations have been noted in medial and lateral septal nuclei, olfactory tubercle, a region anterodorsal to the hypothalamic supraoptic nucleus, and a region tentatively identified as the medial island of Calleja (Larsell, 1950). The contents and courses of the NT in rodents have been revealed recently by immunocytochemistry. LH-RH-positive NT neuron cell bodies are seen at: (a) the rostral end of the MOB, (b) near the branches of the anterior cerebral artery, and (c) just caudal to the MOB. These NT neurons and their processes are spatially closely associated with both CSF (in the arachnoid layer) and cerebral blood supply (Schwanzel-Fukuda and Silverman, 1980). It remains unknown whether

the NT system contributes to neuroendocrine or other kinds of regulatory mechanisms.

(5) Some of the branches of the trigeminal (Vth cranial) nerve innervate the respiratory epithelium of the nasal chambers. They have been shown to transmit chemosensory information and to respond to some molecules at a sensitivity level comparable to, or greater than, that of the main olfactory epithelium (Tucker, 1963; Moulton and Tucker, 1964). There is some experimental support for the idea that the trigeminal fibers contribute to the regulation of the excitability of neurons within the MOB (Stone et al., 1968). Their course to the brain is within the Vth cranial nerve, and thus distant from the olfactory bulbs.

OLFACTORY BULBECTOMY - A MULTITARGET ABLATION

Information provided above is sufficient to show that experimental olfactory bulbectomy, as it is usually performed in rodents, destroys not only the MOB, but also the AOB, vomeronasal nerve, NT and the connections of the septal olfactory organ. Remaining outside of the surgical field is the Vth nerve's innervation of the respiratory epithelium. Thus the operation takes out more than the main olfactory sensory input, but spares a fifth "olfactory" or chemosensory input, that carried by the trigeminal nerve.

Olfactory bulbectomy can be expected also to ablate part of the anterior olfactory nucleus, since the pars intrabulbaris of this nucleus extends rostrally as a network of cells far up within the MOB. Furthermore, olfactory bulbectomy will interrupt or truncate a number of centrifugal fiber systems from higher centers to the olfactory inputs and their relay nuclei (Davis and Macrides, 1981; Macrides et al., 1981; Price and Powell, 1970). Although this aspect of the effects of the lesion might be thought inconsequential with the main olfactory input destroyed, possible repercussions on the net activities of the cells of origin within higher centers still merit critical consideration.

The complexity of the effects of olfactory bulbectomy increases with time after surgery. This in large measure derives from the cellular and synaptic plasticity of the olfactory system, even extending to higher centers of the brain. An experimental demonstration of this has been provided recently by Wright and Harding (1982). These investigators found recovery of olfactory function in mice that had been bilaterally bulbectomized 75 or 410 days earlier. Behavioral discrimination of certain scent cues returned to control levels, and reconstituted olfactory connections, but directly to the forebrain, were supported by experimental findings of several kinds. As will be more apparent later in this review, CNS plasticity at other sites within the olfactory brain is likely

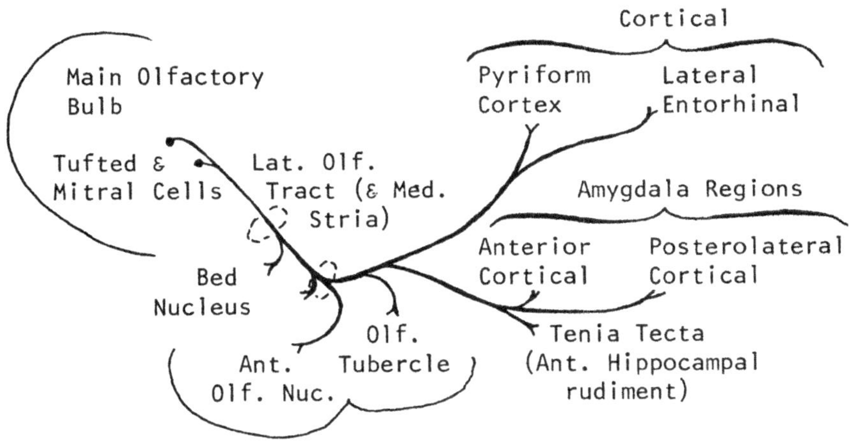

Fig. 1. Major neural projections of the main olfactory bulbs (MOB) in mammals.

to lead to additional modifications in response or behavior after intervals of time following surgery. Some of these appear to be abnormal behaviors and are probably not acceptable models for study of some kinds of mechanisms governing normal responses, at least without additional kinds of supporting information than are usually provided.

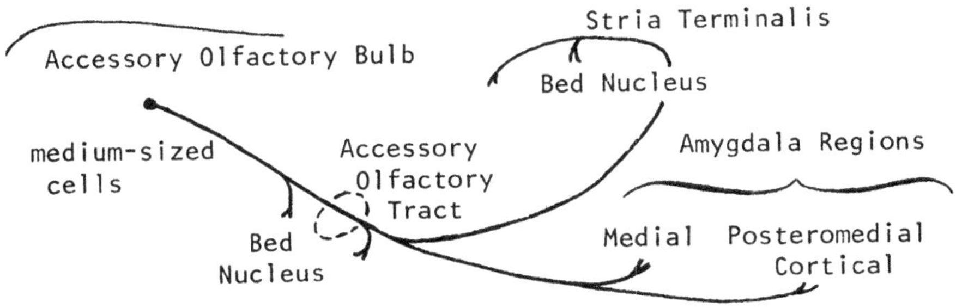

Fig. 2. Major neural projections of the accessory olfactory bulbs (AOB) in mammals.

DIRECT CONNECTIONS OF THE MOB AND AOB

The major afferent and direct neural connections of the main and accessory olfactory bulbs (MOB and AOB) in mammals are summarized in Figures 1 and 2. These are lateral view diagrams of the connections on one side of the brain, with anterior at the left and posterior at the right. Among the usually studied species, there is consistency in the major projections within this evolutionarily primitive and conservative system of the forebrain (Broadwell, 1975; Mascitti and Ortega, 1969; Meyer, 1981; Scalia, 1968; Scalia and Winans, 1975; Skeen and Hall, 1977; White, 1965). The major terminal projections of the MOB and AOB are seen to be separate, not overlapping in any major field, thereby suggesting that these anatomically discrete systems can be expected to be mutually distinctive in their control mechanisms and functional relations (Scalia and Winans, 1975).

POLYSYNAPTIC PROJECTIONS OF THE MOB

Beyond the primary projection areas diagrammed in Figure 1, the MOB has polysynaptic connections with a wide array of regions. These include: dorsomedial and ventromedial thalamic nuclei, lateral habenular nucleus, lateral hypothalamic region, lateral preoptic region and nucleus of the diagonal band of Broca (Powell et al., 1965; Price and Powell, 1971). The primary and polysynaptic projections of the MOB provide a complex entrance for olfactory information to limbic and neuroendocrine systems, as well as to others of secondary concern to us in this review.

TRANSNEURONAL AND CYBERNETIC EFFECTS OF OLFACTORY BULBECTOMY

Interpretations of mechanisms of effects of olfactory bulbectomy on behavioral and neuroendocrine patterns have tended to emphasize phasic and stimulus-related deficits. There are likely to be as well some more complex kinds of mechanisms. These depart from the paradigm of simple "anosmia". One kind of such mechanism is one of general tonic stimulation, or alternatively, of inhibition. Herrick (1933) proposed that the olfactory cortex has a general activating function for "all cortical activities". This concept has been expanded subsequently, but only recently has it been investigated in any depth and in relation to different sensory systems (Wenzel and Ziegler, 1977). It has been concluded that the alteration of many behaviors following a greatly diminished olfactory input is not entirely explicable as a consequence of perceptual loss per se (Wenzel, 1974). Available data, however, generally do not permit an answer to the question whether it is a change in level of tonic stimulation or inhibition, or a change in some kind of neurotrophic influence, that engenders such widespread and diverse effects following deafferentiation (Doty, 1977). Among possible interpretations for many of the behavioral and neuroendocrine effects of olfactory bulbectomy is one proposing that increased tonic activity in some amygdaloid centers is a frequent consequence of olfactory deafferentiation. However, as might be expected, experiments testing the effects of electrical stimulation of different parts of the amygdala reveal far more complex sequences and kinds of results (Edinger et al., 1975; Ono and Oomura, 1975; Zbrozyna, 1963).

Transneuronal degeneration in regions receiving major input from projections of the olfactory bulbs should be considered also as a component in the later effects of olfactory bulbectomy. Transneuronal degenerations within the olfactory bulbs themselves have been demonstrated following destruction of the olfactory mucosa (Le Gros Clark, 1957; Matthews and Powell, 1962). Degenerative changes in mitral, tufted and periglomerular cells started within a month. With ablation of the olfactory bulbs it is conceivable that both anterograde and retrograde transneuronal degenerations can occur, and that these will contribute to diversity and complexity within the resulting behavioral and neuroendocrine changes. Retrograde and retrograde transynaptic degenerations should be looked for inside neuronal systems efferent with respect to the olfactory bulbs and mucosa. Among model studies of this type, but involving other CNS connections, is one on the effects of limbic deafferentiation (Bleier, 1969). A more general account of retrograde degenerative phenomena in the CNS has been provided by Beresford (1965).

BEHAVIORAL EFFECTS OF OLFACTORY BULBECTOMY

Behavioral effects of bulbectomy of laboratory small mammals have been appreciated for over 75 years, and their number, diversity and complexity have been demonstrated extensively in recent years. Ten of these effects, or categories of effects, as seen in rats are: (1) increased irritability; (2) increased or modified muricidal (mouse-killing) behavior; (3) increased cannibalism of pups; (4) increased intermale aggression (species or strain dependent); (5) inhibition of territorial aggression; (6) blockade/increase/ decrease of sexual behaviors; (7) increased locomotor activity and "exploratory" behavior; (8) decreased avoidance learning; (9) modified pattern of daily feeding (decreased intake, increased frequency) ; and (10) decreased REM sleep. References for these may be found in Leonard and Tuite (1981). It is of interest to note that with the possible exceptions of the first and last of these behavioral effects, the loss of pheromonal input can be at least suspected of having a role. Comparative studies with employment of contrastingly responsive or adaptive species or strains in relation to specific pheromonal communications should aid in the determination of the importance of pheromonal deficits in the etiology of these behavioral modifications.

NEUROENDOCRINE EFFECTS OF OLFACTORY BULBECTOMY

Olfactory bulbectomy in the laboratory rat, particularly the immature rat, has been found to lead to a number of hormonal changes that reflect neuroendocrine involvement. Some of the better studied of these are: (1) delay of onset of puberty (Kling, 1964; Reiter, 1969; Reiter and Ellison, 1970); (2) blockade of pheromonal (presence of male) facilitation of ovulation in mice (Zarrow et al., 1970); (3) decreased growth and pituitary levels of growth hormone (Reiter et al., 1971; Sorrentino et al., 1971a,b); (4) increased nocturnal levels of FSH and LH (Rønnekleiv and McCann, 1975a); (5) decreased pituitary and plasma prolactin levels (Rønnekleiv and McCann, 1975b; Pieper and Gala, 1979). The emphasis of most of this research has been on neuroendocrine mechanisms per se, rather than on probable pheromonal inputs whose blockade may be responsible for some of the effects. Therefore, while at least certain of these and other neuroendocrine effects of "anosmia" or olfactory bulbectomy may depend upon interruption of pheromonal input, this has in fact rarely been specifically demonstrated.

Significant changes within the pineal gland are generally not noted after olfactory bulbectomy, but do occur consistently after blinding. Nevertheless, pinealectomy can block or significantly reduce some of the effects of olfactory bulbectomy (Blask and Nodelman, 1980; Reiter, 1969; Reiter et al., 1971; Sorrentino et al., 1971a,b), but not others (Rønnekleiv and McCann, 1975a,b). Olfactory bulbectomy also enhances the sensitivity of the reproductive system to the inhibitory actions of the pineal hormone, melatonin, administered in the late afternoon (Blask and Nodelman, 1980; Reiter et al., 1980). Possible mechanisms of this phenomenon will be discussed in the next section.

Stimulation of some parts of the amygdala can reproduce some of the effects of olfactory bulbectomy on the neuroendocrine-reproductive system axis (Velasco and Taleisnik, 1969). Bilateral lesions of certain parts of the amygdala can produce opposite effects, such as precocious rather than delayed puberty (Döcke et al., 1976; Elwers and Critchlow, 1960; Relkin, 1971; Velasco and Taleisnik, 1971). The specific neuroanatomical locations and physiological circumstances of these experimental manipulations are consistent with the hypothesis that olfactory bulbectomy can increase the tonic sensitivity or level of activity of parts of the amygdaloid complex affecting control of the neuroendocrine-reproductive system axis. However, lesions and stimulations in some of the other parts of the limbic system can have similar or related effects to these.

The question often remains in the interpretation of this work, whether or not the observed effects of olfactory bulbectomy are due to anosmia itself or to some secondary neurological involvement. A more conclusive and specific answer may be obtained by production of anosmia through reversible destruction of peripheral receptors rather than by permanent and gradually extending modification within the central nervous system. Techniques for this approach include nasal irrigation with zinc sulfate, surgical removal of the olfactory epithelium and occlusion of the external nares (Leonard and Tuite, 1981). None of these techniques is ideal for usual purposes however.

MECHANISM OF "PINEAL POTENTIATION" BY OLFACTORY BULBECTOMY

Olfactory bulbectomy of male and female rats leads to what has been interpreted as a potentiation of the neuroendocrine-reproductive system axis to the inhibitory actions of the pineal hormone melatonin (Blask and Nodelman, 1980; Reiter et al., 1980). The currently favored mechanism for this effect is increased sensitivity of the pineal's neural or neuroendocrine targets following anosmia. But a more comprehensive view of present information concerning pineal physiology (Kappers and Pevet, 1979; Quay, 1974) suggests other possibilities as well.

Ten categories of possible kinds of factors can be hypothesized for the linkage between olfactory bulbectomy and apparent potentiation of the pineal gland's neuroendocrine effects. These are not mutually exclusive and are sometimes obviously interrelated. These categories of possible kinds of etiologic factors can themselves be placed in three groups: (A) those factors that follow or depend upon anosmia; (B) those that represent a modification in the balance between various ones of the humoral and neural inputs to the pineal; and (C) those that are derived from modification of sensitivity to pineal hormone(s) by other neurological and/or behavioral mechanisms. The categories of factors within these groups are: (A): (1) increased sensitivity in target cells (hypothalamic neurons, etc.) to pineal hormone, (2) increased activation of stress-related responses, (3) modification of daily feeding pattern and food intake, (4) blockade of particular pheromonal inputs; (B): (5) decreased inhibitory control over particular CNS centers, especially those relating to the limbic system, (6) modified input to the pineal via the habenular nuclei, (7) modified input to the pineal via the sympathetic system, (8) complex neurological imbalances stemming at least in part from transneuronal degenerative changes; (C): (9) complex modification and expression of circadian timing mechanisms, and (10) modified balance in neural and/or humoral input to the hypothalamo-hypophyseal system.

It is probable that some combination of these factors is responsible for "pineal potentiation" following olfactory bulbectomy. One such combination can be conceptualized from the known interactions of the mammalian pineal with biorhythm control systems (circadian, reproductive, circannual). Another can be derived from the known interrelation of the pineal and the sympathetico-adrenal complex. Not yet well accepted, although supported by results from many laboratories, are direct CNS to pineal neural connections, including such from the habenular nuclei (Dafny et al., 1975; Korf and Wagner, 1980; McClung and Dafny, 1975; Nielsen and Møller, 1975; Pazo, 1981; Pfister et al., 1978; Rønnekleiv and Møller, 1979; Rønnekleiv et al., 1980). Although there were some early suggestions of olfactory and habenular influences on pineal activity in mammals (Miline et al., 1962), greatly improved and more diverse kinds of techniques should now enable more detailed and convincing studies concerning these influences.

CONCLUSIONS

(1) Olfactory ablation in the usually studied laboratory rodent models of the mammalian system leads to diverse and complex behavioral and neuroendocrine effects. In many of these a deficit in pheromonal input could play a role.

(2) However, olfactory bulbectomy, the usual method of olfactory ablation in such studies, has serious shortcomings that compromise interpretations of results. The three most serious ones of these are: (a) not all olfactory sensory structures or routes are removed; (b) olfactory regeneration with formation of functional forebrain connections can occur; and (c) other CNS systems are likely to be increasingly involved in the lesion, both directly and indirectly.

(3) Reversible olfactory ablations at the peripheral sensory epithelial level, with critically selected species, olfactory cues and responses, provide superior research designs for study of olfactory influences on behavioral and neuroendocrine systems.

REFERENCES

Adey, W. R., and Tokizane, T., 1967, "Structure and Function of the Limbic System," Prog. Brain Res., 27:1.

Bargmann, W., and Schadé, J. P., 1963, "The Rhinencephalon and Related Structures, "Prog. Brain Res., 3:1.

Beresford, W. A., 1965, A discussion on retrograde changes in nerve fibers, Prog. Brain Res., 14:33.

Blask, D. E., and Nodelman, J. L., 1980, An interaction between the pineal gland and olfactory deprivation in potentiating the effects of melatonin on gonads, accessory sex organs, and prolactin in male rats. J. Neurosci. Res., 5:129.

Bleier, R., 1969, Retrograde transsynaptic cellular degeneration in mammillary and ventral tegmental nuclei following limbic decortication in rabbits of various ages, Brain Res., 15:365.

Broadwell, R. D., 1975, Olfactory relationships of the telencephalon and diencephalon in the rabbit. I. An autoradiographic study of the efferent connections of the main and accessory olfactory bulbs, J. Comp. Neurol., 163:329.

Dafny, J., McClung, R., and Strada, S. J., 1975, Neurophysiological properties of the pineal body. 1. Field potentials, Life Sci., 16:611.

Davies, B. J., and Macrides, F., 1981, The organization of centrifugal projections from the anterior olfactory nucleus, ventral hippocampal rudiment, and piriform cortex to the main olfactory bulb in the hamster: An autoradiographic study, J. Comp. Neurol., 203:475.

Döcke, F., Lemke, M., and Okrasa, R., 1976, Studies on the puberty-controlling function of the mediocortical amygdala in the immature female rat, Neuroendocrinology, 20:166.

Doty, R. W., 1977, Behavioral effects of deafferentiation, Ann. New York Acad. Sci., 290:366.

Edinger, H. M., Siegel, A., and Troiano, R., 1975, Effect of stimulation of prefrontal cortex and amygdala on diencephalic neurons, Brain Res., 97:17.

Elwers, M., and Critchlow, V., 1960, Precocious ovarian stimulation following hypothalamic and amygdaloid lesions in rats, Am. J. Physiol., 198:381.

Gloor, P., 1975, Physiology of the limbic system, Adv. Neurol., 2:1.

Herrick, C. J., 1933, The functions of the olfactory parts of the cerebral cortex, Proc. Nat. Acad. Sci., 19:7.

Huber, G. C., and Guild, S. R., 1913, Observations on the peripheral distribution of the nervus terminalis in mammalia, Anat. Rec., 7:253.

Isaacson, R. L., 1974, "The Limbic System," Plenum, New York and London.

Kappers, J. A., and Pévet, P., 1979, "The Pineal Gland of Vertebrates Including Man," Prog. Brain Res., 52:3.

Kling, A., 1964, Effects of rhinencephalic lesions on endocrine and somatic development in the rat, Am. J. Physiol., 206:1395.

Korf, H.-W., and Wagner, U., 1980, Evidence for a nervous connection between the brain and the pineal organ in the guinea pig, Cell Tissue Res., 209:505.

Larsell, O., 1950, The nervus terminalis, Ann. Otol. Rhinol. Laryngol., 59:414.

Le Gros Clark, W. E., 1957, Inquiries into the anatomical basis of olfactory discrimination, Proc. Roy. Soc. B, 146:299.

Leonard, B. E., and Tuite, M., 1981, Anatomical, physiological, and behavioral aspects of olfactory bulbectomy in the rat, Internat. Rev. Neurobiol., 22:251.

MacLean, P. D., 1959, The limbic system with respect to two basic life principles, in: "The Central Nervous System and Behavior," M. A. B. Brazier, ed., Josiah Macy, Jr., Foundation, New York.

MacLean, P. D., 1960, Psychosomatics, in: "Handbook of Physiology, Sec. 1: Neurophysiology, vol. III," J. Field, H. W. Magoun, and V. E. Hall ed., Am. Physiol. Soc., Washington, D. C.

Macrides, F., Davis, B. J., Youngs, W. M., Nadi, N. S., and Margolis, F. L., 1981, Cholinergic and catecholaminergic afferents to the olfactory bulb in the hamster: A neuroanatomical, biochemical, and histochemical investigation, J. Comp. Neurol., 203:495.

Mascitti, T. A., and Ortega, S. N., 1969, Efferent connections of the olfactory bulb in the cat, an experimental study with silver impregnation methods, J. Comp. Neurol., 127:121.

Matthews, M. R., and Powell, T. P. S., 1962, Some observations on transneuronal cell degeneration in the olfactory bulb of the rabbit, J. Anat., 96:89.

McClung, R., and Dafny, N., 1975, Neurophysiological properties of the pineal body. 2. Single unit recording, Life Sci., 16:621.

Meyer, R. P., 1981, Central connections of the olfactory bulb in the american oppossum (*Didelphys virginiana*): A light microscopic degeneration study, Anat. Rec., 201:141.

Miline, R., Deceverski, V., and Krstic, R., 1963, Influence d'excitations olfactives sur le système habénulo-épiphysaire, Ann. Endocrinol., 24:377.

Moulton, D. G., and Tucker, D., 1964, Electrophysiology of the olfactory system, Ann. New York Acad. Sci., 116:380.

Nielsen, J. T., and Møller, M., 1975, Nervous connections between the brain and the pineal gland in the cat (*Felis catus*) and the monkey (*Cercopithecus aethiops*), Cell Tissue Res., 161: 293.

Ono, T., and Oomura, Y., 1975, Excitatory control of hypothalamic ventromedial nucleus by basolateral amygdala in rats, Pharmacol. Biochem. Behav., 3(Suppl. 1):37.

Papez, J. W., 1937, A proposed mechanism for emotion, Arch. Neurol. Psychiat., 38:725.

Pazo, J. H., 1981, Electrophysiological study of evoked electrical activity in the pineal gland, J. Neural Transmission, 52:137.

Pfister, A., Muller, J., Leffray, P., Guerillot, C., Vendrely, E., and Lage, C. D., 1978, Investigations on a possible extraorthosympathetic innervation of the pineal in rat and hamster, J. Neural Transmission, (Suppl.) 13:390.

Pieper, D. R., and Gala, R. R., 1979, Influence of the pineal gland, olfactory bulbs and photoperiod on surges of plasma prolactin in the female rat, J. Endocr., 82:279.

Powell, T. P. S., Cowan, W. M., and Raisman, G., 1965, The central olfactory connections, J. Anat., 99:791.

Pribram, K. H., and Kruger, L., 1954, Functions of the "olfactory brain," Ann. New York Acad. Sci., 58:109.

Price, J. L., and Powell, T. P. S., 1970, An experimental study of the origin and course of the centrifugal fibres to the olfactory bulb in the rat, J. Anat., 107:215.

Price, J. L., and Powell, T. P. S., 1971, Certain observations on the olfactory pathway, J. Anat., 110:105.

Quay, W. B., 1974, "Pineal Chemistry in Cellular and Physiological Mechanisms," Charles C Thomas, Springfield.

Reiter, R. J., 1969, Antigonadotropic activity of the pineal gland in blinded anosmic female rats, Fed. Proc., 28:318.

Reiter, R. J., and Ellison, N. M., 1970, Delayed puberty in blinded anosmic female rats: Role of the pineal gland, Biol. Reprod., 2:216.

Reiter, R. J., Petterborg, L. J., Trakulrungsi, C., and Trakulrungsi, W. K., 1980, Surgical removal of the olfactory bulbs increases sensitivity of the reproductive system of female rats to the inhibitory effects of late afternoon melatonin injections, J. Exp. Zool., 212:47.

Reiter, R. J., Sorrentino, S., Jr., Ralph, C. L., Lynch, H. J., Mull, D., and Jarrow, E., 1971, Some endocrine effects of blinding and anosmia in adult male rats with observations on pineal melatonin, Endocrinology, 88:895.

Relkin, R., 1971, Relative efficacy of pinealectomy, hypothalamic and amygdaloid lesions in advancing puberty, Endocrinology, 88:415.

Rønnekleiv, O. K., and McCann, S. M., 1975a, Effects of pinealectomy, anosmia and blinding alone or in combination on gonadotropin secretion and pituitary and target gland weight in intact and castrated male rats, Neuroendocrinology, 19:97.

Rønnekleiv, O. K., and McCann, S. M., 1975b, Effects of pinealectomy, anosmia and blinding on serum and pituitary prolactin in intact and castrated male rats, Neuroendocrinology, 17:340.

Rønnekleiv, O. K., and Møller, M., 1979, Brain-pineal nervous connections in the rat: An ultrastructure study following habenular lesion, Exp. Brain Res., 37:551.

Rønnekleiv, O. K., Kelly, M. J., and Wuttke, W., 1980, Single unit recordings in the rat pineal gland: Evidence for habenulo-pineal neural connections, Exp. Brain Res., 39:187.

Scalia, F., 1968, A review of recent experimental studies on the distribution of the olfactory tracts in mammals, Brain Behav. Evol., 1:101.

Scalia, F., and Winans, S. S., 1975, The differential projections of the olfactory bulb and accessory olfactory bulb in mammals, J. Comp. Neurol., 161:31.

Schwanzel-Fukuda, M., and Silverman, A.-J., 1980, LHRH neurons in the nervus terminalis of the guinea pig, Anat. Rec., 196: 168A.

Skeen, L. C., and Hall, W. C., 1977, Efferent projections of the main and accessory olfactory bulb in the tree shrew (*Tupaia glis*), J. Comp. Neurol., 172:1.

Sorrentino, S., Jr., Reiter, R. J., and Schalch, D. S., 1971a, Pineal regulation of growth hormone synthesis and release in blinded and blinded-anosmic male rats, Neuroendocrinology, 7:210.

Sorrentino, S., Jr., Reiter, R. J., Schalch, D. S., and Donofrio, R. J., 1971b, Role of the pineal gland in growth restraint of adult male rats by light and smell deprivation, Neuroendocrinology, 8:116.

Stone, H., Williams, B., and Carregal, E. J. A., 1968, The role of the trigeminal nerve in olfaction, Exp. Neurol., 21:11.

Tucker, D., 1963, Olfactory, vomeronasal and trigeminal receptor responses to odorants, in: "Olfaction and Taste," Y Zotterman, ed., Pergamon, Oxford.

Velasco, M. E., and Taleisnik, S., 1969, Release of gonadotropins induced by amygdaloid stimulation in the rat, Endocrinology, 84:132.

Wenzel, B. M., 1974, The olfactory system and behavior, in: "Limbic and Autonomic Nervous Systems Research," L. V. DiCara, ed., Plenum, New York.

Wenzel, B. M., and Ziegler, H. P., 1977, Introduction, Ann. New York Acad. Sci., 290: 1.

White, L. E., Jr., 1965, Olfactory bulb projections of the rat, Anat. Rec., 152:465.

Wright, J. W., and Harding, J. W., 1982, Recovery of olfactory function after bilateral bulbectomy, Science, 216:322.

Yutzey, D. A., Meyer, D. R., and Meyer, P. M., 1967, Effects of simultaneous septal and neo or limbic-cortical lesions upon emotionality in the rat, Brain Res., 5:452.

Zarrow, M. X., Estes, S. A., Debenberg, V. H., and Clark, J. H., 1970, Pheromonal facilitation of ovulation in the immature mouse, J. Reprod. Fert., 23:357.

Zbrozyna, A. W., 1963, The anatomical basis of the patterns of autonomic and behavioural response effected vis the amygdala, Prog. Brain Res., 3:50.

THE NEUROENDOCRINOLOGY OF SCENT MARKING

Pauline Yahr and Deborah Commins

Department of Psychobiology
University of California, Irvine

How animals behave socially depends on their sex, reproductive condition, and social status, as well as on their previous experience with each other and the location of their encounter. In vertebrates, many of these behavioral biases are caused or accompanied by hormonal changes, such as the secretion of sex- or stress-related steroids. If a vertebrate wanted to forecast the behavior of a conspecific it was approaching, it could scarcely do better than to get the animal's hormonal profile.

In mammals, hormonal profiles are often available in the form of chemicals released from the body. The same steroids that affect a mammal's social behavior also affect the chemical composition of its urine, feces, sweat, and skin gland secretions. Animals that can detect these chemical indicators benefit because they can adjust their own behavior to suit the behavioral dispositions of conspecifics with which they interact. Or, forewarned, they may avoid certain interactions entirely. Inducing behavioral adjustments in conspecifics can, in turn, benefit the animal that revealed its hormonal state.

Such benefits have no doubt contributed to the evolution of mammalian scent marking. Specialized marking behaviors allow animals to restrict their chemical signals to certain parts of their home range. They also allow animals to increase the prominence or half-life of their signals by placing them on particular objects or substrates (Regnier and Goodwin, 1977). In keeping with the fact that the chemical signals mammals lay down are often controlled by steroid hormones, so too are many marking behaviors (Yahr, 1982). Bringing scent marking under the control of steroid hormones only enhances the accuracy of the olfactory communication.

Our goal is to understand how steroid hormones, particularly those secreted by the gonads, act on the mammalian brain to control social behaviors such as scent marking communication. In approaching this goal, we have focused on the neuroendocrine control of ventral scent marking in Mongolian gerbils (Meriones unguiculatus). Gerbils scent mark by rubbing objects with a ventral sebaceous gland. Usually they do this by lowering their bellies onto an object as they pass over it, but they can and do use other postures as needed to scent mark vertical or oblique surfaces.

Both male and female gerbils display scent marking and the behavior is controlled by gonadal steroids in both sexes (Thiessen and Yahr, 1977). Castration decreases scent marking in males. Ovariectomy decreases scent marking in females. These changes can be prevented or fully reversed by supplying gonadectomized animals of either sex with either testosterone or estradiol.

There are, nonetheless, pronounced sex differences in scent marking. Male gerbils scent mark considerably more often than females and are more sensitive to the stimulatory effects of testosterone in adulthood (Thiessen et al., 1969; Turner, 1975). This sex difference in sensitivity to testosterone develops as a result of sex differences in androgen secretion during early development. For females to become as sensitive to testosterone as males are, they must be exposed to testosterone briefly around the time of birth. Conversely, neonatally castrated males develop like females and do not respond as well to testosterone in adulthood as normal males.

In both sexes, gonadal steroids activate scent marking by acting on cells in or near the medial preoptic area (MPOA) of the brain (Yahr, 1977). Implanting testosterone directly into the MPOA reinstates scent marking in castrated males or females. Implants of estradiol, a metabolite of testosterone, are equally effective. Lesions in or near the MPOA can also impair scent marking in male gerbils despite their exposure to testosterone (Yahr et al., 1982).

The MPOA also controls sexually dimorphic patterns of communication in other species. For example, the MPOA mediates androgen control of the mate-calling vocalizations of male frogs (Wada and Gorbman, 1977). In lizards, the MPOA mediates androgen control of male courtship and aggressive displays (Wheeler and Crews, 1978; Morgantaler and Crews, 1978). In dogs and cats, it is directly involved in the control of urinary scent marking (Hart, 1974; Hart and Voith, 1978). In rats, the lateral projections of the MPOA are necessary for androgens to reinstate urinary marking behavior in castrated males (Scouten et al., 1980). Of course, the MPOA also controls male copulatory behavior and regulates ovulation in many vertebrates (Gorski, 1979; Larsson, 1979).

Because the MPOA plays such an important role in reproduction and the control of sexually dimorphic behavior, many efforts have been made to identify within it the neural correlates of these sexually differentiated functions. Several interesting sex differences have been noted in the MPOA using neuroanatomical techniques. In rats, for example, sex differences in synaptic patterning have been observed with the electron microscope (Raisman and Field, 1971, 1973). With the light microscope, sex differences can be observed in the size of cell nuclei (Dorner and Staudt, 1968) and in the volume of a distinctive nuclear group (Gorski et al., 1980). When viewed with a Nissl stain, the MPOA of male rats contains a darkly staining nucleus that is five to eight times the size of the comparable area in females (Gorski et al., 1978). It has therefore been named the sexually dimorphic nucleus (SDN) of the rat MPOA.

In the research summarized here, we used several neuroanatomical techniques to study sex differences and hormone action in the gerbil MPOA. Sex differences in the structure of the MPOA are clearly visible with a Nissl stain. Moreover, the appearance of the MPOA changes in response to changes in gonadal steroid levels in adults. We show here that the distribution of hormone accumulating cells in the male MPOA corresponds closely to the shape of the sexually dimorphic area. We also show that sex differences and hormonal effects on the gerbil MPOA are visible histochemically with a stain that may help to identify the neurotransmitter system controlling marking behavior.

THE SEXUALLY DIMORPHIC AREA OF THE GERBIL MPOA

The sex difference in MPOA structure is illustrated in Figures 1 and 2. Thionin-stained sections through the MPOA of a gonadally intact male (see Fig. 1) reveal a darkly staining hook-shaped structure that flares out on either side of the third ventricle. This structure first appears where the fibers of the anterior commissure begin to close across the midline. At this point, only the medial portion of the structure is visible above the suprachiasmatic nuclei. In more posterior sections, the darkly staining region extends dorsolaterally toward the anterior commissure and assumes its hooked appearance. The structure ends where the anterior commissure ends, its lateral component joining the bed nucleus of the stria terminalis at this point. In contrast, the darkly staining region of the female MPOA is more diffuse and/or ovoid (see Fig. 2). Based on these differences in overall appearance, we can correctly identify the sex of gerbils about 90% of the time simply by observing the MPOA. We therefore refer to this darkly staining structure as the sexually dimorphic area (SDA) of the gerbil MPOA. Within the medial component of the male SDA we usually also see a cluster of cells that stain as intensely as the suprachiasmatic nuclei. We rarely see a cell cluster that stains this intensely in the female SDA.

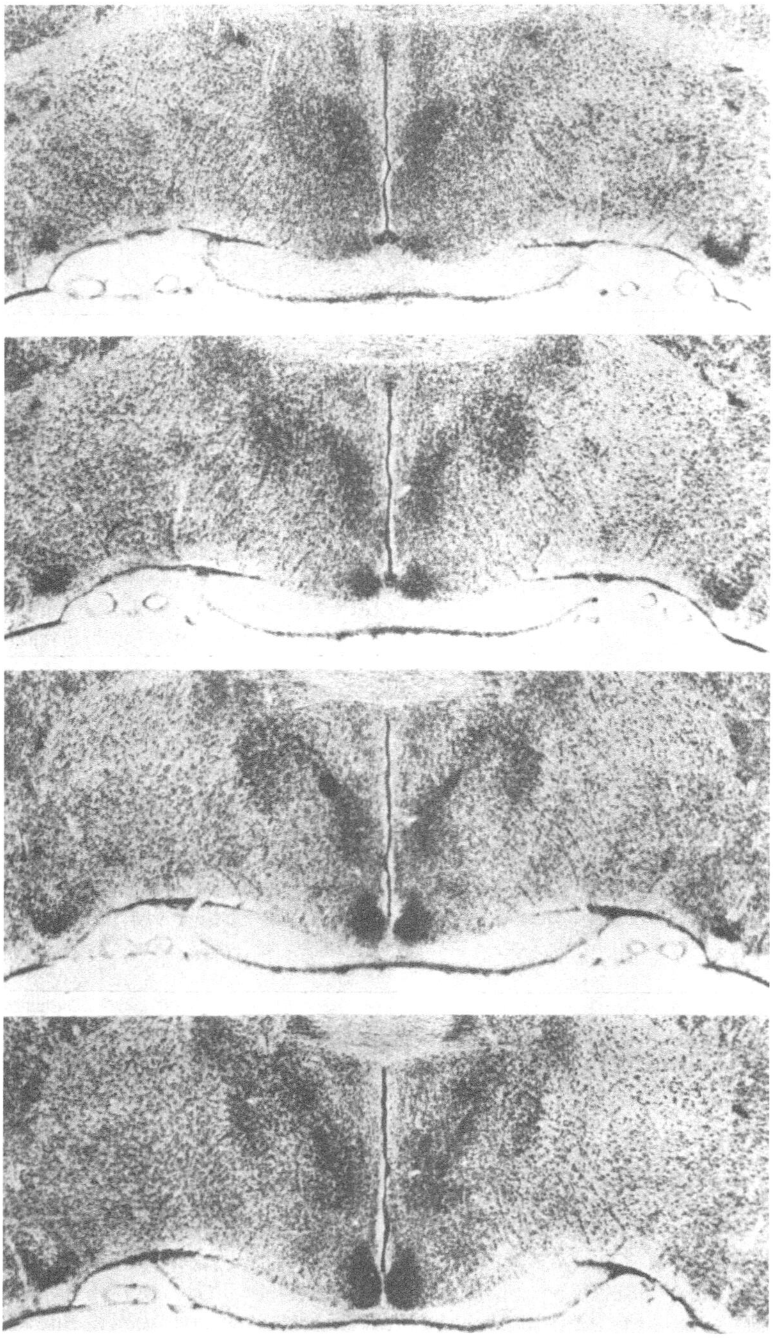

Fig. 1. Sequential 60 μm coronal sections through the medial preoptic area of a gonadally intact male gerbil.

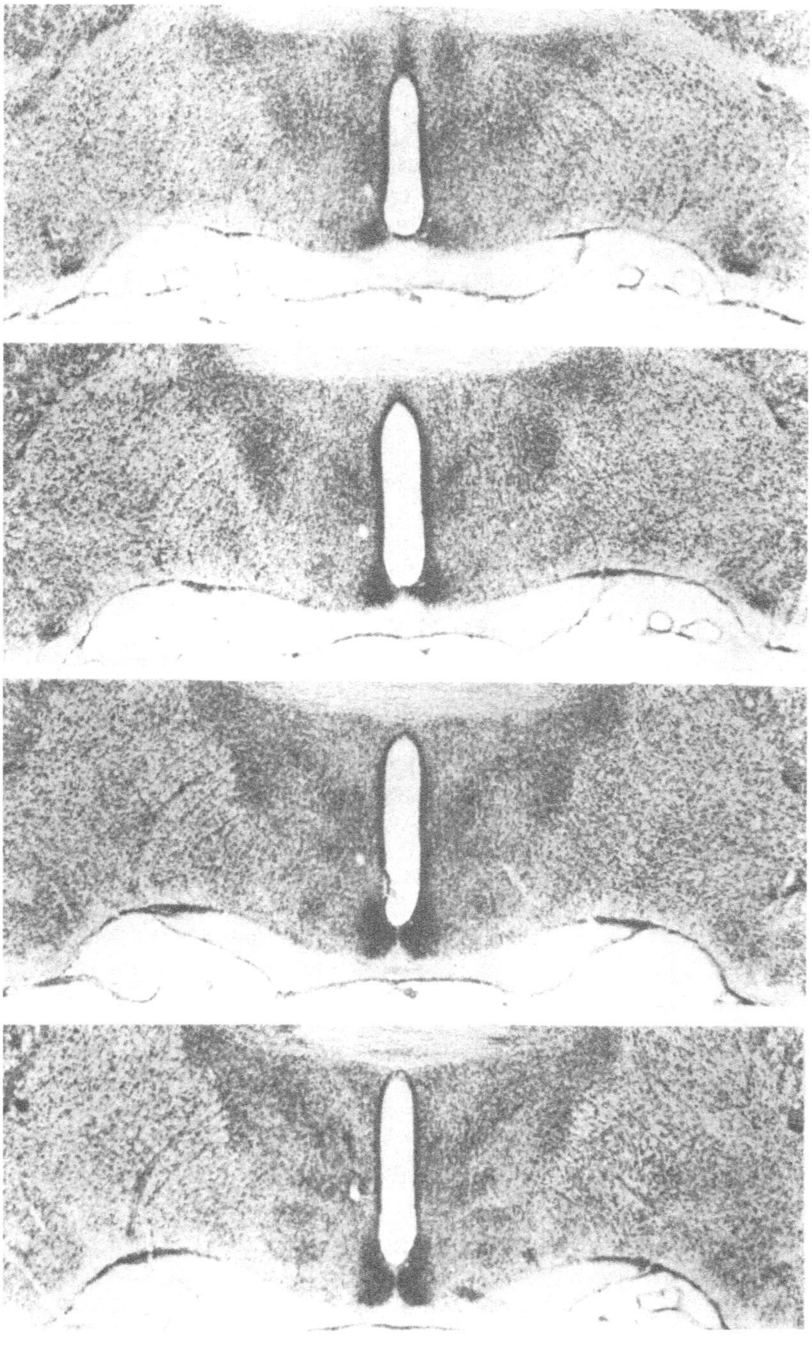

Fig. 2. Sequential 60 μm coronal sections through the medial preoptic area of a gonadally intact female gerbil.

We quantified the sex difference in the gerbil SDA by tracing its outline with the aid of a camera lucida and then retracing this outline with an electronic planimeter. The planimeter computes the area/section of the SDA and calculates the similarity of its outline to a circle on a scale from 0.0 (a line) to 1.0 (a circle). From the area/section, we can derive SDA volume. In area/section, the female SDA is 25% larger than the male SDA, confirming our impression that the female SDA is more diffuse. The female's SDA is also rounder than the male's as reflected in the sex difference in form scores. The form scores of nine gonadally intact males ranged from 0.40 to 0.55, whereas six of eight females scored above this range.

HORMONE-DEPENDENT ASPECTS OF THE SDA

Perhaps the most intriguing aspect of the gerbil SDA is that its appearance changes in the adult when gonadal steroids are removed or supplied in excess. As illustrated in Figure 3, gonadectomy causes the SDA of both sexes to fade and/or become more diffuse. The fading or diffuseness does not occur if the animals receive subcutaneous implants of testosterone at the time of gonadectomy. This is also shown in Figure 3. In this study, some castrated males and ovariectomized females received 10-mm silastic capsules of testosterone, a dose that fully maintains sexual activity and scent marking in castrated males (Yahr et al., 1979) and that can stimulate mounting behavior and scent marking in ovariectomized females (Ulibarri and Yahr, unpublished data). Other subjects, including those shown in Figure 3, received higher doses (25-mm capsules) of testosterone.

In "blind" tests, we could not identify the sex of gonadectomized gerbils by observing the MPOA. We still identified females correctly, but we misclassified nearly all of the males. The objective measures indicated that gonadectomy decreased the volume of the SDA in both sexes and made the SDA rounder in males (33% of the castrates had form scores above 0.55), thus eliminating the sex difference in form. Testosterone treatment made the SDA large and prominent again. Castrated males implanted with testosterone had SDAs that were hook-shaped (no form scores above 0.55) and that covered even a larger area/section and volume than those of gonadally intact males. Ovariectomized females treated with testosterone also had prominent SDAs that increased in area/section and volume relative to gonadally intact animals.

This is the first example of a gross anatomical change in the mammalian brain as a result of steroid hormone changes in adulthood. The only other system in which such changes have been observed is the motor system controlling birdsong, another androgen-dependent pattern of communication. In canaries, four brain areas involved in song are larger in males than females (Nottebohm and Arnold, 1976). In female canaries, testosterone causes two of these areas to enlarge and

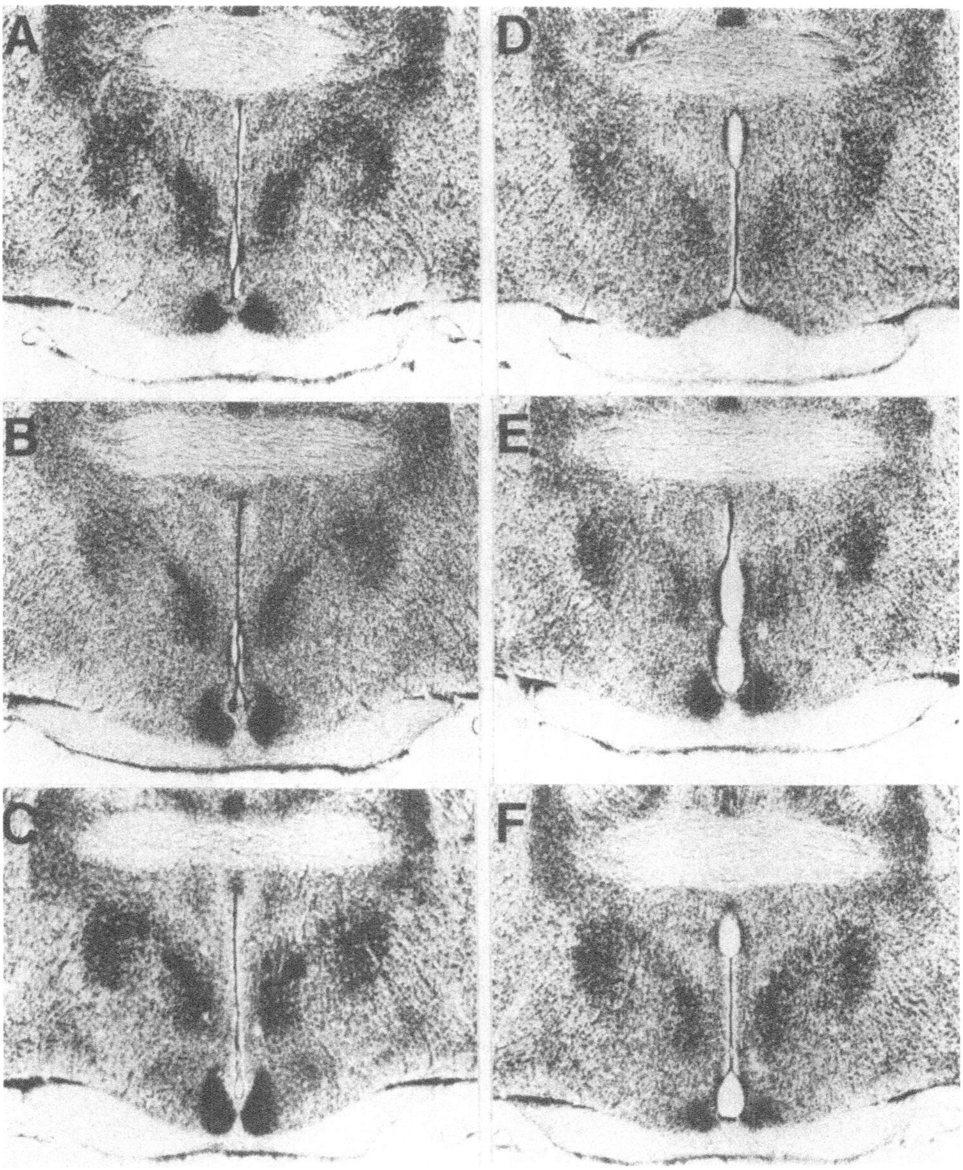

Fig. 3. The effects of gonadectomy and testosterone treatment on the sexually dimorphic area (SDA) of the gerbil MPOA. The photographs show thionin-stained coronal sections through the SDA of (A) a gonadally intact male, (B) a castrated male, (C) a castrated male treated with testosterone, (D) a gonadally intact female, (E) an ovariectomized female, (F) an ovariectomized female treated with testosterone.

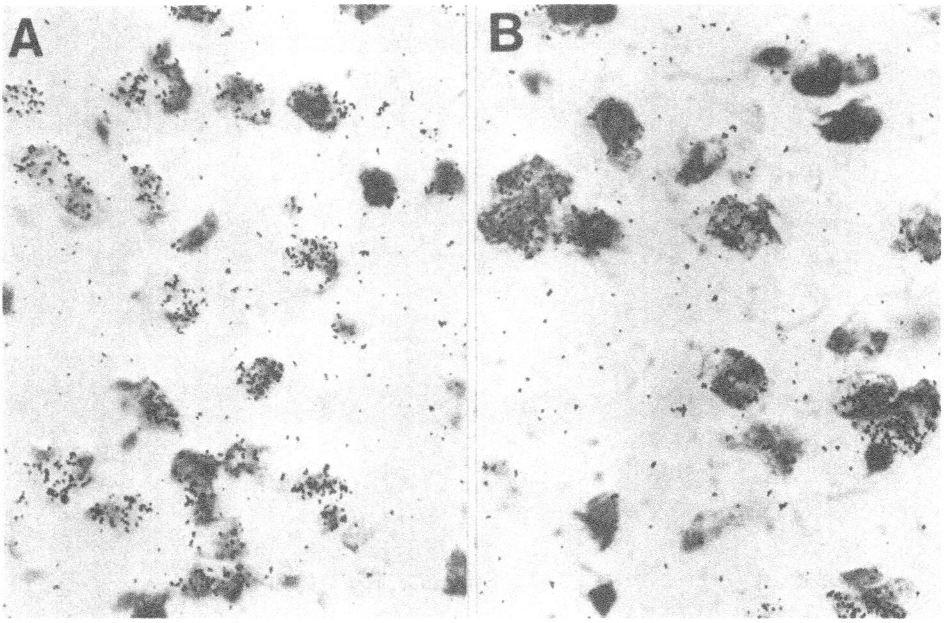

Fig. 4. Cells from the (A) medial and (B) lateral components of the SDA of a male gerbil injected intravenously with tritiated estradiol shortly after castration. The cells are stained with thionin. The dark grains over the cells show the accumulated steroid.

enables the females to sing (Nottebohm, 1980). In males, the same regions show seasonal changes in size. These song-control areas actually enlarge in spring when testosterone secretion is high and the males begin to sing (Nottebohm, 1981). In the brains of mammals, however, anatomical changes have been reported only when gonadal steroids were manipulated during the perinatal period of sexual differentiation (Gorski et al., 1978; Jacobson et al., 1981; Raisman and Field, 1973).

We do not know the cellular nature of the sex differences in the SDA. They could reflect differences in cell number, size, shape or stainability. The sex differences we see probably are not related to the sex difference in cell nuclear size reported by Clancy (1978). He found that MPOA cells of females have larger cell nuclei than MPOA cells of males; however, he sampled cells from sections 2.5 mm anterior to Bregma (Thiessen and Yahr, 1977), i.e., anterior to the SDA.

The hormone-dependent changes in adults are most likely due to changes in the amount of Nissl substance in the cytoplasm of cells, leading to changes in staining intensity. Nissl granules are arrays

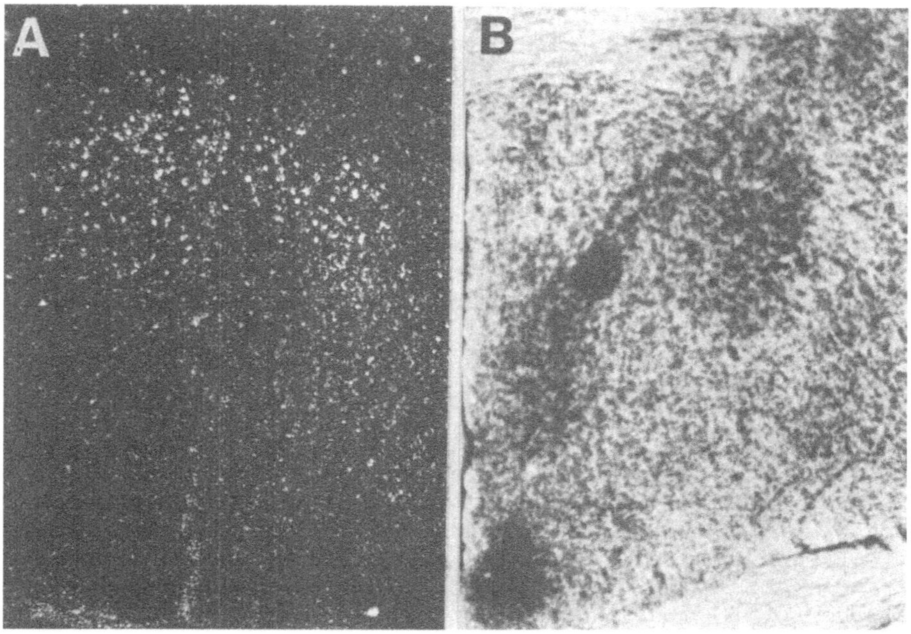

Fig. 5. (A) Darkfield autoradiography of estrogen-concentrating cells in the SDA of a male gerbil. (B) For comparison, a comparable lightfield photograph of a Nissl-stained section through the SDA of castrated male treated with testosterone.

of parallel cisternae of rough endoplasmic reticulum with attached ribosomes (Palay and Palade, 1955). These parallel arrays indicate that the cell body is synthesizing protein for transport to the axon. Some of these proteins are used to package the synaptic transmitters released from the axon terminals (Jacobson, 1978). The amount of Nissl substance in cells is related to the amount of nucleolar RNA and the ribosomes it forms (LaVelle, 1951, 1956). Since gonadal steroids alter the rate of ribosomal RNA synthesis in other target tissues (Mainwaring, 1977; O'Malley and Means, 1974), it is likely that they do so in the brain. Thus the paling or diffuseness of the SDA in animals deprived of gonadal steroids may resemble the chromatolysis, a dissolution of Nissl granules, that can occur in neurons when they are deprived of synaptic input (Globus, 1979).

Cells in the gerbil SDA do accumulate steroids from the blood. Figure 4 shows neurons from both the lateral and medial components of the male SDA that are labelled with tritiated estradiol. We have not yet developed autoradiograms from males injected with radiolabelled testosterone or its metabolite, dihydrotestosterone, or autoradiograms from females given estradiol. Thus we do not yet know what aspects of hormone accumulation are sexually dimorphic or which are specific for estradiol. We would, however, expect neurons involved in either scent marking or male sexual behavior to be labelled by estradiol since this steroid can stimulate both of these behaviors in castrated males (Nyby and Thiessen, 1971; Yahr, 1981). The possibility that males and females differ in their patterns of hormone accumulation also seems promising given that the pattern of estradiol accumulation within the male MPOA corresponds so closely to the distribution of cells in the SDA. As shown in Figure 5, the hook-shaped structure of the male SDA is clearly visible in the steroid hormone autoradiograms. Sex differences in hormone accumulation do occur in the songbird brain (Arnold and Saltiel, 1979).

HISTOCHEMICAL ANALYSIS OF THE SDA

As suggested earlier, gonadal steroids may alter the appearance of the SDA by altering the metabolic activity of neurons in this area. More specifically, testosterone may increase the synthesis of proteins involved in either the production or release of neurotransmitters by presynaptic cells or the sensitivity to neurotransmitters in postsynaptic cells. Analyzing sex differences and hormone-dependent changes in the brain in terms of neurotransmitter metabolism brings us one step closer to identifying their significance in terms of neural function and the control of behavior. We have begun such an analysis of the gerbil SDA by studying sex differences and the effects of gonadectomy on acetylcholinesterase histochemistry.

The cholinergic systems of the gerbil MPOA are particularly interesting to us because they are implicated in the control of scent marking (Yahr, 1977). Exposing the MPOA to drugs that antagonize or mimic the action of acetylcholine can facilitate or inhibit, respectively, the reinstatement of scent marking by testosterone (Yahr, 1977). Atropine sulfate, for example, a cholinergic antagonist, potentiates testosterone's behavioral effects in castrates when both the drug and the hormone are applied simultaneously to the MPOA. Conversely, pilocarpine nitrate, a cholinergic agonist, blocks testosterone's behavioral effects. Recently, we gathered pilot data indicating that biweekly infusions of atropine methyl nitrate into the lateral component of the SDA facilitate the recovery of scent marking in castrated male gerbils given low doses (1-mm capsules) of testosterone systemically.

To test the possibility that testosterone controls scent marking by regulating acetylcholine metabolism in the MPOA, we examined the gerbil MPOA with an acetylcholinesterase stain. The pattern of acetylcholinesterase activity in the gerbil SDA reveals even more clearly the sex differences and hormonal effects originally seen with Nissl stain. In males, reaction product made from the acetylcholinesterase enzyme forms a clear hooked pattern flaring out on either side of the third ventricle. This is illustrated in Figure 6A. In

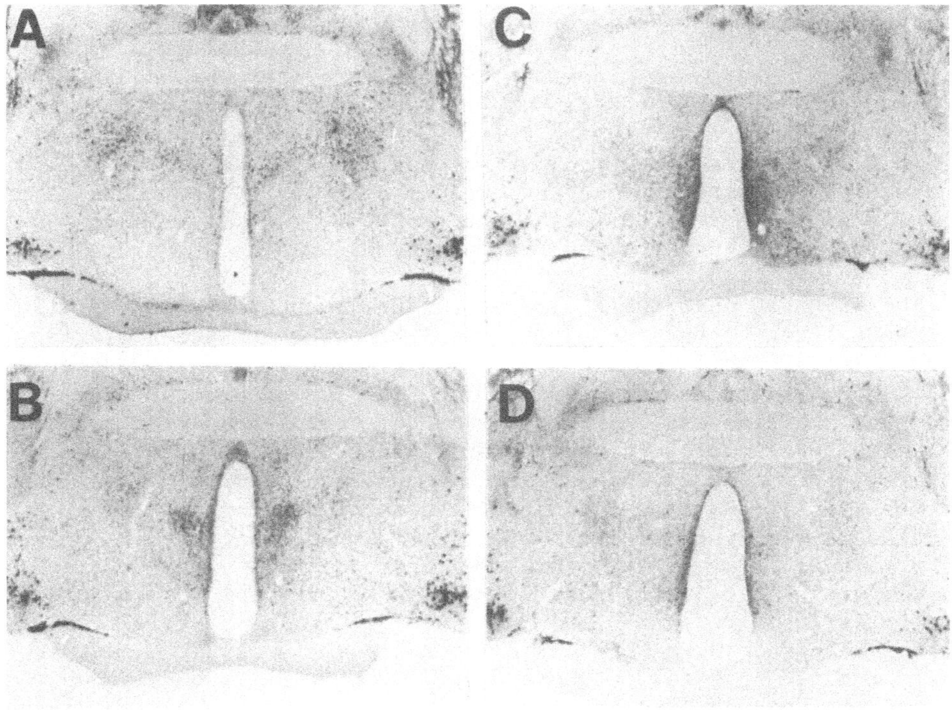

Fig. 6. Sex differences and the effects of gonadectomy on acetylcholinesterase activity in the gerbil MPOA. The photographs show coronal sections through the SDA of (A) a gonadally intact male, (B) a castrated male, (C) a gonadally intact female, (D) an ovariectomized female. The animals were treated with DFP (diisopropylfluorophosphate), an irreversible inhibitor of acetylcholinesterase, nine hours before sacrifice to limit acetylcholinesterase staining to cholinergic and cholinoceptive cells.

females, the familiar ovoid pattern is seen (see Fig. 6C). In both sexes, gonadectomy decreases the intensity of the staining, particularly in the lateral SDA, suggesting that the acetylcholinesterase is being synthesized there at a slower rate. The absolute intensity of the stain varies, of course, from one batch to another; however, we process the brains in matched sets representing each experimental group and always see decreased acetylcholinesterase activity in the gonadectomized animals relative to the controls. We hope that optical densitometry will prove useful for quantifying these differences. It is interesting to note that in zebra finches the song control nuclei in the brain also stain positively for acetylcholinesterase (Ryan and Arnold, 1981) and that in the syrinx and syringeal nerves this enzyme is androgen-dependent (Luine et al., 1980).

A WORKING HYPOTHESIS

Using a neuroanatomical approach to study hormonal control of scent marking has already proved fruitful and we anticipate that it will continue to do so in the future. Our discovery of the gerbil SDA has given us a fresh look at our earlier data on the neuroendocrine basis of scent marking and sexual behavior in this species and has enabled us and to formulate a more explicit hypothesis of how androgens activate these behaviors in adult males.

We propose that the medial and lateral components of the male SDA mediate androgen control of sexual behavior and scent marking, respectively. This hypothesis can explain why various manipulations of the MPOA exert different effects on the two behaviors. For example, MPOA lesions that eliminated or severely impaired sexual activity usually produced only transient deficits in scent marking (Yahr et al., 1982). These lesions often destroyed the medial SDA. We now suspect that marking behavior was not as severely impaired because we rarely destroyed the lateral component of the SDA bilaterally. Involvement of the medial SDA in copulatory behavior also agrees with data showing that infusions of cordycepin (a protein synthesis inhibitor) into the MPOA, near the third ventricle, blocks reinstatement of sexual behavior but not scent marking in castrated males treated with testosterone (Yahr and Ulibarri, unpublished). We predict that cordycepin will inhibit scent marking but not sexual behavior if infused into the lateral component of the SDA. Our hypothesis can also explain why implants of testosterone into the MPOA reinstated scent marking in castrated males (Thiessen et al., 1973; Yahr, 1976) even when they were relatively far from the midline (1.0-1.7 mm lateral) and were often just below the anterior commissure (6.0-6.9 mm below the surface of the skull). Such implants were often right over the lateral component of the SDA.

We do not suggest that the medial SDA is the only area of the MPOA mediating androgenic effects on male copulatory behavior. Indeed, our data indicate that medial aspects of the MPOA anterior, and probably posterior, to the SDA are also involved. Nor do we suggest that androgenic influences on scent marking are limited to the lateral SDA. Again, our data (Yahr et al., 1982) indicate otherwise. We do, however, propose that the SDA is directly involved in both of these behaviors and that its medial and lateral components affect the two behaviors differentially. In particular, we propose that the lateral component of the SDA is critical for scent marking. As a working hypothesis, we also propose that gonadal steroids work within the lateral SDA by altering the synthesis of proteins that eventually affect cholinergic neurotransmission.

In addition to helping us analyze the neuroendocrine basis of scent marking, our neuroanatomical analyses have provided data and ideas pertinent to other issues in behavioral neurobiology. Our data suggest, for example, that the plasticity of the adult brain is greater than is often supposed. Of four species (rats, gerbils, canaries, zebra finches) in which sex differences in brain structure have been studied after hormonal manipulations in adulthood (Arnold, 1980; Gorski et al., 1978; Nottebohm, 1980, 1981; this chapter), two (gerbils and canaries) show neuroanatomical changes. Our data also suggest that neuroanatomical techniques may be more useful than biochemical assays in addressing localization of function issues pertaining to the neuroendocrine functions of the hypothalamus. Finally, our neuroanatomical analyses have nicely revealed the differences that occur between species, even in the cytoarchitecture of a brain region that serves such similar functions throughout vertebrates. The fact that the MPOA of the gerbil shows so much more heterogeneity than the MPOA of the rat may make gerbils particularly useful in studying hypothalamic function.

ACKNOWLEDGEMENT

This research was supported by NIMH Grant MH 26481.

REFERENCES

Arnold, A. P., 1980, Effects of androgens on volumes of sexually dimorphic brain regions in the zebra finch, Brain Res., 185:441.

Arnold, A. P., and Saltiel, A., 1979, Sexual difference in pattern of hormone accumulation in the brain of a songbird, Science, 205:702.

Clancy, A. N., 1978, "Hormonal Differentiation of Ventral Gland Marking in the Mongolian Gerbil (Meriones unguiculatus), Unpublished Ph.D. Dissertation, University of Texas at Austin.

Dorner, G., and Staudt, J., 1968, Structural changes in the preoptic anterior hypothalamic area of the male rat following neonatal castration and androgen substitution, Neuroendocrinol., 3:136.

Globus, A., 1979, Brain morphology as a function of presynaptic morphology and activity, in: "The Developmental Neuropsychology of Sensory Deprivation," A. H. Riesen, ed., Academic Press, New York.

Gorski, R, A., 1979, The neuroendocrinology of reproduction: an overview, Biol. Reprod., 20:111.

Gorski, R. A., Gordon, J. H., Shryne, J. E., and Southam, A. M., 1978, Evidence for a morphological sex difference within the medial preoptic area of the rat brain, Brain Res., 148:333.

Gorski, R. A., Harlan, R. E., Jacobson, C. D., Shryne, J. E., and Southam, A. M., 1980, Evidence for the existence of a sexually dimorphic nucleus in the preoptic area of the rat, J. Comp. Neurol., 193:529.

Hart, B. L., 1974, Medial preoptic-anterior hypothalamic area and the sociosexual behavior of male dogs: a comparative neuropsycho-

logical analysis, J. Comp. Physiol. Psychol., 86:328.
Hart, B. L., and Voith, V. L., 1978, Changes in urine spraying, feeding and sleep behavior of cats following medial preoptic-anterior hypothalamic lesions, Brain Res., 145:406.
Jacobson, C. D., Csernus, V. J., Shryne, J. E., and Gorski, R. A., 1981, The influence of gonadectomy, androgen exposure, or a gonadal graft in the neonatal rat on the volume of the sexually dimorphic nucleus of the preoptic area, J. Neurosci., 1:1142.
Jacobson, M., 1978, "Developmental Neurobiology," Plenum Press, New York.
Larsson, K., 1979, Features of the neuroendocrine regulation of masculine sexual behavior, in: "Endocrine Control of Sexual Behavior," C. Beyer, ed., Raven Press, New York.
LaVelle, A., 1951, Nucleolar changes and development of Nissl substance in the cerebral cortex of fetal guinea pigs, J. Comp. Neurol., 94:453.
LaVelle, A., 1956, Nucleolar and Nissl substance development in nerve cells, J. Comp. Neurol., 104:175.
Luine, V., Nottebohm, F., Harding, C., and McEwen, B. S., 1980, Androgen affects cholinergic enzymes in syringeal motor neurons and muscle, Brain Res., 192:89.
Mainwaring, W. I. P., 1977, "The Mechanism of Action of Androgens," Springer-Verlag, New York.
Morgantaler, A., and Crews, D., 1978, Role of the anterior hypothalamus-preoptic area in the regulation of reproductive behavior in the lizard, Anolis carolinensis: implantation studies, Horm. Behav., 11:61.
Nottebohm, F., 1980, Testosterone triggers growth of brain vocal control nuclei in adult female canaries, Brain Res., 189:429.
Nottebohm, F., 1981, A brain for all seasons: cyclical anatomical changes in song control nuclei of the canary brain, Science, 214:1368.
Nottebohm, F., and Arnold, A., 1976, Sexual dimorphism in vocal control areas of the songbird brain, Science, 194:211.
Nyby, J., and Thiessen, D. D., 1971, Singular and interactive effects of testosterone and estrogen on territorial marking in castrated male Mongolian gerbils, Horm. Behav., 2:279.
O'Malley, B. W. and Means, A. R., 1974, Female steroid hormones and target cell nuclei, Science, 183:610.
Palay, S. L., and Palade, G. E., 1955, The fine structure of neurons J. Biophys. Biochem. Cytol., 1:69.
Raisman, G., and Field, P. M., 1971, Sexual dimorphism in the preoptic area of the rat, Science, 173:731.
Raisman, G., and Field, P. M., 1973, Sexual dimorphism in the neuropil of the preoptic area of the rat and its dependence on neonatal androgen, Brain Res., 54:1.
Regnier, F. E., and Goodwin, M., 1977, On the chemical and environ-environmental regulation of pheromone release from vertebrate scent marks, in: "Chemical Signals in Vertebrates," D. Müller-

Schwarze and M. M. Mozell, eds., Plenum Press, New York.
Ryan, S. M., and Arnold, A. P., 1981, Evidence for cholinergic participation in the control of birdsong: acetylcholinesterase distribution and muscarinic receptor autoradiography in the zebra finch brain, J. Comp. Neurol., 202:211.
Scouten, C. W., Burrell, L., Palmer, T., and Cegavske, C. E., 1980, Lateral projections of the medial preoptic area are necessary for androgenic influence on urine marking and copulation in rats, Physiol. Behav., 25:237.
Thiessen, D. D., Blum, S. L., and Lindzey, G., 1969, A scent marking response associated with the ventral sebaceous gland of the Mongolian gerbil (Meriones unguiculatus), Anim. Behav., 18:26.
Thiessen, D. D., Friend, H. C., and Lindzey, G., 1968, Androgen control of territorial marking in the Mongolian gerbil, Science, 160:432.
Thiessen, D. D., and Yahr, P., 1977, "The Gerbil in Behavioral Investigations: Mechanisms of Territoriality and Olfactory Communication," University of Texas Press, Austin, Texas.
Thiessen, D. D., Yahr, P. I., and Owen, K., 1973, Regulatory mechanisms of territorial marking in the Mongolian gerbil, J. Comp. Physiol. Psychol., 82:382.
Turner, J. W., 1975, Influence of neonatal androgen on the display of territorial marking behavior in the gerbil, Physiol. Behav., 15:265.
Wada, M., and Gorbman, A., 1977, Relation of mode of administration of testosterone to evocation of male sexual behavior in frogs, Horm. Behav., 8:310.
Wheeler, J. M., and Crews, D., 1978, Role of the anterior hypothalamus-preoptic area in the regulation of male reproductive behavior in the lizard, Anolis carolinensis: lesion studies, Horm. Behav., 11:42.
Yahr, P., 1976, The role of aromatization in androgen stimulation of scent marking, Horm. Behav., 7:259.
Yahr, P., 1977, Central control of scent marking, in: "Chemical Signals in Vertebrates," D. Müller-Schwarze and M. M. Mozell, eds., Plenum Press, New York.
Yahr, P., 1981, Scent marking, sexual behavior and aggression in male gerbils: comparative analysis of endocrine control, Amer. Zool., 21:143.
Yahr, P., 1982, Hormonal influences on territorial marking behavior, in: "Hormones and Aggressive Behavior," B. B. Svare, ed., Plenum Press, New York, in press.
Yahr, P., Commins, D., Jackson, J. C., and Newman, A., 1982, Independent control of sexual and scent marking behaviors of male gerbils by cells in or near the medial preoptic area, Horm. Behav., in press.
Yahr, P., Newman, A., and Stephens, D. R., 1979, Sexual behavior and scent marking in male gerbils: comparison of changes after castration and testosterone replacement, Horm. Behav., 13:175.

PRIMING PHEROMONES IN MICE

Anna Marchlewska-Koj

Institute of Zoology
Jagiellonian University
30-060 Krakow, Poland

INTRODUCTION

When we consider the present knowledge of pheromonal systems in mammals, undoubtedly the best documented are behavioral and physiological activities involved in reproduction of mice. More is known about priming pheromonal action in this species than in any other, but all the information comes from the laboratory while observations in wild populations are very limited. The investigators were more successful working with mice than with other laboratory rodents, e.g., rats, because *Mus musculus* is particularly sensitive to the pheromonal factors. Another possible explanation is that during the long term breeding in laboratory conditions selection between various strains occurred and some of them have accumulated genes which condition special sensitivity to pheromonal stimulants. However, the influence of priming pheromones on reproduction physiology was also described in other rodents: *Peromyscus maniculatus* of the Cricetidae family (Bronson and Dezell, 1968; Eleftheriou et al., 1962; Vandenbergh, 1980) and in a number of Microtidae: *Microtus pennsylvanicus* (Clulow and Langford, 1971), *Microtus agrestis* and *Clethrionomys glareolus* (Clulow and Clarke, 1968; Clarke and Clulow, 1973), *Microtus ochrogaster* (Hasler and Nalbandov, 1974) and *Dicrostonyx groenlandicus* (Hasler and Banks, 1975; Mallory and Brooks, 1980). In all investigated rodents similar reactions to the olfaction stimulants were found. This points out that in Rodentia the same or similar pheromonal activity can influence reproduction physiology. Of course it cannot be generalized that all pheromonal phenomena observed in *Mus musculus* are also characteristic for other species, but the mouse reproduction system can be used as a model system for macrosmatic small animals.

This paper gives a brief summary of recent information about the possible sources and hormonal control of priming pheromones in mice. Also the physiological effects and their genetical background will be considered.

PHEROMONAL INTERACTION BETWEEN FEMALES

Female pheromones strongly influence the hormonal activity of mice during their whole life, starting from maturation and extending through the whole reproduction period.

Young females caged separately attain the first vaginal estrus earlier than mice caged in groups (Vandenbergh et al., 1972; Drickamer, 1974). Using as the indicator the age of the first vaginal estrus McIntosh and Drickamer (1977) found that females exposed to external urine of grouped females matured later than control ones, or than females treated with urine collected from singly caged females. However, there was no difference if the urine was obtained from bladders. This indicates that puberty-delaying pheromone is synthesized by both grouped and singly caged females, but in the last case the activity disappears before urine is excreted. Homogenates of urethras of singly caged females with urine containing the pheromone did not delay puberty, while urethras tissue from grouped females did not inhibit the pheromonal effect. It is possible that urethras of singly caged females contain a factor which blocks capacity to delay puberty.

In the experiments of Drickamer and coworkers (1978) the presence of puberty-delaying pheromone was not affected by ovariectomy. Similarly to gonad-intact females, pheromone was present in bladder urine of singly caged and grouped females, but released with external urine only by grouped females. Also ovariectomy did not influence the blocking effect associated with urethras in singly caged females.

Since the ovariectomy did not affect the synthesis of puberty-delaying pheromone the adrenal glands were considered as another possible source of hormone. Results of the later experiments (Drickamer and McIntosh, 1980) showed that after adrenalectomy the pheromone disappears from bladder and external urine from singly caged or grouped females. This indicates that the presence of some substance(s) from the adrenal glands is necessary for the production of this pheromone. The experiments with homogenates of urethras with urine containing the puberty-delaying pheromone produced results which were not different from those obtained by using urethras from intact females. Hence the phenomenon of blocking or deactivating the pheromone by some substance associated with the urethra is not affected by adrenal glands. There is no more information about hormonal control of puberty-delaying factor(s), but the experiments described above indicate that apart from the gonadal system also some other factors are involved in inhibition of the reproductive function.

Pheromonal interaction between mature female mice appears as suppression of the estrous cycle. Housing females in groups causes a prolongation of diestrous phase of cycles (Lee and Boot, 1955; Bronson and Chapman, 1968). A small number of females caged together showed some disturbances in normal (4-5 day) cycles, and the frequency of estrus decreased in response to increase of density in cages. This effect can be mimicked by housing a single female in the cage recently soiled by a group of females (Champlin, 1971). Caging females in large groups caused anestrus in grouped females accompanied with decrease of ovarian weight and absence of corpora lutea (Whitten, 1959). Grouping females also increased frequency of incidence of spontaneous pseudopregnancy (Lee and Boot, 1956; De War, 1959).

Ryan and Schwartz (1977) working with white Swiss mice found that grouped females (20 females per cage) show suppression of estrous cycle in a very high percentage of animals. There was also a high incidence of 10 to 12 day cycles, but autopsies did not indicate the occurrence of anestrous phase. Further investigation of this group effect revealed that on the 4th or 5th day of the cycle uteri of grouped females were sensitive to trauma which caused decidual cell response (DCR), and consequently increased the uterus weight. The grouped females showed DCR reaction very similar to that formed in sterile-mated females. The differences between the groups of females (mated and grouped) concerned the time of sensitivity of DCR formation for at least 72 hours while identical determinations in grouped females indicated shorter period of 24-36 hours. The hormonal mechanism which is involved in pseudopregnancy induced by grouping females will not be discussed here, although it should be mentioned that the function of corpora lutea which condition the uterus reaction is probably stimulated by prolactin.

The suppressive effect of grouping on estrous cycle was abolished by ovariectomy while estradiol treatment of ovariectomized females restored this ability (Clee et al., 1975). Also in our laboratory the experiments were designed to determine the influence of density on the course of estrous cycle in outbred females. Animals were reared on a constant lighting schedule (lights on 7 am to 10 pm) and fed a standard pelleted diet and water. Ten females per polyethylene cage (18 cm x 20 cm x 10 cm) were maintained from the 24th day of life and after 8 to 10 weeks they were divided into five experimental groups (for detail see Table 1). The influence of a presence of ovariectomized (ovx) females on the length of estrous cycle was tested. In this experiment female mice were spayed at least 4 weeks before using.

Table 1. The effect of different grouping of female mice on number of estrous phases in 20 days.

Group	Density	No. of tested females	No. of estrus per female
A	2	20	4.0 ± 0.24[1]
B	5	20	3.0 ± 0.21
C	10	20	2.8 ± 0.34
D	20	20	2.4 ± 0.17
E	2 + 3 ovx[2]	20	3.2 ± 0.27
F	2 + 18 ovx	18	2.8 ± 0.25

Comparison of Groups by Student's t test.

A versus B	$p<0.01$	A versus E	$p<0.05$
A versus C	$p<0.01$	A versus F	$p<0.01$
A versus D	$p<0.001$	B versus D	$p<0.05$

[1]Mean ± s.e.m.

[2]Ovariectomized 4 weeks earlier.

The assay procedure for the estrous cycle was based on vaginal smears. Frequency of estrus was determined by counting the episodes of cornified smears observed in each mouse over the 20 day period which followed grouping.

As indicated by the results summarized in Table 1 the frequency of the estrus was significantly higher when two females were reared per cage than when 5, 10 or 20 cycling females were together. The estrous cycle was also affected by ovariectomized females. The presence of even three ovariectomized females inhibited estrus in two cycling animals. The decrease of estrous frequency was more pronounced when 2 cycling females were reared together with 18 ovariectomized females.

Inhibition of the estrous phase evoked by presence of gonadally intact or ovariectomized female mice can be provoked by various factors accompanying density increase. This phenomenon has to be further investigated but olfactory factors cannot be completely ruled out. A similarity between the inhibition of estrous cycle and puberty-delaying pheromone release by ovariectomized mice (Drickamer et al., 1978) should be considered.

PHEROMONAL INTERACTION BETWEEN FEMALES AND MALES

Influence of Females on Males

Comparatively little is known about female pheromone action on the male reproduction system. This is probably due to the fact that we have no adequate and simple bioassay for estimation of hormonal activity in male mice. Practically the only way is to measure androgens and gonadotropins by RIA which is an excellent technique but still limited to certain laboratories.

At present only very few data are available on the hormonal excitation of males evoked by female odor in hamsters (Macrides et al., 1974) and in rats (Kamel et al., 1977). Some more information has been completed for house mice. Vandenbergh (1971) reported that acceleration of sexual maturation in young male mice can be evoked by cohabitation with adult females from weaning until 36-37 days life. Continued exposure after that age yielded adults that were indistinguishable from isolated controls. Hence the age limits the sensitivity of juvenile males to the presence of mature females (Maruniak et al., 1978). Acceleration of puberty which appears as increase of testes and accessory glands weights was observed only when males were exposed to females between 20 and 36 days of life. At this age a short period of contact with female urine stimulated LH release.

Also in adult males the urine of females can elicit gonadotropin secretion in the absence of other types of cues (Maruniak and Bronson, 1976). The response to urine was more pronounced in LH than FSH discharge. The most intriguing is the fact that increase of LH plasma level was evoked not only by cycling and pregnant females but also by urine of ovariectomized females. The effect was connected with female genetical sex because exposure to gonadally intact or castrated males did not affect hormonal excitation of mature males.

Similarly to adult males, juvenile males release LH gonadotropin when exposed for a short period to a female factor. A single short exposure of 36-day-old males to female urine elevated LH plasma level after 30 minutes. Prolonged contact did not elicit increase of LH discharge (Maruniak et al., 1978). When an adult male contacted a female for a short time, LH gonadotropin discharge occurred, but caging together the same female and male for longer periods gradually diminished this hormone response. A subsequent exposure of this male to a novel female dramatically stimulated luteinizing hormone release (Coquelin and Bronson, 1979). Also testosterone was elevated after a short exposure to the females while in males paired for one week testosterone level remained similar to that in all-male groups (Macrides et al., 1975).

Discharge of luteinizing hormone after introduction of a novel female appears at the same period as a mounting behavior which is commonly observed in similar experimental conditions. But since the gonadotropin stimulating effect is evoked to the same extent in males by gonadally intact and by ovariectomized females a direct correlation between olfactory stimulants delivered by females and sexual behavior of a male is questionable.

Influence of Males on Females

One of the first pheromonal effects which was described in mammals is associated with the influence of male olfaction stimulants on reproduction physiology of female mice. Exposure to a male induces normal ovarian cyclicity in previously anestrous females (Whitten, 1956a). Sexually immature females reared in the presence of adult males attained puberty significantly earlier than females housed in the absence of a male (Vandenbergh, 1967). Recently inseminated females housed with a strange male during the first four days of pregnancy blocked the first pregnancy and could be inseminated by the strange male (Bruce, 1959). Further investigations demonstrated that all these phenomena are also evoked by a contact of females with urine of adult males (Dominic, 1964; Bronson and Whitten, 1968; Colby and Vandenbergh, 1974). The presence of substances which accelerated puberty (Vandenbergh, 1969), synchronize estrous cycle (Bronson and Whitten, 1968) or block pregnancy (Bruce, 1965) is dependent on circulating testosterone. These olfaction stimulants are produced only by mature males and their release can be inhibited by castration but restored by testosterone treatment. It is possible that three priming pheromonal effects associated with the urine are mediated by the same chemical messenger(s) which evoke a different reaction depending on the hormonal status of recipients.

Pheromonal activity of males decreases rather slowly after castration. The ability to accelerate puberty disappeared between the 10th and 15th day (Lombardi et al., 1976) and pregnancy blocking pheromone was still released on 10th day after tested extirpation: from 12 tested 5 females blocked pregnancy (unpublished data). Injection of 1 mg testosterone propionate into castrated males stimulated release of pheromones: the puberty accelerating factor appeared after 24 hours and reached maximum level by 60 hours. After similar treatment the pregnancy blocking pheromone was released in the period between the 2nd and 7th day.

Despite extensive studies during the last few years the chemical nature of pheromones is still unknown. A direct correlation between circulating testosterone and synthesis of priming pheromones indicates that these biologically active substances are either the products of catabolism of androgens or their synthesis is stimulated by androgens. The urine of male mice contains a high level of protein. A major urinary protein complex is synthesized in the liver, depends on androgen hormones and occurs in mature males and androgen-treated females (Rümke and Thung, 1964; Finlayson et al., 1965). Vandenbergh et al. (1975), using simple and precise bioassay for ascertaining accelerated sexual development of juvenile females based on uterine weight increase, demonstrated that exposure of immature females to urinary protein fraction of male advances the puberty. Also the proteins salted out with ammonium sulphate from male urine synchronized the estrous cycle similarly as the presence of a male behind a net (Marchlewska-Koj and Biaty, 1978) and evoked the typical pregnancy block effect (Marchlewska-Koj, 1977). The females exposed to these urinary proteins blocked pregnancy in 90% of the tested animals. The effect was similar to the reaction elicited by the presence of non-stud males (80%).

During further experiments on the identification of the puberty-accelerating pheromone Vandenbergh and coworkers (1976) detected the biological activity in urinary peptide fraction with molecular weight of approximately 860. This fraction, kindly supplied by Dr. Vandenbergh was also tested for pregnancy block effect and all 8 tested females interrupted pregnancy and returned to the estrous on 5th or 6th day after coitus.

The source of male priming pheromones is still unknown. All three male pheromonal effects can be evoked by urine collected directly from bladder and thus free of any accessory glands secretion (Bronson and Whitten, 1968; Marchlewska-Koj, 1977; Drickamer and Murphy, 1978). Also the hemizygous Tabby J males with inherited lack of preputial glands evoked the pregnancy block effect in a very high percentage of tested females (Hoppe, 1975) and preputialectomized males accelerated first estrus in young females as well as intact animals (Colby and Vandenbergh, 1974). Since several metabolic processes of the kidney are under androgen control (Pettengill and Fishman, 1962; Kochkian et al., 1963) this organ may represent the site of pheromone synthesis. As discussed above the biologically active substance appears to be strongly bound to the urinary protein fraction produced in the liver so the liver is the next candidate for testing.

However, the role of the preputial glands cannot be definitely ruled out. These glands are known as the source of androgen-dependent olfaction stimulants such as aggression-promoting pheromone (Mugford and Nowell, 1971) or sex-attractant pheromone (Bronson and Caroom, 1971). Also the biologically active substances accelerating the estrous cycle (Gaunt, 1968; Chipman and Albrecht, 1974) and evoking the Bruce effect (Marchlewska-Koj, 1977) were found in these glands. In conclusion - at the present state of knowledge it is quite possible that several organs are involved in the synthesis of male pheromones.

It is still disputable whether urinary pheromones are volatile or not. In the initial work of Whitten and coworkers (1968) it was demonstrated that the estrus-inducing pheromone can be transported in a tunnel if the air flow is 6 m/min. Also Hoppe (1975) observed differences in the gas chromatography pattern of urine from "blocking" and "non-blocking" mouse males and suggested that the Bruce effect is evoked by a volatile pheromone. However, the persistence of pheromonal activity in the residue after evaporation of urinary peptide fraction indicates that the puberty-accelerating pheromone (Vandenbergh et al., 1976) and the pregnancy-blocking pheromone (Marchlewska-Koj, 1981) are non-volatile substances.

As suggested by the results presented by Johns et al. (1978) the olfaction stimulants released by rat males belong to non-volatile substances. The reflex ovulation in female rats reared under constant light can be provoked by the male urine but only when the females had a direct contact with urine-soiled bedding. The lack of volatility is also the property of the aggression-reducing signal which is released by female mice (Evans et al., 1978). One can hypothesize that quite different olfaction stimulants, the synthesis of which depends on steroid hormones, are all low-volatile compounds. The low-volatile substances from urine after reaching vomeronasal epithelial cells can be easily transposed into the organ undoubtedly involved in the transmission of chemical signals (Wysocki et al., 1980).

GENETICAL ASPECTS OF PRIMING PHEROMONE EFFECTS

It is well known and generally accepted that the prerequisite of a successful bioassay is selection of a proper genotype used in the experiments. Very soon after Bruce had described the pregnancy block effect in female mice Marsden and Bronson (1965) criticized these results because they were unable to evoke a similar effect when testing four different strains of mice. This could indicate that the olfaction-dependent pregnancy termination is a characteristic feature for a mutation present only in albino Parkes stock females. Later Godowicz (1968) and Chapman and Whitten (1969) confirmed previous results obtained by Bruce but they found great variations between the tested strains of mice. These differences concerned both the sensitivity of females and ability of pheromone production by males.

The sensitivity or non-sensitivity of females to olfaction stimulants which evoke the pregnancy block effect appears to be a very stable feature. Marsden and Bronson (1965) failed to reproduce the pregnancy block effect when using the C57BL/6J females from the Jackson Laboratory (USA) colony. Also Godowicz (1968) found the C57BL/kw females much less sensitive than CBA/kw females from a colony reared in Poland. On the other hand, the females from DBA strain very easily terminate pregnancy due to olfactory stimulants. They were used as the model of Bruce effect by Dominic in 1967 (colony from Banaras University), and also in our laboratory until now.

A low susceptibility of C57BL females to male olfaction stimulants concerns also other male effects. The presence of mature males facilitated ovulation in the PMS-induced ovulation in immature mice. Among a few tested strains of mice the least sensitive were C57BL/6J females (Zarrow et al., 1973; Ho and Wilson, 1980).

Genetical differences between female mice concern not only sensitivity to a male factor but also interaction between females. Champlin (1971) reported strain differences in estrous response to grouping females. In the C57BL/6 strain four females per cage appears to be the grouping maxímum beyond which there was no additional effect on frequency of estrus. Different reactions were found in BALB females, where the increase of density from two females per cage through four, six or eight was accompanied by consecutive prolongation of estrous cycles.

Variations between genotypes were found not only in sensitivity of female mice but also in pheromonal production of males. Previously mentioned C57BL males blocked pregnancy in a very low percentage of tested females (Chipman and Bronson, 1968; Hoppe, 1975; Marchlewska-Koj, 1977) while the CBA males are known as the producers of pheromones which frequently block pregnancy (Dominic, 1967). Eleftheriou and coworkers (1972, 1973) who have been working on the genetic control of male pheromonal facilitation of ovulation in juvenile female mice, found that the presence of C57BL/6J males did not result in a significant enhancement of the number of released ova. When females were exposed to SWR/J males the number of ovulated eggs significantly increased. Taking into account the results of experiments with hybrids between C57BL/6J and SWR/J males, the authors suggest that pheromone production is controlled by a single gene. They propose to denote as Ph^h the gene determining high pheromonal production and Ph^l as its allele responsible for low production.

During long-term laboratory selection the different strains of mice were established and now they can be used as an excellent model for pheromonal investigations. For instance, CBA mice represent typical animals for which olfaction stimulants play an important role in the reproductive physiology. The females are sensitive - they easily block pregnancy, and the males produce the pheromone which is effective in blocking pregnancy. Mice from C57BL strain belong to the opposite group. The females are not very sensitive, both as juvenile and during pregnancy. Also the males from this strain do not produce strong pheromones. A correlation between the production of pheromones and sensitivity to the olfaction stimulants is not a general rule for genotypes. Experiments with the KE/kw strain showed that males from this strain were able to block pregnancy in different genotype females (e.g., in CBA/kw) but they did not evoke any effect when tested with KE/kw pregnant females. However, these females were sensitive and easily terminated pregnancy when exposed to CBA/kw males (Godowicz, 1970). Lack of the stimulating effect of KE/kw males on the females from their own strain was also confirmed when the Whitten effect was examined, while other genotype males increased frequency of estrus (Krzanowska, 1964).

The results summarized above point out that the reaction of female mice is strongly determined by the genotype and can be regarded as an inherited feature concerning variation in sensitivity of the olfactory system (Whitten, 1956b; Gaudelman et al. 1972; Vandenbergh, 1973) and responsiveness of the hormonal system (Hoppe and Whitten, 1972; Champlin, 1977).

We have little information about genetical control of pheromone synthesis in males. Since the synthesis of pregnancy-blocking pheromone depends directly on the presence of testosterone, all the loci which control the synthesis of this androgen have some influence on pheromonal production. This idea is supported by the observations made on C57BL males. They are characterized by a low level of plasma testosterone (Bartke, 1974) and by low pheromonal activity (as described above). However, the lethal yellow (A^y) gene in the heterozygous state affects the reproductive function of male mice, probably due to the influence on cholesterol metabolism, but it does not decrease the ability to evoke the Bruce effect. Yellow (A^ya) males blocked pregnancy in a similar percentage as non-yellow (aa) males (Bartke, 1968; Kakihana et al., 1974).

It is a pity that genetical aspects of pheromonal functions represent such an under-investigated area. Any information about the causes of inheritance responsible for the sensitivity and production of olfaction stimulants could be very helpful in further investigations of the role of pheromonal communication in the reproduction physiology and in understanding of the mechanisms involved in population selection.

POSSIBLE ADAPTIVE FUNCTION OF PRIMING PHEROMONES

Current evidence indicates that pheromones are important factors in the regulation of density population. Unfortunately, most of the data concerning hormonal and physiological background of priming pheromones were collected for mice reared in the laboratory. Hence a question arises which of the described effects could fulfil some adaptive function in a wild population.

The first evidence that wild female mice living under natural conditions produce a urinary component that delays the onset of puberty in juvenile female mice was recently reported by Massey and Vandenbergh (1980). These authors found that only urine collected from females in a population at its maximum density delayed puberty in tested females. Urine collected when the population was less dense, or from a population which remained sparse failed to delay puberty.

The density increase of free-living rodent populations is followed by decrease of reproductive rate, increase of neonatal mortality, or both. A number of investigators have reported a delay or inhibition of puberty in laboratory or natural populations of *Mus musculus* (Christian, 1956), *Microtus agrestis* (Clarke, 1955) and *Microtus pennsylvanicus* (Christian, 1971). The results obtained by Massey and Vandenbergh (1980) suggest that a urinary factor present at high densities may delay puberty and thus help to slow further population growth.

The pregnancy block effect described in a few rodent species in laboratory conditions is sometimes classified as a typical laboratory artifact. But there is also some evidence that female mice (Chipman and Fox, 1966) and female meadow voles (Mallory and Clulow, 1977) from wild populations blocked pregnancy too, but the experiments with these animals were done in laboratories. However, in the ovaries of wild trapped *M. pennsylvanicus* females, two or three sets of corpora lutea can indicate the number of copulations in consequence of pregnancy blockage. This is the only experimental evidence that the pregnancy block effect appears in a wild population but several researchers have already attempted to explain pregnancy block effects in terms of ecological concepts.

The Bruce effect can act at the peaks of density when females are living in a high density population. They can probably meet strange males easily and frequently. It may be that in such conditions the olfactory block of pregnancy reduces reproduction and suppresses population growth (Chipman et al., 1966; Rogers and Beauchamp, 1976), or as Bruce and Parrot (1966) suggested, this is a mechanism which prevents inbreeding in a small deme. Even if this adaptation could be explained only by group selection, it is still unknown how this feature is fixed in the course of natural selection.

The pregnancy block may be a product of post-copulatory male-male competition. In reproduction procedures one sex generally invests more than the other. In mammals this is usually the female since she suffers the costs of gestation and lactation. As a result females cannot produce and rear the offspring at the rate at which males can father them. Consequently while the reproductive success of females will usually be limited by the number of young they can produce and raise, the males will often be limited by factors affecting the number of females they can fertilize. In this relationship between the male and female, the olfaction stimulants delivered in a proper period can give an advantage to the male to inseminate a female, even if she was already fertilized. There may be some analogies between the prevention of pregnancy and killing of unrelated young. Infanticide of unrelated neonates was reported in lions (Bertram, 1975) and primates (Hrdy, 1977) and also in a few species of rodents: wild mice (Labov, 1980), several voles (Boonstra, 1980) and collared lemmings (Mallory and Brooks, 1980). Infanticide behavior may have a selective advantage only for males, because females invested much time and energy. The interruption of pregnancy before implantation in mammals which produce microlecithal ova practically does not cost females any energy. This can explain why selection did not avoid susceptibility to male stimulants. Labov (1981) minimized the consequences of male infanticide and hypothesized that the Bruce effect is a possible advantage for females because, "those individuals which are capable of utilizing the process may actually produce more offspring which will survive past infancy and thus increase their potential individual fitness." The hypothesis that the pregnancy-block effect is the infanticide of unborn offspring explains in a simple way the role of the Bruce effect in the mechanism of population control. But this hypothesis has to be proved by evidence that the male which interrupted pregnancy has a chance to fertilize this particular female. Even now, when we have still practically no information about the Bruce effect in a feral population, a possible function of this phenomenon cannot be completely ruled out.

The possible function of puberty-delaying pheromone and pregnancy-blocking pheromone in feral population was discussed above. Probably also other priming pheromonal effects play important roles in reproduction of wild populations. The main question now is how to transfer a successful experimental approach from the laboratory to natural conditions.

REFERENCES

Bartke, A., 1968, Pregnancy blocking by yellow ($A^y a$) and non-yellow (aa) male mice, Genetics, 60:161.

Bartke, A., 1974, Increased sensitivity of seminal vesicles to testosterone level, J. Endocr., 60:145.

Bertram, B. C. R., 1975, Social factors influencing reproduction in wild lions, J. Zool. (Lond.), 177:463.

Boonstra, R., 1980, Infanticide in microtines: importance in natural populations, Oecologia, 46:262.

Bronson, F. H., and Caroom, D., 1971, Preputial gland of the male mouse: attractant function, J. Reprod. Fert., 25:279.

Bronson, F. H., and Chapman, V. M., 1968, Adrenal-oestrous relationships in grouped or isolated female mice, Nature, 218:483.

Bronson, F. H., and Dezell, H. E., 1968, Studies on the estrus-inducting (pheromonal) action of male deermouse urine, Gen. Comp. Endocr., 10:339.
Bronson, F. H. and Whitten, W., 1968, Oestrus-accelerating pheromones of mice: assay, androgen dependency and presence in bladder urine, J. Reprod. Fert., 15:131.
Bruce, H. M., 1959, An exteroceptive block to pregnancy in the mouse, Nature, 184:105.
Bruce, H. M., 1965, The effect of castration on the reproductive pheromones of male mice, J. Reprod. Fert., 10:141.
Bruce, H. M. and Parrott, D. M. V., 1960, Role of olfactory sense in pregnancy block by strange males, Science, 131:1526.
Champlin, A. K., 1971, Suppression of oestrus in grouped mice: the effects of various densities and the possible nature of the stimulus, J. Reprod. Fert., 27:233.
Champlin, A. K., 1977, Strain differences in estrous cycle and mating frequencies after centrally produced anosmia in mice, Biol. Reprod., 16:513.
Chapman, V. M. and Whitten, W. K., 1969, Occurrence and inheritance of pregnancy block in inbred mice, Genetics, 61:S9.
Chipman, R. K., and Albrecht, E. D., 1974, The relationship of the male preputial gland to the acceleration of oestrus in the laboratory mouse, J. Reprod. Fert., 38:91.
Chipman, R. K. and Bronson, F. H., 1968, Pregnancy blocking capacity and inbreeding in laboratory mice, Experientia, 24:199.
Chipman, R. K., Holt, J. A., and Fox, K. A., 1966, Pregnancy failure in laboratory mice after multiple short-term exposures to strange males, Nature, 210: 653.
Chipman, R. K., and Fox, K. A., 1966, Oestrus synchronization and pregnancy blocking in wild house mice (*Mus musculus*), J. Reprod. Fert., 12:233.
Christian, J. J., 1956, Adrenal and reproductive responses to population size in mice from freely growing populations, Ecology, 37:258.
Christian, J. J., 1971, Fighting, maturity, and population density in *Microtus pennsylvanicus*, J. Mammal., 52:556.
Clarke, J. R., 1955, Influence of numbers on reproduction and survival in two experimental vole populations, Proc. R. Soc. Lond. ser. B., 144:68.
Clarke, J. R., and Clulow, F. K., 1973, The effect of successive matings upon bank vole (*Clethrionomys glareolus*) and vole (*Microtus agrestis*) ovaries, in: "The development and maturation of the ovary and its functions," H. Peters, ed., Experta Medica, Amsterdam.
Clee, M. D., Humphreys, E. M., and Russell, J. A., 1975, The suppression of ovarian cyclical activity in groups of mice, and its dependence on ovarian hormones, J. Reprod. Fert., 45:395.
Clulow, F. K., and Clarke, J. R., 1968, Pregnancy-block in *Microtus agrestis*, an induced ovulation, Nature, 219:511.
Clulow, F. K., and Langford, P. E., 1971, Pregnancy-block in the

meadow vole, Microtus pennsylvanicus, J. Reprod. Fert., 24:275.
Colby, D. R. and Vandenbergh, J. G., 1974, Regulatory effects of urinary pheromones on puberty in the mouse, Biol. Reprod., 11: 268.
Coquelin, A., and Bronson, F. H., 1979, Release of luteinizing hormone in male mice during exposure to females: habituation of the response, Science, 206:1099.
De War, A. D., 1959, Observation on pseudopregnancy in the mouse, J. Endocr., 18:186.
Dominic, C. J., 1964, Source of the male odour causing the pregnancy block in mice, J. Reprod. Fert., 8:266.
Dominic, C. J., 1967, Effect of exogenous prolactin on olfactory block to pregnancy in mice exposed to urine of alien males, Ind. J. Exp. Biol., 5:47.
Drickamer, L. C., 1974, Sexual maturation of female house mice: social inhibition, Devel. Psychobiol., 7:257.
Drickamer, L. C., and McIntosh, T. K., 1980, Effects of adrenalectomy on the presence of a maturation-delaying pheromone in the urine of female mice, Horm. Behav., 14:146.
Drickamer, L. C., McIntosh, T. K., and Rose, E. A., 1978, Effects of ovariectomy on the presence of a maturation-delaying pheromone in the urine of female mice, Horm. Behav., 11:131.
Drickamer, L. C., and Murphy, R. X., 1978, Female mouse maturation: effects of excreted and bladder urine from juvenile and adult males, Devel. Psychobiol., 11:63.
Eleftheriou, B. E., Bailey, D. W., and Zarrow, M. X., 1972, A gene controlling male pheromonal facilitation of PMSG induced ovulation in mice, J. Reprod. Fert., 31:155.
Eleftheriou, B. E., Bronson, F. H., and Zarrow, M. X., 1962, Interaction of olfactory and other environmental stimuli on implantation in the deer mouse, Science, 137:764.
Eleftheriou, B. E., Christenson, C. M., and Zarrow, M. S., 1973, The influence of exteroceptive stimuli and pheromonal facilitation of ovulation in different strains of mice, J. Endocr., 57:363.
Evans, C. M., Mackintosh, J. H., Kennedy, J. F., and Robertson, S. M., 1978, Attempts to characterise and isolate signals from the urine of mice, Mus musculus, Physiol. Behav., 20:129.
Finlayson, J. S., Asofsky, R., Potter, M., and Runner, C. C., 1963, Major urinary protein complex of normal mice: origin, Science, 149:981.
Gaudelman, R., Zarrow, M. X., and Denenberg, V. H., 1972, Reproductive and maternal performance in the mouse following removal of the olfactory bulbs, J. Reprod. Fert., 28:453.
Gaunt, S. L., 1968, Studies on the preputial gland as a source of a reproductive pheromone in the laboratory mouse (Mus musculus), Ph.D. Thesis, Burlington, University of Vermont.
Godowicz, B., 1968, Pregnancy block in inbred mice and the F_1 crosses, Folia Biol. (Crac.), 16:199.
Godowicz, B., 1970, Influence of the genotype of males on pregnancy-block in inbred mice, J. Reprod. Fert., 23:237.

Hasler, J. F., and Banks, E. M., 1975, The influence of mature males on sexual maturation in female collared lemmings (Dicrostonyx groenlandicus), J. Reprod. Fert., 42:583.
Hasler, M. J., and Nalbandov, A. V., 1974, The effect of weaning and adult males on sexual maturation in female voles (Microtus ochrogaster), Gen. Comp. Endocrinol., 23:137.
Ho, H., and Wilson, J. R., 1980, Genetic and hormonal aspects of male facilitation of PMSG-induced ovulation in immature mice, J. Reprod. Fert., 59:57.
Hoppe, P. C., 1975, Genetic and endocrine studies of the pregnancy blocking pheromones of mice, J. Reprod. Fert., 45:109.
Hoppe, P. G., and Whitten, W. K., 1972, Pregnancy block: initiation by administered gonadotropin, Biol. Reprod., 7:254.
Hrdy, S. B., 1977, "The langurs of Agu: female and male strategies of reproduction," Harvard Univ. Press, Cambridge.
Johns, M. A., Feder, H. H., Komisaruk, B. R., and Mayer, A. D., 1978, Urine-induced reflex ovulation in anovulatory rats may be a vomeronasal effect, Nature, 272:446.
Kakihana, R., Ellis, L. B., Gerling, S. A., Blum, S. L., and Kessler, S., 1974, Bruce effect competence in yellow-lethal heterozygous mice, J. Reprod. Fert., 40:483.
Kamel, F., Wright, W. W., Mock, E. J., and Frankel, A. J., 1977, The influence of mating and related stimuli on plasma levels of luteinizing hormone, follicle stimulating hormone, prolactin and testosterone in the male rats, Endocrinology, 101:421.
Kochakian, C. D., Hill, J., and Aonuma, S., 1963, Regulation of protein biosynthesis in mouse kidney by androgens, Endocrinology, 72:354.
Krzanowska, H., 1964, Studies on heterosis. III, The course of the sexual cycle and the establishment of pregnancy in mice, as affected by the type of mating, Folia Biol. (Crac.), 12:415.
Labov, J. B., 1980, Factors influencing infanticidal behavior in wild male house mice (Mus musculus), Behav. Ecol. Sociobiol., 6:297.
Labov, J. B., 1981, Pregnancy blocking in rodents: adaptive advantages for females, Am. Nat., 118:361.
Lee, S. van der, and Boot, L. M., 1955, Spontaneous pseudopregnancy in mice, Acta Physiol. Pharm. neerl., 4:442.
Lee, S. van der, and Boot, L. M., 1956, Spontaneous pseudopregnancy in mice. II, Acta Physiol. Pharm. neerl., 5:213.
Lombardi, J. R., Vandenbergh, J. G., and Whitsett, J. M., 1976, Androgen control of the sexual maturation pheromone in house mouse urine, Biol. Reprod., 15:179.
Macrides, F., Bartke, A., and Dalterio, S., 1975, Strange females increase plasma testosterone levels in male mice, Science, 189: 1104.
Macrides, F., Bartke, A., Fernandez, F., and D'Angelo, W., 1974, Effects of exposure to vaginal odor and receptive females on plasma testosterone in the male hamster, Neuroendocr., 15:355.

Mallory, F. F., and Brooks, R. J., 1980, Infanticide and pregnancy failure: reproductive strategies in the female collared lemming (Dicrostonyx grenlandicus), Biol. Reprod., 22:192.
Mallory, F. J., and Clulow, F. V., 1977, Evidence of pregnancy failure in the wild meadow vole, Microtus pennsylvanicus, Can. J. Zool., 55:1.
Marchlewska-Koj, A., 1977, Pregnancy block elicited by urinary proteins of male mice, Biol. Reprod., 17:729.
Marchlewska-Koj, A., 1981, Pregnancy block elicited by urinary peptides in mice, J. Reprod. Fert., 61:221.
Marchlewska-Koj, A., and Biata, E., 1978, Modification of the oestrous cycle by urinary proteins of male mice, Folia Biol., (Crac.) 26:311.
Marsden, H. H., and Bronson, F. H., 1965, Strange male block to pregnancy: its absence in inbred mouse strains, Nature 207:878.
Maruniak, J. A., and Bronson, F. H., 1976, Gonadotropic responses of male mice to female urine, Endocrinology, 99:963.
Maruniak, J. A., Coquelin, A., and Bronson, F. H., 1978, The release of LH in male mice in response of female urinary odors: characteristics of the response in young males, Biol. Reprod., 18:251.
Massey, A., and Vandenbergh, J. G., 1980, Puberty delay by a urinary cue from female house mice in feral populations, Science, 209: 821.
McIntosh, T. K., and Drickamer, L. C., 1977, Excreted urine, bladder urine, and the delay of sexual maturation in female house mice, Anim. Behav., 25:999.
Mugford, R. A., and Nowell, N. W., 1971, The preputial glands as a source of aggression-promoting odors in mice, Physiol. Behav., 6:247.
Pettengill, O. S., and Fishman, W. H., 1962, Influence of testosterone on glycine incorporation into mouse kidney-β-glucuronidase, Exp. Cell. Res., 28:218.
Rogers, J. G., Jr., and Beauchamp, G. K., 1976, Some ecological implications of primer chemical stimuli in rodents, in: "Mammalian olfaction reproductive processes and behavior," R. L. Doty, Ed., Academic Press, New York.
Rümke, P., and Thung, P. J., 1964, Immunological studies on the sex-dependent prealbumin in mouse urine and its occurrence in the serum, Acta endocrinol., 47:156.
Ryan, K. D., and Schwartz, N. B., 1977, Grouped female mice: demonstration of pseudopregnancy, Biol. Reprod., 17:578.
Vandenbergh, J. G., 1967, Effect of the presence of a male on the sexual maturation of female mice, Endocrinology, 81:345.
Vandenbergh, J. G., 1969, Male odor accelerates female sexual maturation in mice, Endocrinology, 84:658.
Vandenbergh, J. G., 1971, The influence of the social environment on sexual maturation in male mice, J. Reprod. Fert., 24:383.
Vandenbergh, J. G., 1973, Effects of central and peripheral anosmia on reproduction of female mice, Physiol. Behav., 10:257.

Vandenbergh, J. G., 1980, The influence of pheromones on puberty in rodents, in: "Chemical signals in vertebrates and aquatic invertebrates," D. Müller-Schwarze and R. M. Silverstein, eds., Plenum Press, New York.

Vandenbergh, J. G., Drickamer, C. C. and Colby, W. R., 1972, Social and dietary factors in the sexual maturation of female mice, J. Reprod. Fert., 28:397.

Vandenbergh, J. G., Finlayson, J. S., Dobrogosz, W. J., Dills, S. S., and Kost, T. A., 1976, Chromatographic separation of puberty accelerating pheromone from male mouse urine, Biol. Reprod., 15: 260.

Vandenbergh, J. G., Whitsett, J. M., and Lombardi, J. R., 1975, Partial isolation of pheromone accelerating puberty in female mice, J. Reprod. Fert., 43:515.

Whitten, W. K., 1956a, Modifications of the oestrous cycle of the mouse by external stimuli associated with the male, J. Endocr., 13:399.

Whitten, W. K., 1956b, The effect of removal of the olfactory bulbs on the gonads of mice, J. Endocr., 14:160.

Whitten, W. K., 1959, Occurrence of an oestrus in mice caged in groups, J. Endocr., 18:102.

Whitten, W. K., Bronson, F. H., and Greenstein, J. A., 1968, Estrus-inducting pheromone of male mice: transport by movement of air, Science, 161:584.

Wysocki, C. J., Wellington, J. L., and Beauchamp, G. K., 1980, Access of urinary nonvolatiles to the mammalian vomeronasal organ, Science, 207:781.

Zarrow, M. X., Eleftheriou, B. E., and Denenberg, V. H., 1973, Sex and strain involvement in pheromonal facilitation of gonadotrophin-induced ovulation in the mouse, J. Reprod. Fert., 35:81.

PHEROMONAL CONTROL OF THE BOVINE OVARIAN CYCLE

J. G. Vandenbergh and M. K. Izard

Department of Zoology
North Carolina State University
Raleigh, NC 27650

INTRODUCTION

The ovarian cycle of several species of rodents is influenced by priming pheromones from both males and females (Bronson, 1979; Vandenbergh, 1973; 1980). A chemical cue contained in the urine of male housemice accelerates the onset of puberty in juvenile females (Vandenbergh, et al., 1976) and influences the rythmicity of the ovarian cycle of the adult female (Whitten, 1956). In rats, synchronization of estrus occurs as a result of airborne chemical cues produced by other females (McClintock, 1978; 1981).

The economic implications of using priming pheromones to control ovarian function in animals are considerable. Research has focused on two areas. Priming pheromones, particularly those involved in estrus suppression, could be an important, density dependent regulating factor in the control of population growth in some rodents (Bronson, 1979; Massey and Vandenbergh, 1980; 1981). Research in this area is in a very preliminary stage and, at this time, we can only speculate on the regulation of rodent populations. The second area that has received attention recently is the possible role of priming pheromones in regulating ovarian function in domestic farm animals. In this paper we will describe some recent studies demonstrating the strong likelihood that priming pheromones can be useful tools to aid in the regulation of the ovarian cycle of the cow. First, I will describe our attempts at accelerating the onset of puberty in heifers by chemical stimuli and second, we will report on our work showing that a chemical cue from the estrous cow can influence ovarian cyclicity in herdmates.

CONTROL OF PUBERTY

The first study was designed to test whether puberty could be advanced in beef heifers by exposure to bull urine (Izard and Vandenbergh, in press). Further we hypothesized that if more heifers treated with bull urine than control heifers reached puberty early in the breeding season then these heifers would calve earlier in the calving season.

Fifty-two crossbred beef heifers were divided into two equal groups on the basis of age, weight, and breed of sire. The groups were then randomly assigned to receive either bull urine or water as a treatment. The heifers ranged on two separate pastures of approximately equal size and quality. Both pastures were isolated from pastures containing bulls. Puberty in heifers is usually defined as the age at which the heifer will first stand for a mount, either by a bull or by another heifer. We could not use a bull to test for puberty since bull urine was our treatment variable so we used ovarian palpation to assess pubertal status. Within two weeks after assignment to a group the heifers were palpated via the rectum and classified as pubertal or prepubertal based on the presence or absence of a palpable ovarian follicle or corpus luteum. In a few cases where the heifers were too small to palpate or the ovarian structure was questionable, blood was collected for progesterone assay.

At the time of palpation the heifers received the first of eight weekly treatments with either bull urine or water. The heifers were briefly restrained in a shute and sprayed on the nose and mouth with 3 ml of urine previously collected from a bull of proven fertility or, in a similar manner with water. At the eighth treatment the heifers were again weighed and palpated for ovarian structures.

At the time of the first palpation, 35% of the heifers destined to be in the urine treatment group had already attained puberty as had 27% in the water treatment group. These animals were left with their assigned groups to keep group size equivalent. The focus of the experiment, however, was on the remaining heifers, i.e., those that were prepubertal at the first palpation. When palpated for a second time, 67% of these previously prepubertal heifers had attained puberty in the group treated with bull urine but only 32% of their counterparts in the water treated group had reached puberty. This difference is significant at the <0.05 level on a chi square test. Thus, eight weekly exposures to bull urine resulted in a higher proportion of heifers attaining puberty during a fixed interval of time.

After being palpated for the second time, the two herds were split in half randomly and combined with half of the other group.

An adult Angus bull was released into each pasture. The heifers were left with the bulls for a 90-day breeding period from January through March. In June all heifers were palpated for pregnancy and the calving date was designated as day 1 of the calving season. No differences were found in comparing the incidence of pregnancy between the two groups. Seventy-nine per cent of the heifers treated with water had palpable fetuses in their uteri. Forty-one of the heifers used in the experiment calved. Of these, the originally prepubertal heifers treated with bull urine calved significantly earlier than those treated with water.

These data suggest that bull urine contains a priming pheromone which can accelerate the onset of puberty. This effect is much like that reported in housemice (Colby and Vandenbergh, 1974). These results must be considered preliminary. The magnitude of the effect on sexual maturation is relatively small, though statistically significant, and replication is necessary to confirm the effect. As yet we have no information on endocrine changes in the heifer following exposure to bull urine. Work similar to that of Bronson and Desjardins (1974) measuring circulating concentrations of reproductively significant hormones in the female mice after male stimulation is necessary now in the cow. A further goal of this work is to identify the compound or compounds serving as priming pheromones. To accomplish this elusive goal requires the development of a rapid and reliable bioassay as used by Vandenbergh, et al. (1975) in studies on housemice. We have tested whether bull urine had an effect on puberty in housemice. If it did, a proven assay would be available to us. Unfortunately, female mice exposed to bull urine matured at the same rate as those exposed to water.

Further research to confirm and extend our finding that bull urine hastens the onset of puberty in heifers may result in husbandry techniques that can significantly improve the reproductive efficiency of cattle. One of the important benefits of earlier puberty is that a higher proportion of heifers will attain puberty early in the birth season. This pattern of earlier breeding can be expected to continue throughout an individual's lifetime (Burris and Priode, 1958; Short and Bellows, 1971; Leimeister et al., 1973).

CONTROL OF ADULT CYCLE

The ovarian functions of other domestic farm animals may also be susceptible to male stimuli. Izard (in press) reviews this literature extensively. Here we only have time to mention the recent work on the pig. In 1969, Brooks and Cole reported that the presence of the boar has a strong effect on the age of puberty in gilts. Kirkwood and Hughes (1979) have recently confirmed and amplified this finding. Preliminary tests indicate that 5α-androstenone, the signalling pheromone produced by the boar is not the active priming pheromone. A chemical cue from the boar does seem to be important to the gilt since olfactory bulbectomy of gilts eliminated their ability to respond to the boar with accelerated puberty (Kirkwood et al., 1981).

Now that we have evidence that priming pheromones can help to bring a cow into sexual maturity, we can examine whether priming pheromones play a role in scheduling the ovarian cycle of the adult. This work grew from a report that cows whose estrous cycles are synchronized by injections of synthetic progestagens or prostaglandin $F_{2\alpha}$ ($PGF_{2\alpha}$) appear to enhance cycling in herdmates so that untreated cows in the same herd display a sympathetic synchronized estrus shortly after the treated cows (Weston and Ulberg, 1976; Zimbelman, Lauderdale and Moody, 1977). Weston and Ulberg (1976) suggested a possible role for pheromones in this phenomenon. Two studies have been conducted to follow-up on this suggestion (Izard and Vandenbergh, in press) and we will summarize these studies here.

The overall design of these studies was to pretreat dairy heifers with $PGF_{2\alpha}$. This prostaglandin causes regression of the corpora lutea and thus places all of the heifers in a similar early follicular phase of the cycle. Heifers treated with $PGF_{2\alpha}$ are then exposed to secretions from estrous cows to determine if such exposure reduces the variance in the onset of estrus among the heifers.

In the first test, 47 $PGF_{2\alpha}$-treated dairy heifers were exposed to either a mixture of urine and cervical mucus from estrous cows or to water as a control. The putative pheromone source was sprayed in the nostrils and mouths of the cows once daily for three days after $PGF_{2\alpha}$ injection. The heifers were then observed for estrus behavior and bred by artificial insemination (A.I.) 12 hours after estrus was detected. Exposure to the mixture of urine and cervical mucus resulted in a significant reduction in the variance of estrus onset compared to the control group exposed to water.

A second study was conducted utilizing a similar design to test each of the putative pheromone sources, urine and cervical mucus, separately. This work revealed that cervical mucus alone was capable of enhancing the synchronization of estrus following $PGF_{2\alpha}$ administration. Urine from an estrous cow did not differ from water.

The availability of a technique to enhance estrus synchronization after $PGF_{2\alpha}$ administration could improve the success of set-time A.I. and could be useful in establishing a rhythm breeding program in large herds.

One surprising result from this study was the finding that the conception rate in heifers treated with cervical mucus was only 25 per cent compared to 64 per cent in water-treated heifers and 58 per cent in urine-treated heifers. Exposure to cervical mucus may have shortened the follicular phase of the ovarian cycle and thus the time of A.I. may not have been appropriate. At this time we have no supportable explanation for this apparent reduction in fertility of the heifers treated with cervical mucus.

Without additional research it would be premature to draw firm and generalizable conclusions from the studies on the influences of pheromones on the ovarian cycle of the cow that we have summarized in this paper. The results do have the weight of precedence behind them. That is, the acceleration of puberty in heifers by bull urine was preceded by the studies showing that puberty is accelerated in juvenile mice by male mouse urine. Similarly, the synchronization of estrus in a herd by chemical signals from group-mates was preceded by the work of McClintock on rats. These similarities provide reassurance that we might be on the right track in extending our work to cows. But, it will require a great deal more experimentation on cows to verify these findings.

REFERENCES

Bronson, F. H., 1979, The reproductive ecology of the house mouse, Quart. Rev. Biol., 54:265.

Bronson, F. H., and Desjardins, C., 1974, Circulating concentrations of FSH, LH, estradiol and progesterone associated with acute, male-inudced puberty in female mice, Endocrinol., 94:1658.

Brooks, P. H., and Cole, D. J. A., 1970, The effect of the presence of a boar on the attainment of puberty in gilts, J. Reprod. & Fertil., 23:435.

Burris, M. J., and Priode, B. M., 1958, Effect of calving date on subsequent calving performance, J. Anim. Sci., 17:527.

Colby, D.R., and Vandenbergh, J. G., 1974, Regulatory effects of urinary pheromones on puberty in the mouse, Biol. Reprod., 11: 268.

Izard, M. K., Pheromones and reproduction in domestic animals, in: "Pheromones and Reproduction in Mammals," J. G. Vandenbergh, ed., Academic Press, NY (in press).

Izard, M. K., and Vandenbergh, J. G., 1982, The effects of bull urine on puberty and calving date in crossbred beef heifers, J. Anim. Sci., (in press).

Izard, M. K., and Vandenbergh, J. G., 1982, Priming pheromones from oestrous cows increase synchronization of oestrous in dairy heifers after PGF-$_{2\alpha}$ injection, J. Reprod. Fertil., (in press).

Kirkwood, R. N., and Hughes, P. E., 1979, The influence of age at first boar contact on puberty attainment in the gilt, Anim. Prod., 29:231.

Kirkwood, R. N., Forbes, J. M., and Hughes, P. E., 1981, Influence

of boar contact on attainment of puberty in gilts after removal of the olfactory bulbs, J. Reprod. Fert., 61:193.
Lesmeister, J. L., Burfening, P. J., and Blackwell, R. L., 1973, Date of first calving in beef cows and subsequent calf production, J. Anim. Sci., 36:1.
Massey, A., and Vandenbergh, J. G., 1980, Puberty delay by a urinary cue from female house mice in feral populations, Sci., 209:821.
Massey, A., and Vandenbergh, J. G., 1981, Puberty acceleration by a urinary cue from male mice in feral populations, Biol. Reprod., 24:523.
McClintock, M. K., 1978, Estrous synchrony in the rat and its mediation by airborne chemical communication (Rattus norvegicus), Horm. Behav., 10:264.
McClintock, M. K., 1981, Social control of the ovarian cycle and function of estrous synchrony, Am. Zool., 21:243.
Short, R. E., and Bellows, R. A., 1971, Relationship among weight gains, age at puberty and reproductive performance in heifers, J. Anim. Sci., 32:127.
Vandenbergh, J. G., 1973, Acceleration and inhibition of puberty in female mice by pheromones, J. Reprod. Fert., 19:411.
Vandenbergh, J. G., 1980, The influence of pheromones on puberty in rodents, in: "Chemical Signals," D. Müller-Schwarze and R. M. Silverstein (eds.), Plenum Publishing Corp.
Vandenbergh, J. G., Whitsett, J. M., and Lombardi, J. R., 1975, Partial isolation of a pheromone accelerating puberty in female mice, J. Reprod. Fertil., 43:515.
Vandenbergh, J. G., Finlayson, J. S., Dobrogosz, W. J., Dills, S. S., and Kost, T. A., 1976, Chromatographic separation of puberty accelerating pheromone from male mouse urine, Biol. Reprod., 15:260.
Weston, J. S., and Ulberg, L. C., 1976, Responses of dairy cows to the behavior of treated herdmates during the postpartum period, J. Dairy Sci., 59:1985.
Whitten, W. K., 1956, Modifications of the oestrous cycle of the mouse by external stimuli associated with the male, J. Endocrinol., 13:399.
Zimbelman, R. G., Lauderdale, J. W., and Moody, E. L., 1977, Beef A.I. and prostaglandin $F_{2\alpha}$, Proc. 11th Confr. A.I. Beef Cattle, Denver, Col., p. 66.

SYNCHRONIZING OVARIAN AND BIRTH CYCLES BY FEMALE PHEROMONES

Martha K. McClintock

Department of Behavioral Sciences
The University of Chicago
Chicago, Illinois 60637

OVARIAN SYNCHRONY

The timing of ovulation can be altered by olfactory signals from other females in the social environment. In the rat, olfactory communication results in the synchronization of estrous cycles within a female social group: the majority of females are likely to be at the same phase of their estrous cycles on the same day (McClintock, 1978; McClintock, 1981). This may be a mechanism to coordinate fertility and infant care with an appropriate social and physical environment (McClintock, 1981; McClintock, In press)

It is known that a common air supply is sufficient for the development of estrous synchrony in rats (McClintock, 1978). However, the number of olfactory signals that are produced, the times at which females are sensitive to these signals and how these factors interact to generate synchrony in the context of a social group have not been clear. In order to address these questions, it is useful to conceptualize the ovarian cycle as an oscillator or simple clock (Campbell, 1964; Best, 1975; Winfree, 1980). This is an appropriate model because the same sequence of events is repeated regularly and spontaneously in the ovarian cycle. Ovarian synchrony can then be viewed as coupling between oscillators with olfactory signals serving as the coupling mechanism. We have been able to use this coupled oscillator model to examine the olfactory signals that generate estrous synchrony in the rat, demonstrating its power for investigating the coupling mechanisms of ovarian synchrony in other species (McClintock, 1971; McClintock, In press).

Not all coupled oscillators become synchronized. For synchrony to occur in a population, both the period and phase relationships of the individual oscillators must be similar. When the phase and period of each oscillator in a group are exactly the same, perfect synchrony occurs, i.e. 100% of the population is at the same phase of the cycle at the same time. As the variance in period length increases, the level of synchrony is reduced even though the cycles may still be coupled and relatively coordinated (von Holst, 1969; Winfree, 1967). In the rat, estrous synchrony is rarely perfect across the entire group; some individuals may occasionally come into heat just before or just after the rest of the group. When the estrous cycles of a group range in length from 4 to 6 days, the level of estrous synchrony fluctuates between 50% and 85%. Furthermore, the degree of synchronization reaches a stable level only after three to four cycles.

Synchronization of coupled oscillators is seen in a variety of biological systems. In some cases synchrony is produced by an external synchronizer. In mice, a male pheromone reinstates cyclicity across a group of anestrous or pseudopregnant females (Whitten, 1958). The capacity for this form of inertial synchrony is enhanced by the coupling between females that keeps them in the same phase by prolonging one phase of the cycle. Synchrony is then generated when each female resumes short cycles at the same time, having been started in phase by the same male signal. Likewise, the synchrony of cell division cycles in yeast can be triggered by a change in the culture medium and then sustained by coupling between the cells (von Meyenburg, 1973).

Synchrony can also develop without an external zeitgeber if the oscillators are coupled to each other. This is a form of mutual entrainment. A spectacular example is found in the synchronous flashing of an entire swarm of male fireflies (Pteroptyx). In this case, synchrony results from a response to the averaged interactions among all members of the group (Buck and Buck, 1968). In others, (e.g. in the case in the synchronization of circadian leaf movements in soybean and oil perilla plants), one member of the group is dominant (in this case the higher leaf) and the other members are entrained by the dominant rhythm of the zeitgeber (Gurevich, 1967; Kübler, 1969). Because ovarian synchrony does not require an external zeitgeber and results instead from interactions within the social group (McClintock, 1971; McClintock, 1978), estrous synchrony is an example of the mutual entrainment of coupled oscillators.

In many cases of external and mutual entrainment, synchrony is generated by the coordinated action of signals with two opposing effects on the rhythm that is entrained: one that produces a phase delay and a second that produces a phase advance. This is a property common to the entrainment of circadian activity cycles by the daily light/dark cycle (DeCoursey, 1961), relative coordination of fish fin movements (von Holst, 1969), synchrony of cAMP pulsing in slime mold suspensions (Malchow, Nanjundiah and Gerisch, 1978), neural entrainment of electrical synchrony in cells of the sinoatrial pacemaker of the heart (Jalife and Moe, 1979) and synchronized emergence cycles in 13-year cicadas (Magicicada spp.; Hoppensteadt and Keller, 1976). In some systems, the coupling mechanism that produces synchrony relies on one signal that has opposite effects on the endogenous rhythm, effects which are determined by the phase, Ø, that the signal is received. This is often demonstrated by positive and negative phase changes in a phase resetting curve (Ø vs. ΔØ; although others prefer Ø vs. Ø' to avoid conceptal ambiguity; Winfree, 1980, p.383). In other systems, there are two different signals that alter the entrained rhythm in opposite ways: one signal phase advances and the other phase delays (e.g. the estrous synchrony that results from the interaction of female and male pheromones in mice; Marsden and Bronson, 1965).

One way of demonstrating the effects of an entraining signal is to present the signal as a constant stimulus and measure the change in the period of the endogenous rhythm. This is a technique that has been used to study the entrainment of circadian activity rhythm by daily fluctuation in light intensity. For example, when a diurnal animal is exposed to constant bright light, the animal's activity cycle becomes shorter than 24 hours. This shortened cycle reflects a series of phase advances in the endogenous rhythm. When the intensity of constant light is reduced, the cycle is longer than 24 hours, reflecting a series of phase delays in the endogenous rhythm (Aschoff, 1960). Bright light produces a phase advance; dim light produces a phase delay. A similar strategy has also been used to study the mutual entrainment of cricket chirps (Heiligenberg, 1969) and neural pacemakers (Perkel, Schulman, Bullock, Moore and Segundo, 1964). This method is particularly useful because it allows one to assess the effects of a putative entraining stimulus without knowing the time of maximum sensitivity to the signal. Once a set of entraining stimuli have been thus characterized, they can be presented in discrete pulses at different times in the cycle in order to assess the type of response at each phase of the cycle.

OLFACTORY MECHANISMS OF OVARIAN SYNCHRONY

Effect of Estrous Cycle Odors

The first step in the identification of the olfactory coupling mechanism was to partition the estrous cycle into three hormonally distinct phases that could potentially produce different signals: the follicular phase (lights-on of diestrus to lights-on of proestrus), the ovulatory phase (lights-on of proestrus to lights-on of estrus) and the luteal phase (lights-on of metestrus to lights-on of diestrus). See Figure 1. Note that progesterone is produced in the luteal phase, although this phase is not fully functional in the rat (Long and Evans, 1922).

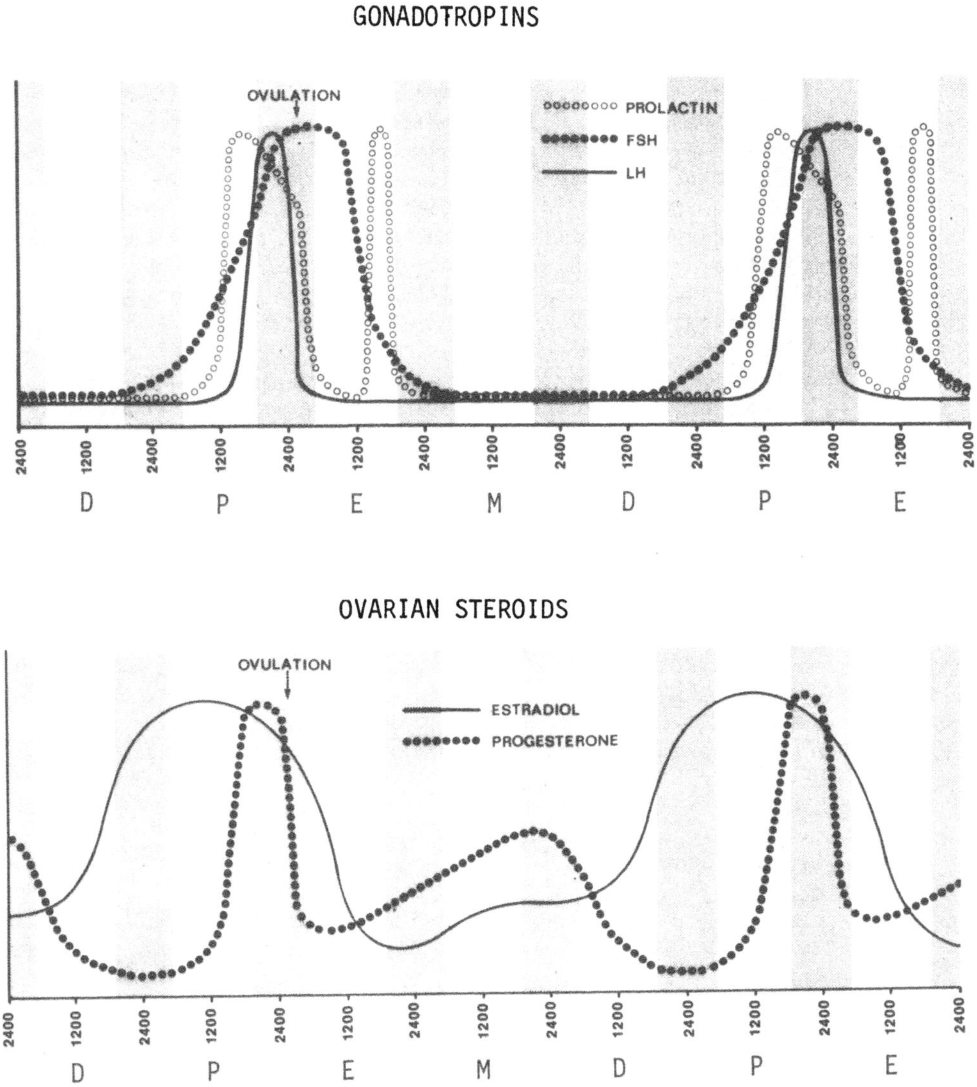

Fig. 1. A schematic of gonadotropin and ovarian steroid levels during a 4-day estrous cycle in the rat. D = diestrus, P = proestrus, E = estrus, M = metestrus. Figure redrawn from data in Butcher, Collins and Fugo (1974), Smith, Freeman and Neill (1975), and Nequin, Alvarez and Schwartz (1979).

In order to generate constant odors from each of these phases, six 4-month-old Sprague-Dawley rats that were expected to enter a particular phase of the estrous cycle were put in an odor source box. (Similar equipment is described in McClintock and Adler, 1978). They remained in the source box for 24 hours and were then removed and replaced with another set of animals that were just then entering the same phase of the estrous cycle. Six different females lived downwind from the odor source box in separate compartments of the downwind box and a control group of six females lived in a similar box upwind (McClintock, In prep.).

The data shown in Figure 2A suggest that at least two different pheromones are produced during the rat's estrous cycle. These pheromones have opposing effects on the timing of the estrous cycle. Odors from the follicular phase shorten or phase advance the estrous cycle of females living downwind, increasing the probability of four day cycles (see Figure 2A). Ovulatory odors have the opposite effect; they lengthen or phase delay the cycle, increasing the probability of longer cycles ($\geq$ 5 days). Luteal odors do not have a significant effect on estrous cycle length when presented against the background odors of this study.

Because anovulatory cycles are relatively rare in the young rat (Bingel and Schwartz, 1969), and changes in vaginal cytology are a reasonably good indicator of ovulation (Everett, 1964), the length of the vaginal smear cycle is a good measure of the interval between successive ovulations. This enables us to examine the effect of estrous cycle odors on the rate or probability of ovulation by replotting the same cycle distributions presented in the lower three histograms of Figure 2A as a log survivor plot (see Figure 2B).

In log survivor analysis, the probability of ending a cycle, in this case the probability or rate of ovulating, is proportional to the slope of the log survivor curve at a particular time, t (Fagen and Young, 1978). Therefore, the constant but different slopes of each of these log survivor curves indicates that females ovulate at constant but significantly different rates in each condition (Figure 2B). (The rate of ovulation during exposure to luteal phase odors can serve as the referent because the distribution of cycles is not significantly different from that during exposure to background odors.) When females are exposed to ovulatory odors, the rate of ovulation is significantly lower than during exposure to luteal phase odors ($p < 0.0001$, Mantel-Cox). Nonetheless, the rate is still constant and each cycle length is equiprobable. On the other hand, follicular odors increase both the rate of ovulation and the dependence of that rate on time since the last ovulation ($p < 0.0001$, Mantel-Cox). Once four days have passed since the last ovulation, the probability of ovulating again is 1.00. The endogenous rhythm of the ovarian cycle does limit the phase advances produced by follicular odors; cycles are not shortened below

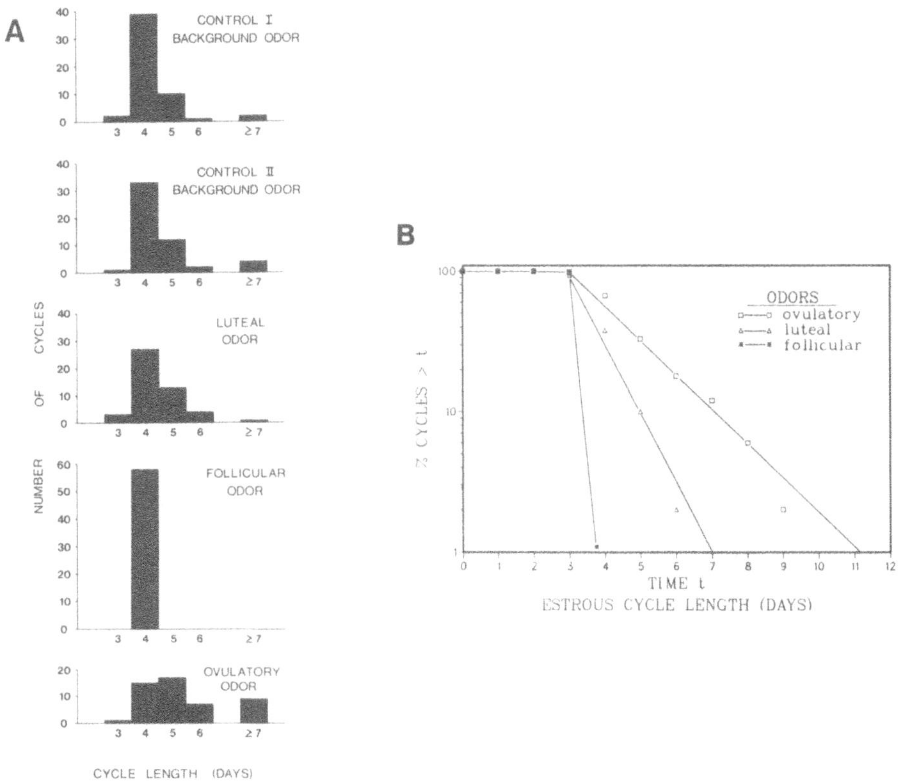

Fig. 2. (A) Histograms and (B) a log survivor plot depicting the effect of different estrous cycle odors on estrous cycle length.

the 4-day limit which normally marks the lower bound of the length of the rat's ovarian cycle (Everett, 1948). (In another replication of this experiment, the effect was only slighly less dramatic; 85% of cycles were four days long).

Follicular odors also advance the time of ovulation within the circadian time frame, an inference that is based on a detailed analysis of the timing of changes in vaginal cytology. The transition from a predominantly cornified vaginal epithelium to the emergence of leukocytes in the vaginal smear indicates that ovulation has occurred (Schwartz, 1964; Weick, Smith, Dominguez, Dhariwal and Davidson, 1971). If ovulation is delayed (e.g. late in the morning of estrus), then the transition to leukocytes will also be delayed. Therefore, the vaginal smear taken at the standard time (midway through the dark phase) will be taken before the transition to leukocytes and will still contain cornified cells. However, if ovulation is advanced, then the transition will also be advanced ahead of the time that vaginal smears are taken and the smear will contain leukocytes. This method of interpreting the pattern of vaginal smears has been validated by assessing the circadian timing of ovulation by examination of the oviducts through laparotomy (Matthew, pers. comm.).

The proportion of cell types in the vaginal smear was recorded for each day of the estrous cycle. All seven animals that had a large proportion of four day cycles during exposure to both background and follicular odors had a pattern of vaginal smear types that was consistent with a phase advance in the time of ovulation ($p = 0.008$; binomial test). Thus follicular odors also advance the circadian timing of ovulation during the night between proestrus and estrus. It was not possible to assess the effect of ovulatory odors on the circadian timing of ovulation, because the number of cycles of any one length was dramatically reduced by ovulatory odors (see Figure 2). Nonetheless, these findings are consistent with other reports that circadian rhythmicity is affected by ovarian cycle events (Morin, Fitzgerald and Zucker, 1977).

Mechanisms for Changing Ovarian Cycle Length

There are two processes that affect the timing of ovulation and the length of the ovarian cycle in the rat (Schwartz, 1969; Nequin et al., 1979). The first is the rate of follicular growth, which can be accelerated by follicle stimulating hormone and retarded by progesterone (Buffler and Roser, 1974). The second process is the timing of the LH surge that precedes ovulation. The neural prerequisite for the LH surge is a circadian event. In addition, estrogen from the ripening follicles is necessary to stimulate the LH surge. Progesterone, secreted by the preceding corpus luteum, can inhibit neural events in the hypothalamus, delay the subsequent LH surge, and lengthen the cycle (Everett, 1964; Naftolin, Brown-Grant and Corker, 1972; Nequin et al., 1979).

Therefore there are a variety of excitatory and inhibitory mechanisms through which estrous cycle pheromones could regulate cycle length. Because cornified epithelial cells in the vaginal smear indicate estrogen produced by ripening follicles (Long and Evans, 1922; Nequin et al., 1979) and the appearance of leukocytes is associated with progesterone produced by the corpus luteum (Parkes, 1929), a detailed record of the proportion of epithelial cells and leukocytes in the vaginal smear can be used to evaluate the alternative regulatory mechanisms.

Follicular odors could shorten the cycle by accelerating follicular development and estrogen production; threshold levels of estrogen would be attained sooner and the time of the LH surge would be advanced. If this were the case, there should be an increase in the number of cycles that have a large proportion of cornified cells, particularly in the follicular phase (late diestrus and early proestrus) and an increase in 3-day cycles (because of an increase in estrogen priming; Krey, Tyrey and Everett, 1973). Alternatively, follicular odors could shorten cycle by shortening the lifespan of the corpus luteum and reducing progesterone which would delay the LH surge; in the absence of the delaying effect of progesterone, the LH surge would be advanced. If this were the case, the time course of follicular development would be normal.

The latter hypothesis is favored because follicular odors do not produce cycles with a higher proportion of days with cornified cells ($p = 0.008$) nor an increase in 3-day cycles (see Figure 2A). This suggests a normal follicular and a short luteal phase. However, odors may in addition act centrally to change the sensitivity to feedback from ovarian steroids.

Ovulatory odors may lengthen the cycle through this same set of alternative mechanisms. Ovulatory odors could act centrally to delay the neural events necessary for the LH surge even when follicular development occurs on time. If this were the case, the longer cycles would have extended follicular phases indicated by an increase in the number of days of vaginal cornification in the vaginal smear. Alternatively, ovulatory odors could lengthen the cycle by prolonging the lifespan of the corpus luteum. Progesterone produced by the corpus luteum would retard follicular development and delay the gonadotropin surge. If this were the case, the longer cycles would have only one or two days of cornified cells, indicating a follicular phase of normal length, and an extended number of days with leukocytes in the vaginal smear.

Ovulatory odors produced cycles with extended periods of leukocytes, indicating an extended luteal phase ($p \leq 0.05$). The length of the follicular phase (indicated by cornified cells) was not prolonged. Because progesterone produced by the corpus luteum can retard follicular development, it is likely that ovulatory odors prolong the cycle by increasing the lifespan of the corpus luteum, increasing the duration of progesterone secretion, and delaying follicular development and ovulation.

Thus the most parsimonious hypothesis is that estrous cycle odors alter the length of the estrous cycle by modulating the lifespan of the corpus luteum rather than the length of the follicular phase. Nonetheless our data do not allow us to rule out the additional possibilities that follicular odors act centrally to facilitate the LH surge and that ovulatory odors also inhibit follicular development directly as well as through progesterone production.

Effects on Other Components of the Ovarian Cycle: The Lordosis Reflex

When female rats are socially isolated, the lordosis reflex can be elicited by palpation on every day of the estrous cycle (McClintock, 1981). Normally, when females live in groups, the lordosis reflex can only be elicited when the vaginal smear contains cornified epithelial cells and is therefore temporally coordinated with vaginal cytology. A mixture of odors from all phases of the estrous cycle failed to reinstate the cyclic lordosis pattern normally found in grouped females (McClintock, 1981). Furthermore, in this study, odors presented singly from each phase of the estrous cycle also failed to affect the lordosis reflex (see Table 1). There was also no significant correlation within individuals between the temporal disruption of the lordosis reflex in response to social isolation of the downwind boxes living compartments and the change in the vaginal smear cycle length in response to estrous cycle odors (diestrous odor, r = .27, NS; ovulatory odor, r = .35, NS).

Table 1. Effect of estrous cycle odors on the average daily strength of lordosis reflex in solitary females (ordinal scale 0 - 9). The differences between control groups probably reflects differences in cohorts or season at the time of testing; a later replication using follicular odors found levels that were similar to those of the ovulatory and luteal groups.

ODORS	UPWIND CONTROL	DOWNWIND EXPERIMENTAL	Matched-pairs signed-ranks test
Follicular	3.5	3.4	N.S.
Ovulatory	5.4	6.4	N.S.
Luteal	6.4	5.7	N.S.

A coupled oscillator model predicts that odors will alter the timing and cyclicity of the lordosis reflex only if the mechanism of the lordosis reflex is tightly coupled to the mechanism underlying the vaginal smear cycle. For example, in yeast cells, cell budding is normally temporally correlated with the excretion of a substance into the medium. The two activites are, however, dependent on each other physiologically. Changes in the culture medium that phase shift and synchronize cell budding do not phase shift this excretion, resulting in temporal dissociation of the two events (Kraeplin, 1973). Similarly, estrous cycle pheromones do not alter the temporal pattern of the lordosis reflex, although cyclicity of vaginal cytology is affected. This provides additional evidence that the lordosis reflex (as measured by palpation) is not tightly coupled to the vaginal smear component of the ovarian cycle (Gerall and McCrady, 1970; McClintock, 1981) and appears to be regulated by social signals other than estrous cycle pheromones from other females.

Time of Sensitivity

In the rat, olfactory sensitivity is greatest just prior to ovulation (Curry, 1971; Pietras and Moulton, 1974), suggesting that this may be the time when the rat is most responsive to pheromonal cues. Considering that the corpus luteum forms immediately after ovulation (Nikitovitch-Winer and Everett, 1958), sensitivity at this time would be consistent with the hypothesis that estrous cycle pheromones change cycle length by altering the lifespan of the corpus luteum and the neural thresholds underlying the LH surge.

The next step in testing the coupled oscillator model will be to determine the phase response curves for each of the estrous cycle odors. Because the response is a change in the length of the cycle, the phase response curve will be a phase resetting curve where increases and decreases in phase angle are plotted as a function of estrous cycle phase. After the phase resetting curve is generated, it will be possible to model the development and maintenance of synchrony in a variety of group sizes. It is possible that the same coupling mechanisms will produce synchrony 'in phase' in some populations and 'out of phase' in others.

OTHER FORMS OF SYNCHRONY: THE BIRTH CYCLE

In many species, females spend a larger portion of their reproductive life spans in birth cycles of conception, pregnancy and lactation than they do in spontaneous unfertilized ovarian cycles (Short, 1976; Altmann, Altmann, Hausfater and McCuskey, 1977). Females can influence the fertility of other females in their social group at a variety of points in this birth cycle. For example, female gerbils will not be fertile if their mother is pregnant or lactating, although they will become fertile when she is having spontaneous ovarian cycles (Payman and Swanson, 1980). Because pregnancy and lactation have neuroendocrine mechanisms in common with the ovarian cycle, it is possible that the birth cycle may also generate pheromones that regulate fertility (McClintock, 1981).

To test this hypothesis, four pregnant female rats were placed in separate compartments of a wind tunnel's odor source box. Each female in the source box had mated on the same day so that the sequence of birth cycle odors from the source box were synchronized. This was designed to provide a strong signal that would be relatively unaffected by individual variation. Six female Sprague-Dawley rats with regular spontaneous estrous cycles were placed in separate compartments of a box downwind. A control group of six females lived upwind. The timing of the estrous cycles of the downwind and upwind females were measured as the females in the source box went through a birth cycle of pregnancy, birth and lactation (44 days). In addition, a second set of females was placed in another wind tunnel. The six downwind females were also exposed to odors from four females over the course of a birth cycle: pregnancy, birth, and lactation. However, the downwind females in this second wind tunnel were first exposed to follicular odors for 20 days by placing a series of diestrous females in the source box (see the previous section; McClintock, In prep.). This additional exposure was designed to delay the time of exposure to birth cycle odors and thus control for the effects of living in the wind tunnel for a long time and chance variations in the laboratory environment. It also provided another replication of the effect of follicular odor and established a basis of comparison for the effect of birth cycle odors.

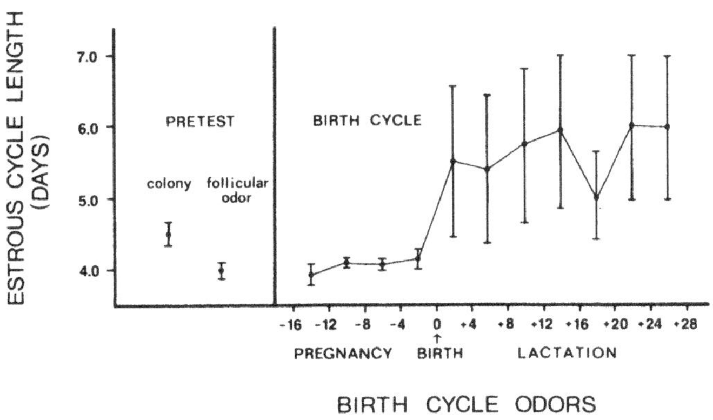

Fig. 3. The effect of birth cycle odors on the length of the estrous cycle. Each point represents the current cycle length for each successive four day block of time (mean ± S.E.M.).

Odors from pregnant females shortened the estrous cycle ($p \leq 0.033$, sign test, $N = 12$; see Figure 3). Although this effect was not quite as dramatic as that of follicular odors ($p \leq 0.030$, sign test, $N = 6$; see Figure 3), there was no significant difference between the changes produced by pregnancy and follicular odors ($p > 0.050$, sign test). The control females living upwind did not show either of these changes ($p > 0.050$, sign test, $N = 12$).

Odors from lactating females and their pups lengthened the estrous cycle and increased its variance ($p < 0.004$, sign test, $N = 12$), whereas upwind controls were not significantly affected ($p > 0.050$, sign test, $N = 12$). These changes occurred soon after the odor source females gave birth whether or not the downwind females had been previously exposed to follicular odors. The effects were thus independent of time spent in the wind tunnel.

These findings suggest a mechanism for the incidental observation that females have irregular estrous cycles when pregnant and lactating females are in a colony room (Aron, 1973), particularly because lactation odors appear to be more potent than pregnancy odors. However, it remains to be determined whether the source of the pheromone is the lactating female or her pups. One candidate is the maternal pheromone which is produced by lactating mothers and serves as an attractant for her pups (Leon and Moltz, 1971). However, the maternal pheromone is not fully effective until 14 days postpartum when pups begin to eat solid food; the estrous cycle can be disrupted by odors from females and pups immediately after birth. Nonetheless, it is possible that the estrous cycle is more sensitive to this same pheromone or perhaps another pheromone is produced during lactation. Another candidate is odors from the pups. Contact with newborn pups can induce maternal behavior and disrupt the estrous cycle of virgin females (Marinari and Moltz, 1978). Our data suggest that these changes are caused by a pheromone that is either carried or produced by the pups.

In the mouse, the effect of lactation and pup odors is the opposite of that in the rat: estrus is prolonged and fertility is enhanced, although the length of the estrous cycle is not changed (Hoover and Drickamer, 1979). These contrasting effects of birth cycle odors are in keeping with the contrasting effects of estrous cycle odors in the two species: ovarian cycles are lengthened by an all female social environment in the mouse (Whitten, 1959) but shortened in the rat (Aron, 1973; McClintock, 1981).

The birth cycle may not only be an appropriate temporal context to evaluate the production of pheromones, it may also be a time when the female is particularly sensitive to cues in her environment (McClintock, 1981). Because the rat is capable of conceiving at the postpartum estrus, 12 hours after giving birth, female rats have the opportunity to carry a second pregnancy while they are caring for their newborn at the nest. This requires a large energy investment and the availability of abundant resources. If the physical and social environments are not appropriate, the female may either delay implantation or not conceive at all at the postparum estrus (Krehbiel, 1941; Woodside, Wilson, Chee, and Leon, 1981). Therefore, females may be particularly sensitive to pheromonal cues at this time if the cues provide reliable information about the availability of resources in the immediate environment or the likelihood of resource competition from other lactating females.

DOMINANCE: AN ADDITIONAL FACTOR

In pairs of female hamsters, the socially dominant female maintains regular ovarian cycles while the socially subordinate female has variable cycles that entrain to those of her dominant partner (Handelmann, Ravizza and Ray, 1980). This is a form of synchrony in which there is a dominant rhythm or zeitgeber which entrains the other rhythms in a population (Buck and Buck, 1968; Winfree, 1980). We also found evidence for estrous entrainment in female hamsters, studying them in groups of five. Variance in cycle length was less within than between groups (McClintock, unpublished observations). However, in our larger groups, we did not find the synchrony which develops between pairs. This may be explained by the difference in group size and therefore the number of oscillators that are coupled in the population, or it may reflect disruption of cyclicity by the fights that are common in large groups of unrelated female hamsters.

Nonetheless, it is very significant that social dominance can correlate with having the dominant estrous cycle in a group. In yellow baboons (Papio cynocephalus), pubescent females reach menarche in synchrony, lead by the dominant female of the cohort (Altmann, in prep.). A similar correlation between the dominant rhythm and social dominance is found in the dwarf mongoose (Helogale parvula; Rood, 1980). The dominant female of a pack produces litters at regular intervals. The subordinate females come into heat in synchrony with the alpha female, but usually do not have offspring; they may fail to conceive or they may have spontaneous abortions.

Although pheromonal mechanisms have yet to be documented for these species, the findings do suggest that social dominance may modulate a pheromonal system and that dominant females may synchronize and/or suppress the estrous cycles of subordinates with pheromonal signals. For example, in mice, all the studies to date of estrous suppression have failed to document complete suppression; there are always some females in the group that maintain regular estrous cycles (van der Lee and Boot, 1955; Lamond, 1959). A comparison with ovarian suppression of subordinates in other species suggests that the few cycling females may be socially dominant

(McClintock, In press). This hypothesis has recently been confirmed by Kennedy and Lenington (In litt.). Eight different groups of wild house mice (Mus musculus) containing 2 to 4 males and 4 to 6 females were placed in a seminatural environment for 20 days. Only 32% of the females mated. Although the number of groups with documented matings was small (6/8), the probability of mating tended to correlate with dominance rank (r = 0.81). The only females to carry a pregnancy to term were the dominant or second ranked females. Furthermore, second ranked females only carried to term if the dominant female in her group also had a successful pregnancy. It may be that subordinate females are suppressed by pheromones from females or that dominant females are more sensitive to male pheromones. In either case, this is the first preliminary evidence that reproductive pheromones are modulated by dominance status in a heterosexual population.

IN SUMMARY: PHEROMONES OR CHEMOSIGNALS?

The term pheromone has been used throughout this discussion primarily because ovarian and birth cycle pheromones operate similarly to the classic set of primer pheromones that regulate ovarian cyclicity and pregnancy in the mouse. However, the application of the term "pheromone" to mammals has raised objections, primarily because it implies a simplicity of action that was once thought to characterize insect pheromones. Therefore, by way of summary, it may be helpful to review the five criteria for a pheromone that were set out by Beauchamp, Doty, Moulton, and Mugford (1976) and evaluate the appropriateness of the term for the signals which produce ovarian synchrony in the rat. It should be borne in mind, however, that many insect pheromones do not meet all of these criteria.

(1) Serve a well defined neuroendocrine function. Estrous cycle odors alter the length of the ovarian cycle and the rate of ovulation, but do not affect the strength or cyclicity of the lordosis reflex. (2) Uniqueness of the compound in producing a response. These compounds can operate against a background of odors from a rat colony and in this sense are relatively unique signals. Nonetheless, similar alterations of ovarian cycle length can be induced by other odors. For example, prolonged estrous cycles are produced both by ovulatory odors and by odors from lactating females and their pups. In addition, ammonia is a nonspecific noxious odor that can also prolong the cycle and produce pseudopregnancy in rats (Takewaki, 1949). The estrous cycle can be shortened both by odors from females in a follicular phase and by odors from males (Chateau, Roos and Aron, 1972). The responses to these different odors may depend on different neuroendocrine mechanisms or they may share a final common path. (3) Species specificity. The estrous cycles of rats are shortened when they are paired with hamsters (Weizenbaum, McClintock and Adler, 1977). Although neither this phenomenon nor

estrous synchrony in hamsters have a documented pheromonal mechanism, these data suggest that rats may also respond to hamster estrous cycle odors. (4) Minimal dependence on learning. Aron (1973) reported that estrous cycles are shortened only if females have been isolated since weaning and not exposed to odors of other rats. We have found that females which have not been isolated will also respond (McClintock, 1981; McClintock, In press), demonstrating that the pheromone system is relatively robust and can develop in a variety of social environments. (5) Relatively few compounds in the signal. The compounds are airborne chemosignals that most probably stimulate receptors in the main olfactory or the vomeronasal system (Johns, 1980). An accurate guess of their number must await isolation and chemical characterization. Given that many insect pheromone systems also meet some but not all five of these criteria, it seems appropriate to treat the estrous and birth cycle odors that modulate and synchronize ovarian cyclicity as pheromones.

ACKNOWLEDGMENTS

I thank T. Butler, R. B. Church, S. Cogswell, C. Hedricks, J. Herman, J. LeFevre, M. Mostert and D. Wolf for data collection; J. Charrow for incisive comments on an earlier draft; and T. Butler for graphics. This work was supported by grants from the National Science Foundation BNS 78-03658 and 80-19496 and the National Institute of Aging PHS 5 R23AG02408.

REFERENCES

Altmann, J., Altmann, S., Hausfater, G. and McCuskey, S. A. 1977. Life history of yellow baboons: physical development, reproductive parameters, and infant mortality. Primates 18: 315-330.

Altmann, J. In preparation. Synchronous menarche in baboons.

Aron, C. 1973. Phéromones et regulation de la durée du cycle oestral chez la ratte. Arch. Anat. Embr. Norm. et Exp. 56: 209-216.

Aschoff, J. 1960. Exogenous and endogenous components in circadian rhythms. Symp. Quant. Biol. 25: 11-28.

Beauchamp, G. K., Doty, R. L., Moulton, D. G. and Mugford, R. A. 1976. The pheromone concept in mammalian chemical communication: a critique. In: Mammalian Olfaction, Reproductive Processes, and Behavior (R. L. Doty, ed.) pp. 144-160. Academic Press, New York.

Best, E. N. 1975. Exploration of a menstrual cycle model. Simulation Today 25: 117-120.

Bingel, A. S. and Schwartz, N. B. 1969. Pituitary LH content and reproductive tract changes during the mouse oestrus cycle. J. Reprod. Fertil. 19: 215-222.

Buck, J. and Buck, E. 1968. Mechanism of rhythmic synchronous flashing of fireflies. Science 159: 1319-1327.

Buffler, G. and Roser, S. 1974. New data concerning the role played by progesterone in the control of follicular growth in the rat. Acta. Endocr. 75: 569-578.

Butcher, R. L., Collins, W. E. and Fugo, N. W. 1974. Plasma concentration of LH, FSH, prolactin, progesterone and estradiol-17 throughout the 4-day estrous cycle of the rat. Endocrinology 94: 1704-1708.

Campbell, A. 1964. The theoretical basis of synchronization by shifts in environmental conditions. In: Synchrony in Cell Division and Growth (E. Zeuthen, ed.) pp. 469-484. Wiley, New York.

Chateau, D., Roos, J. and Aron, C. 1972. Action de l'urine mâle ou femelle provenant de rats normaux ou castrés sur la durée du cycle oestral chez la ratte. C. R. Soc. Biol. 166: 1110-1113.

Curry, J. J. 1971. Effects of estrogen on LH evolved potentials in the olfactory system of female rats. In: Influence of Hormones on the Nervous System (D. H. Ford, ed.) pp. 255-268. Krager, Basel.

DeCoursey, P. 1961. Effect of light on the circadian activity rhythm of the flying squirrel, Glaucomys volans. Z. vergl. Physiol. 44: 331-354.

Everett, J. W. 1948. Progesterone and estrogen in the experimental control of ovulation time and other features of the estrous cycle in the rat. Endocrinology 43: 389-405.

Everett, J. W. 1964. Preoptic stimulative lesions and ovulation in the rat: 'thresholds' and LH-release time in late diestrus and proestrus. In: Major Problems in Neuroendocrinology (E. Bajusz & G. Jasmin, ed.) pp. 346-366. Karger, Basel.

Fagen, R. M. and Young, D. Y. 1978. Temporal patterns of behaviour: durations, intervals, latencies & sequences.. In Quantitative Ethology J. Wiley & Sons, New York. (P. N. Colgan, ed.) pp. 79-114.

Gerall, A. A. and McCrady, D. E. 1970. Receptivity scores of female rats stimulated either manually or by males. J. Endocrinol. 46: 55-59.

Gurevich, B. Kh. 1967. Specific communications in plants on the basis of diurnal physiological rhythms. Dokl. (Proc.) Acad. Sci. USSR, Botanical Sec. 173: 46-48.

Handelmann, G., Ravizza, R. and Ray, W. J. 1980. Social dominance determines estrous entrainment among female hamsters. Horm. Behav. 14: 107-115.

Heiligenberg, W. 1969. The effect of stimulus chirps on a cricket's chirping (Acheta domesticus). Z. vergl. Physiologie. 65: 70-97.

Hoover, J. E. and Drickamer, L. C. 1979. Effects of urine from pregnant and lactating female house mice on oestrous cycles of adult females. J. Reprod. Fert. 55: 297-301.

Hoppensteadt, F. and Keller, J. B. 1976. Synchrony of periodical cicada emergences. Science 194: 335-336.

Jalife, J. and Moe, G. K. 1979. Phasic effects of vagal stimulation of pacemaker activity of the isolated sinus node of the young cat. Circ. Res. 45: 595-608.

Johns, M. A. 1980. The role of the vomeronasal system in mammalian reproductive physiology. In: Chemical Signals in Vertebrates (D. Müller-Schwarze & R. M. Silverstein, eds.) pp. 341-364. Plenum Publishing Corp., New York.

Kraeplin, G. 1973. Physiological rhythms in Saccharomyces cerevisiae populations. In: Biological and Biochemical Oscillators (B. Chance, E. K. Pye, A. K. Ghosh & B. Hess, eds.) pp. 419-427. Academic Press, New York.

Krehbiel, R. H. 1941. The effects of lactation on the implantation of ova of a concurrent pregnancy in the rat. Anat. Rec. 81: 43-63.

Krey, L. C., Tyrey, L. and Everett, J. W. 1973. The estrogen-induced advance in the cyclic LH surge in the rat: dependency on ovarian progesterone secretion. Endocrinology 93: 385-390.

Kübler, F. 1969. Wechselseitige Synchronisation der Blattbewegungen innerhalb einer Pflanze. Z. Pflanzenphysiol. Bd. 61: 310-313.

Lamond, D. R. 1959. Effect of stimulation derived from other animals of the same species on oestrous cycles in mice. J. Endocrin. 18: 343-349.

Leon, M. and Moltz, H. 1971. Maternal pheromone: discrimination by preweaning albino rats. Physiol. Behav. 7: 265-267.

Long, J. A. and Evans, H. M. 1922. The oestrous cycle in the rat and its associated phenomena. Mem. Univ. California 6: 1-148.

Malchow, D., Nanjundiah, V. and Gerisch, G. 1978. pH-oscillations in cell suspensions of Dictystelium discoideum: their relation to cyclic-AMP signals. J. Cell. Sci. 30: 319-330.

Marinari, K. T. and Moltz, H. 1978. Serum prolactin levels and vaginal cyclicity in concaveated and lactating female rats. Physiol. Behav. 21: 525-528.

Marsden, H. M. and Bronson, F. H. 1965. The synchrony of oestrus in mice: relative roles of the male and female environments. J. Endocrin. 32: 313-319.

McClintock, M. K. 1971. Menstrual synchrony and suppression. Nature 229: 244-245.

McClintock, M. K. 1978. Estrous synchrony and its mediation by airborne chemical communication (Rattus norvegicus). Horm. Behav. 10: 264-276.

McClintock, M. K. 1981. Social control of the ovarian cycle and the function of estrous synchrony. Amer. Zool. 21: 243-256.

McClintock, M. K. In press. Pheromonal regulation of the ovarian cycle: enhancement, suppression and synchrony. In: Pheromones and Reproduction in Mammals (J. G. Vandenbergh, ed.) Academic Press, New York.

McClintock, M. K. In preparation. Modulation of the ovarian cycle and estrous synchrony by female chemosignals.

McClintock, M. K. and Adler, N. T. 1978. Induction of persistent estrus by airborne chemical communication among female rats. Horm. Behav. 11: 414-418.

Morin, L., Fitzgerald, K. M. and Zucker, I. 1977. Estradiol shortens the period of hamster circadian rhythms. Science 196: 305-307.

Naftolin, F., Brown-Grant, K. and Corker, C. S. 1972. Plasma and pituitary luteinizing hormone and peripheral plasma oestradiol concentrations in the normal oestrous cycle of the rat and after experimental manipulation of the cycle. J. Endocrin. 53: 17-30.

Nequin, L. G., Alvarez, J. and Schwartz, N. B. 1979. Measurement of serum steroid and gonadotropin levels and uterine and ovarian variables throughout 4 day and 5 day estrous cycles in the rat. Biol. Reprod. 20: 659-670.

Nikitovitch-Winer, M. E. and Everett, J. W. 1958. Comparative study of luteotropin secretion by hypophysial autotransplants in the rat. Effect of site and stages of the estrous cycle. Endocrinol. 62: 522-253.

Parkes, A. S. 1929. The functions of the corpus luteum. II. The experimental induction of plancentomata in the mouse. Proc. Roy. Soc. B. 104: 183-188.

Payman, B. C. and Swanson, H. H. 1980. Social influence on sexual maturation and breeding in the female mongolian gerbil (Meriones unguiculatus). Anim. Behav. 28: 528-535.

Perkel, D. H., Schulman, J. H., Bullock, T. H., Moore, C. P. and Segundo, J. P. 1964. Pacemaker neurons: effects of regularly spaced synaptic input. Science 145: 61-63.

Pietras, R. J. and Moulton, D. G. 1974. Hormonal influences on odor detection in rats: changes associated with the estrous cycle, pseudopregnancy, ovariectomy, and administration of testosterone propionate. Physiol. Behav. 12: 475-491.

Rood, J. P. 1980. Mating relationships and breeding suppression in the dwarf mongoose. Anim. Behav. 28: 143-150.

Schwartz, N. B. 1964. Acute effects of ovariectomy on pituitary LH, uterine weight, and vaginal cornification. Am. J. Physiol. 207: 1251-1259.

Schwartz, N. B. 1969. A model for the regulation of ovulation in the rat. Recent Prog. Horm. Res. 25: 1-55.

Short, R. V. 1976. The evolution of human reproduction. Proc. R. Soc. Lond. B. 195: 3-24.

Smith, M. S., Freeman, M. E. and Neill, J. D. 1975. The control of progesterone secretion during the estrous cycle and early pseudopregnancy in the rat: prolactin, gonadotropin, and steroid levels associated with rescue of the corpus luteum of pseudopregnancy. Endocrinology 96: 219-225.

Takewaki, K. 1949. Occurrence of pseudopregnancy in rats placed in vapor ammonia. P. Japan Acad. 25(7): 38-39.

van der Lee, S. and Boot, L. M. 1955. Spontaneous pseudopregnancy in mice. Acta. Physiol. Pharmacol. Neerlandica 4: 442-444.

von Holst, E. 1969. Zur Verhaltenphysiologie bei Tieren und Menschen: Gesammelte Abhandlungen. Trans. R. Martin, The Behavioral Physioloy of Animals and Man. University of Miami Press, Coral Gables.

von Meyenburg, H. K. 1973. Stable synchrony oscillations in continuous cultures of Saccharomyces cerevisiae under glucose limitation. In Biological and Biochemical Oscillators (B. Chance, E. K. Pye, A. K. Ghosh & B. Hess, eds.) pp. 411-417. Academic Press, New York.

Weick, R. F., Smith, E. R., Dominguez, R., Dhariwal, A. P. S. and Davidson, J. M. 1971. Mechanism of stimulatory feedback effect of estradiol benzoate on the pituitary. Endocrinology 88: 293-301.

Weizenbaum, F., McClintock, M. K. and Adler, N. 1977. Decreases in vaginal acyclicity of rats when housed with female hamsters. Horm. Behav. 8: 342-347.

Whitten, W. K. 1958. Modification of the oestrus cycle of the mouse by external stimuli associated with the male. J. Endocrin. 17: 307-313.

Whitten, W. K. 1959. Occurrence of anoestrus in mice caged in groups. J. Endocrin. 18: 102-107.

Winfree, A. T. 1967. Biological rhythms and the behavior of coupled oscillators. J. Theoret. Biol. 16: 15-42.

Winfree, A. T. 1980. The Geometry of Biological Time. Springer-Verlag, New York.

Woodside, B., Wilson, R., Chee, P. and Leon, M. 1981. Resource partitioning during reproduction in the Norway rat. Science 211: 76-77.

VOLATILE AND NONVOLATILE CHEMOSIGNALS OF FEMALE RODENTS: DIFFERENCES IN HORMONAL REGULATION

John Nyby

Department of Psychology
Lehigh University
Bethlehem, Pa.

INTRODUCTION

Among rodents, males and females produce certain chemosignals that are different. These sexually dimorphic odors serve both to attract conspecifics and to coordinate reproductive physiology and behavior. In fact, among some rodents such as housemice and golden hamsters, chemoreception may constitute the single most important sensory modality for gender recognition and stimulation of sexual arousal.

At the present time a great deal more is known about the physiological control of chemosignals produced by males than those produced by females. In almost all cases of male-specific chemosignals, chemosignal production is androgen dependent. Male castration typically causes production of chemosignals to cease, and chemosignal production can be restored by exogenous androgen therapy. However, considerably less research has examined female-produced chemosignals and the physiological mechanisms that mediate their production.

Some of the documented effects of chemosignals produced by females include attraction of males and offspring, stimulation of male reproductive physiology and reproductive behavior, alteration of male aggression, and inhibition of estrous cycling in other females. While a single cue produced by females could theoretically serve multiple functions, evidence exists that some of these functions of female-specific chemosignals may be subserved by different chemosignals. For example, two separate chemosignals have been postulated to exist in both female mouse urine (Dixon and Mackintosh, 1975) and in female hamster vaginal secretion

(Macrides, Johnson and Schneider, 1977; Johnston, 1977).

Multiple female-specific chemosignals might exist for at least 2 reasons. One possible reason is that female-specific chemosignals may be required to serve very different functions that are best accomplished with separate substances. For example, among rats the "maternal pheromone" that is present in the feces of lactating females is very likely a different chemosignal than the sex attractant present in the urine of nonlactating females. A second possible reason is that multiple chemosignals carrying basically the same information may exist to provide redundant backup systems. For example in our laboratory we have found that male mice will emit ultrasonic courtship vocalizations in response to female urine, female vaginal odors and female facial odors (Nyby, Wysocki, Whitney and Dizinno, 1977). More recently we have found female saliva also to be an effective stimulus for male vocalizations (unpublished). While these various female odors remain to be chemically characterized, the possibility exists that more than one chemosignal promotes courtship behavior in male mice.

As will be described, different laboratories have often reached different and sometimes seemingly contradictory conclusions about the hormonal control of female chemosignals. Differing conclusions could arise for a variety of reasons. First of all, the different bioassays utilized by different laboratories may be specific for different female chemosignals. Different chemosignals that serve very different functions may be physiologically regulated quite differently.

However, even laboratories that are assessing the effects of the same chemosignal could conceivably come to different conclusions about hormonal regulation. For example some bioassays may be less sensitive to cyclicity of chemosignal production than others. Even cyclically low levels of the chemosignal may be sufficient in some cases to stimulate high levels of response (O'Connell, Singer, Stern, Jesmajian and Agosta, 1981). Another potential problem is that different investigators may utilize animals with differing levels of social and sexual experience. Certain infant and adult experiences are sometimes crucial in determining the types of responses that males make to female chemosignals (Nyby and Whitney, 1980). Finally, different behavioral responses are sometimes seen depending upon whether the chemosignal of interest is presented in isolation or an animal is used as a vehicle for chemosignal presentation (Nyby, Wysocki, Whitney, Dizinno, Schneider and Nunez, 1981; Dixon and Mackintosh, 1971).

With these precautions in mind, the remainder of this paper will examine the hormonal regulation of female-specific cues in a variety of different rodent species. The evidence to be presented appears consistent with at least four different classes of female-specific chemosignals among rodents. Two different classes of chemosignals affect male behavior. One is volatile, ovarian dependent and determines male preference at a distance. The other is nonvolatile, shows surprising independence from ovarian physiology, and appears to affect male preference and sexual arousal at close range. The other two classes of chemosignals also appear not to vary in accordance with ovarian cyclicity. One is volatile, serves as a maternal pheromone, and is regulated by prolactin. The other affects female reproductive physiology, is regulated by adrenal hormones, and its volatility is not known. At the end of the paper I will discuss some of the implications of these hypothesized classes of female specific chemosignals.

REGULATION OF FEMALE SPECIFIC CHEMOSIGNALS

In the following section of the paper, research bearing on the hormonal regulation of the chemosignals of female rodents will be examined. To facilitate comparisons, different laboratories that appear to be examining similar phenomena in the same species will be grouped.

Rat Sex Odors

Male rats are attracted to and prefer chemosignals from females over corresponding chemosignals from males. Chemosignals attractive to males are known to emanate from the female's body (Carr, Loeb and Dissinger, 1965) the female's urine (Le Magnen, 1952) and from preputial gland extracts of females (Gawienowski, Orsulak, Stacewicz-Sapuntzakis and Pratt, 1976). However, conflicting evidence bears on the issue of how these female chemosignals are hormonally regulated. Most studies indicate that sexually experienced malesprefer the odors of estrous females over the odors of diestrous (or ovariectomized) females (LeMagnen, 1952; Carr et al., 1965; Pfaff and Pfaffman, 1969; Stern, 1970; Lydell and Doty, 1972; Gawienowski et al., 1976; Landauer, Weise and Carr, 1977). However, one carefully-controlled, comprehensive study (Brown, 1977) found that while sexually-experienced males showed a strong preference for female odors, estrous and diestrous odors were preferred equally. The difference between Brown's results and those of the others can not be explained in terms of the sources of odors utilized. For example Pfaff and Pfaffman (1969), and Lydell & Doty (1972) as well as Brown (1977) all utilized urine from females as the stimulus in their preference tests.

Brown (1977) suggests that one possible explanation for this difference may lie in the past experience of the males whose preferences were being tested. For example Pfaff and Pfaffman (1969), Lydell and Doty (1972), Carr et al. (1965) and Stern (1970) all found that only sexually experienced males showed a preference for odors from estrous females over those of diestrous females. Sexually inexperienced males in these studies and in the study by Brown (1977) did not show a preference. While Brown (1977) found that sexual experience increased the preference for female odors, the preference for both estrous and diestrous odors increased equally. Brown points out that he allowed his sexually-experienced males contact with only sexually-receptive females, whereas in the other studies males may have had experience with both receptive and nonreceptive females. Thus, according to Brown's hypothesis, males normally show a similar preference for estrous and diestrous odors, but can learn to avoid diestrous odors as a result of experience with nonreceptive diestrous females.

While male social/sexual experience is undoubtedly important, I would like to suggest a second plausible hypothesis. Interestingly, in all studies where a preference for estrous odors was demonstrated, the male subjects were not allowed to physically contact the odor source. Furthermore, a decline in preference for estrous urine occurs when the estrous urine is first exposed to air for several days before use (Lydell and Doty, 1972). I would suggest from these findings that males must be discriminating odors from estrous and diestrous females on the basis of a volatile chemosignal whose presence is ovarian-dependent. In contrast, in the study by Brown (1977) that did not demonstrate a preference by sexually experienced males for estrous odors, the males were allowed to physically contact the odor source. Perhaps, if a male is allowed to contact the odor source, his preference is determined by a nonvolatile chemosignal whose production is ovarian-independent. In summary the existing evidence is consistent with two different types of female-specific chemosignals that affect male preference: a volatile, ovarian-dependent, sex attractant which determines preference at a distance, and a hypothesized ovarian-independent, nonvolatile chemosignal which determines preference after physical contact with the odor source. Evidence will be presented that chemosignals with similar properties may also exist for mice, hamsters, and guinea pigs.

There appear to be at least 2 sites of release of the ovarian-dependent, volatile sex attractant of female rats. Lydell and Doty (1972) found that the bladder urine of estrous females is as preferred as externally voided urine. This finding indicates that the volatile component enters the urine via kidney filtration or perhaps might be produced in the kidney itself. Gawienowski et al. (1976), on the other hand, have demonstrated that males also prefer a volatile component of preputial gland extract from estrous females. Whether urinary and preputial chemosignals represent the same chemical substance being excreted in two different ways or whether they are two different chemical substances producing similar preferences remains to be determined. While the hypothesized nonvolatile chemosignal would appear to exist in female urine (Brown, 1977) additional sites of chemosignal release cannot be ruled out.

Rat Aggression Promoting Chemosignal

Taylor (1975, 1980) has described a chemosignal produced by female rats that is sufficient to promote aggression between strange males. The chemosignal is present to a greater extent in estrous than diestrous females and is volatile. The effects of this chemosignal can be observed by drawing air across a female or across female-soiled cage bedding. While the site of release is not known, the degree of volatility and hormonal control mechanisms are consistent with this cue being the same as that which acts as a long-range sex attractant.

Rat Maternal Pheromone

Lactating female rats and nulliparous females that are induced to behave maternally produce a female-specific chemosignal that is present in the feces (Leon, 1974; Leidahl and Moltz, 1975). While there is some disagreement as to the exact mechanisms leading to chemosignal production (see Kilpatrick and Moltz, 1981), the chemosignal itself is highly volatile and dependent upon the presence of prolactin. Since this female-specific chemosignal appears to have a different site of release and is produced only by rats behaving maternally, this chemosignal is probably chemically different from the previous 2 hypothesized classes of female-specific chemosignals.

Urinary Sex Signals in Mice

Female mouse urine elicits male ultrasonic courtship vocalizations (Nyby, Wysocki, Whitney and Dizinno, 1977), promotes the occurrence of male social and sexual behavior towards conspecifics (Davies and Bellamy, 1974), reduces male aggression towards conspecifics (Mugford and Nowell, 1971), and stimulates luteinizing hormone (LH) release in males (Maruniak and Bronson, 1976). Despite somewhat different time courses of response, all four responses appear related to male reproduction and will be discussed together to facilitate comparisons.

The hormonal mechanisms regulating the urinary chemosignals from female mice promoting ultrasonic vocalizations from males (Nyby, Wysocki, Whitney, Dizinno and Schneider, 1979) appear highly similar to the mechanisms which regulate LH release from males (Maruniak and Bronson, 1976). In both cases, the ability of female urine to elicit a response does not show significant variation across the female estrous cycle nor is this ability substantially altered by long-term ovariectomy. Ovariectomy does, however, lead to a rather small, but statistically reliable, depression in the ultrasound eliciting ability of female urine (Nyby et al., 1979). Both chemosignals appear to be present in the urine of pregnant females and absent from the urine of intact males, castrated males and hypophysectomized females (Maruniak and Bronson, 1976; Nyby et al., 1979; Johnston, 1981; Nyby, unpublished). While the ultrasound eliciting chemosignal is of low volatility, the volatility of the LH releasing chemosignal is unknown. Nevertheless, the high degree of similarity in physiological regulation and ovarian independence is consistent with the same chemosignal eliciting both courtship vocalizations and LH release from males.

Female urine is also known to inhibit the aggression of males toward conspecifics (Mugford and Nowell, 1970a). In contrast to the aggression promoting factors in female rat urine (Taylor, 1975) female mouse urine must be applied directly to a stimulus animal; simply putting volatile components of urine into the air is not sufficient to see an effect. This and other findings (Evans, Mackintosh, Kennedy and Robertson, 1978) indicate that the aggression-inhibiting factor, like the factor eliciting courtship vocalizations, is of low volatility.

Conflicting evidence exists as to the role of ovarian hormones in regulating production of the aggression-inhibiting cue. Early work (Mugford and Nowell, 1971a; Haag, Jerhoff and Kirkpatrick, 1974) indicated that urine from either estrous or diestrous females would suppress attack behavior but that urine from ovariectomized females was ineffective. Complicating these original findings, subsequent research (Mugford and Nowell, 1971b) found that ovariectomized females themselves, in contrast to the effects of their urine, were relatively immune to attack. This latter finding suggested that some nonurinary aggression inhibitor might exist. However, more recent evidence (Dixon and Mackintosh, 1975) indicates, in contradiction to the earlier findings, that urine from ovariectomized females was just as effective in suppressing attack as urine from estrous females.

Subsequent research by Dixon and Mackintosh (1976) indicated that, in addition to urine from ovariectomized females, urine from prepubertal females (but not prepubertal males) also inhibits aggression. Thus at all ages females, but not males, possess urine that provides immunity from attack. While the aggression-inhibiting property of prepubertal female urine appears as effective as that of adult females, only subsequent research can confirm whether the same chemosignal is involved. Nevertheless, the most recent research with mice indicates that the aggression-inhibiting properties of female mouse urine are ovarian-independent.

In contrast to the effects of urine described so far, Dixon and Mackintosh (1975) state that the ability of female urine to promote male sexual and social behaviors is ovarian-dependent. However, examination of their data indicates that while ovariectomy depresses the ability of female urine to elicit male sexual and social behaviors, this ability is not altogether eliminated. Thus the ability of female urine to promote male sexual and social behaviors does show some degree of ovarian independence. In fact, the results differ only in degree from our findings (Nyby et al., 1979) that ovariectomy produces a small, but significant, depression in the ability of female urine to elicit courtship vocalizations. Paralleling each other further, neither the ultrasound-eliciting factor (Nyby et al., 1979) nor the factor promoting sexual and social behavior (Dixon and Mackintosh, 1976) is present in female urine prior to puberty.

It should be noted that Dixon and Mackintosh (1975, 1976) measure 14 different behaviors as elements of their index of social and sexual behavior. One possible explanation for their conclusions about hormonal regulation is that perhaps some of their behaviors such as "attend," "approach," "investigate" may in fact be stimulated by volatile chemosignals that are ovarian-dependent while other behaviors such as "attempted mount," "mount," and "genital groom" may be stimulated by ovarian-independent, nonvolatile substances. Unfortunately their published data are not presented in a form which allows a test of this hypothesis.

Preliminary work on the source of production and chemical characterization of the aggression-reducing and sex-promoting properties (Evans et al., 1978) as well as the courtship vocalization eliciting properties (Nyby et al., 1979) of female urine suggests that all three of these male responses may be affected, at least in part, by the same urinary substance.

Maturation-delaying Factor in Mouse Urine

In addition to effects upon males, female mouse urine also affects female conspecifics. Specifically the puberty of young females can be delayed by exposing them to the urine of group-housed females (Drickamer, 1974, 1977). Excreted urine from singly-housed females does not contain the maturation-delaying chemosignal although bladder urine from either group-housed or singly-housed females is effective. Apparently a factor is present in the urethra of singly-housed females which deactivates, during excretion, the puberty-delaying qualities of their urine.

The production of the maturation-delaying factor continues following ovariectomy as does the blocking factor of singly-housed females (Drickamer, McIntosh and Rose, 1978). Thus both factors appear to be ovarian-independent. However, the delaying effect (but not the deactivating factor) can be eliminated by adrenalectomy (Drickamer and McIntosh, 1980). Thus the ability of a urinary chemosignal from group-housed females to delay puberty appears to be regulated by the adrenal gland.

Nonurinary Chemosignals in Mice

Male mice will emit ultrasounds not only to female urine but also to female vaginal odors, skin odors in the facial area (Nyby et al., 1979) and more recently we have demonstrated a response to female saliva (Bernhard, Wysocki, Byatt and Nyby, unpublished). A series of studies in my laboratory was recently completed to determine whether these nonurinary ultrasound-eliciting cues are hormonally regulated in a fashion similar to those found in urine (Byatt and Nyby, unpublished). For each of these three sources of chemosignal collection (vaginal odors, facial odors, and saliva), we obtained cues from intact females, long-term ovariectomized females and long-term hypophysectomized females and tested them for their ability to elicit ultrasounds. In general, the hormonal dependence of odors from these sources were highly similar to the hormonal dependence of the odor found in female urine. Odors from ovariectomized females appeared to be roughly comparable to odors from intact females while those from hypophysectomized females were significantly less effective. Thus the hormonal regulation of cues eliciting vocalization appeared similar regardless of the female body area where the cues were obtained. Whether the similarity in hormonal regulation reflects several different substances that are regulated by similar hormonal mechanisms or the same substance which is spread about the body or excreted via several different routes is unknown.

In contrast, Hayashi and Kimura (1974) found that sexually experienced males preferred the odor of excised vaginal tissue from estrous females over that from diestrous females. Furthermore, diestrous females smeared with the vaginal secretion of estrous females were more likely to be mounted than diestrous females. Also in contrast to some of the work already described, urine from either estrous or diestrous females was without effect in promoting male sexual behavior. At the present time, the work by Hayashi and Kimura appears difficult to reconcile with the findings of others.

Vaginal Chemosignals in Hamsters

Female hamsters produce a vaginal discharge whose amount varies cyclically with the female estrous cycle. The discharge appears to reach its peak either around or slightly after the time of estrus (Brom and Schwartz, 1968; O'Connell et al., 1981). Females show cyclical changes in scent marking with this secretion and Johnston (1972, 1977, 1979) has suggested that the scent marking may serve as an advertisement of approaching estrus.

Hamster vaginal secretion has been demonstrated to have a variety of effects upon males. The odor of the secretion is highly attractive to males (Johnston, 1977) being preferred even over highly attractive food odors. Once in its vicinity, males continue to show a strong preference for the discharge and spend substantial time sniffing and licking it. The discharge, when applied to conspecifics, will reduce aggression and stimulate sex behavior towards the odorized conspecific. This discharge in isolation will also reduce scent-marking behavior (Johnston, 1980) and increase male plasma testosterone levels (Macrides, Bartke, Fernandez and Angelo, 1974).

Recent evidence indicates that some of the different effects of hamster vaginal secretion appear to be accounted for by different components of the secretion. Approximately half of the long-range attractiveness of hamster vaginal secretion can be accounted for by a single compound, dimethyl disulfide (Singer, Agosta, O'Connell, Pfaffman, Bowen and Field, 1976). The remaining components inducing long-range attraction are yet to be identified. Dimethyl disulfide, however, does not appear to account for the sexually arousing properties of hamster vaginal secretion. Rather, nonvolatile compounds appear to be important in this regard (Johnston, 1977; Singer, Macrides and Agosta, 1980).

The evidence bearing upon the hormonal control of female sex signals in hamsters is not altogether clear. The female hamster vaginal discharge in general and the dimethyl disulfide component in particular (O'Connell et al., 1981) show pronounced variations in production which vary systematically with the female cycle. In contrast, much research seems to indicate that odors (cage shavings, volatile body odors, vaginal secretions) from estrous females are not responded to differently then those from diestrous females (Landauer and Banks, 1973; Darby, Devor and Chorover, 1975; Landauer, Banks and Carter, 1978; Kwan and Johnston, 1980; Johnston, 1977). These findings would be consistent with the presence of an ovarian-independent cue affecting a variety of male behaviors. More recent evidence (Johnston, 1980) indicates that sexually experienced males under some circumstances can behave differently in response to estrous and diestrous odors. Sexually inexperienced males failed to show differential responses. This findings would seem to indicate that an ovarian-dependent cue is also available to be responded to by sexually experienced males.

ANOGENITAL ODORS IN GUINEA PIGS

Very recently Ruddy (1980), using a learning task, demonstrated that both male and female guinea pigs could learn to discriminate anogenital odors from estrous and diestrous females. Since the guinea pigs were not allowed to physically contact the odor source, the discrimination must have been made on the basis of volatile chemical cues.

In contrast, other work (Beauchamp, 1973; Beauchamp and Berüter, 1973; Beauchamp, personal communication) indicated that even sexually experienced male guinea pigs do not show a preference for urine from estrous females over that of diestrous females. While there are a variety of methodological differences between the work of Ruddy and

Beauchamp, it is of interest that Beauchamp allowed his males to physically contact the odor source. Thus work with guinea pigs also appears consistent with two different female specific chemosignals: one that is volatile and ovarian dependent, and a hypothesized chemosignal that is of low volatility and ovarian independent. Preference at close range would be determined by the hypothesized ovarian independent chemosignal.

CONCLUDING REMARKS

Many difficulties arise when comparing the research on female-specific chemosignals from different laboratories and different rodent species. Nevertheless the evidence appears compelling that female rodents produce multiple female-specific chemosignals. For example many of the different chemosignals reviewed here exhibit different mechanisms of hormonal regulation, different degrees of volatility, as well as different sites of chemosignal production and release. While many chemically distinct substances may serve as female-specific chemosignals I would tentatively suggest that the chemosignals reviewed here can be grouped into four different classes.

Table 1. Hypothesized classes of chemosignals produced by female rodents.

	Function	Hormonal Control	Volatility	Sources
1.	long-range sex attractant, aggression promotor	ovarian hormones	high	rat urine, rat preputial odors, hamster vaginal secretions, guinea pig anogenital area
2.	sex behavior promotor, LH primer, and aggression inhibitor	pituitary hormones (ovarian independent)	low	mouse, rat, & guinea pig urine, hamster vaginal secretion
3.	puberty inhibitor	adrenal hormones (ovarian independent)	not known but probably low	mouse urine
4.	offspring attractant	prolactin (ovarian independent)	high	rat and mouse feces

Two of these classes of female-specific chemosignals are known to affect male behavior and physiology. One of the classes appears to be released to a greater extent by estrous than diestrous females, is volatile, and male responses to the chemosignal require sexual experience. This class of chemosignal probably serves as a long-range sex attractant and may also promote male aggression. Given its degree of volatility and the experiential component necessary for the male response, I would also suggest that this chemosignal may be perceived by the primary olfactory system (see Wysocki, 1979). Evidence that such a chemosignal exists in rat urine is quite good. Volatile ovarian-dependent chemosignals also exist in the anogenital odors of guinea pigs and in the vaginal secretions of golden hamsters. Some evidence also exists for the presence of this class of chemosignals in mouse vaginal secretions.

A second class of chemosignals that affects males appears to be ovarian-independent, and of low volatility. This class of chemosignals may determine preference for female secretions at close range and also act to stimulate male reproductive behavior and arousal and reduce male aggression. Responses to this chemosignal appear to be less dependent upon male experience. Given low volatility and an intimate relationship to male sex behavior and physiology, I would speculate that this class of chemosignal may be perceived by the vomeronasal system (Wysocki, 1979). This class of chemosignal is known to exist in female mouse urine and can be inferred to exist in hamster vaginal secretion. While this chemosignal has not yet been demonstrated to exist in rat urine or guinea pig anogenital odors, its existence would account for some of the apparent inconsistencies concerning male preference for the odors of estrous versus diestrous females.

A third class of chemosignal appears to inhibit female reproductive physiology. Although its production is independent of ovarian hormones, its presence in the excreted urine of group-housed but not singly-housed females appears inconsistent with the previous class of chemosignal. The volatility of this class of chemosignal has not been explicitly examined. At the present time, the best evidence for this factor comes from research with mice.

A fourth class of chemosignal serves as an attractant for offspring. It appears to be volatile and its production depends upon prolactin. The hormonal control mechanism, the site of release, and chemosignal presence only in maternal females argue strongly for a separate category for this class of chemosignal. At the present time this class of chemosignal has been most extensively studied in rats although evidence also exists for a similar chemosignal among mice (Breen and Leshner, 1977).

As a final note, most investigators have concentrated their own research efforts on a single site of chemosignal release.

However, redundancy seems to be a common feature in many biological systems. From the evidence reviewed here, I would suggest the possibility that a particular type of behavioral or physiological response may occur in some cases in response to more than one female specific chemosignal.

ACKNOWLEDGEMENTS

The writing of this manuscript was supported in part by National Science Foundation Grant BNS-8111344. Thanks to Chuck Wysocki, Gladye Whitney, and Murray Itzkowitz for critically reading earlier versions of this manuscript.

REFERENCES

Beauchamp, G. K., 1973, Attraction of male guinea pigs to conspecific urine. Physiol. Behav., 10:589.

Beauchamp, G. K., and Berüter, J., 1973, Source and stability of attractive components in guinea pig (Cavia porcellus) urine, Behav. Biol., 9:43.

Breen, M. F., and Leshner, A. I., 1977, Maternal pheromone: a demonstration of its existence in the mouse. Physiol. Behav., 18:527.

Brown, R. E., 1977, Odor preference and urine-marking scales in male and female rats: effects of gonadectomy and sexual experience on responses to conspecific odors. J. Comp. Physiol. Psychol., 91:1190.

Carr, W. J., Loeb, L. S., and Dissinger, M. L., 1965, Responses of rats to sex odors. J. Comp. Physiol. Psychol., 59:370.

Darby, E. M., Devor, M., and Chorover, S. L., 1975, A presumptive sex pheromone in the hamster: some behavioral effects, J. Comp. Physiol. Psychol., 88:496.

Davies, V. J., and Bellamy, D., 1972, The olfactory response of mice to urine and effects of gonadectomy, J. Endocrinol., 55:11.

Davies, V. J., and Bellamy, D., 1974, Effects of female urine on social investigations in male mice, Anim. Behav., 22:239.

Dixon, A. K., and Mackintosh, J. H., 1971, Effects of female urine upon the social behavior of adult male mice, Anim. Behav., 19:138.

Dixon, A. K., and Mackintosh, J. H., 1975, The relationship between the physiological condition of female mice and the effects of their urine on the social behavior of adult males, Anim. Behav., 23:513.

Dixon, A. K., and Mackintosh, J. H., 1976, Olfactory mechanisms affording protection from attack to juvenile mice, (Mus musculus, L.) Z. Tierpsychol., 41:225.

Drickamer, L. C., 1974, Sexual maturation of female house mice: social inhibition, Dev. Psychobiol., 7:257.

Drickamer, L. C., 1977, Delay of sexual maturation in female house mice by exposure to grouped females or urine from grouped females, J. Reprod. Fert., 51:77.
Drickamer, L. C., McIntosh, T. R., and Rose, E. A., 1978, Effects of ovariectomy on the presence of a maturation-delaying pheromone in the urine of female mice. Horm. Behav., 11:131.
Drickamer, L. C. McIntosh, T. K., 1980, Effect of adrenalectomy on the presence of a maturation-delaying pheromone in the urine of female mice, Horm. Behav., 14:146.
Evans, C. M., Mackintosh, J. H., Kennedy, J. F., and Robertson, S. M., 1978, Attempts to characterize and isolate aggression reducing olfactory signals from the urine of female mice, Mus musculus, Physiol. Behav., 20:129.
Gawienowski, A. M., Orsulak, P. J., Stacewicz-Sapuntzakis, M., and Pratt, J. J., 1976, Attractant effect of female preputial gland extracts on the male rat, Psychoneuroendocrinol., 1:411.
Haag, Claudia, Jerhoff, B., Kirkpatrick, J. F., 1979, Ovarian hormones and their role of aggression inhibition among male mice, Physiol. Behav., 13:175.
Hayashi, B., and Kimura, T., 1974, A sex attractant emited by female mice, Physiol. Behav., 13:563.
Johnston, R. E., 1977, Sex pheronobes in golden hamsters, in: "Chemical Signals in Invertebrates," D. Müller-Schwarze and M. M. Mozell, eds., Plenum Press, New York.
Johnston, R. E., 1980, Responses of male hamsters to odors of females in different reproductive states, J. Comp. Physiol., 94:894.
Johnston, R., 1981, Hormonal control of odor-elicited LH responses and attraction in male mice, Paper presented at Conference on Reproductive Behavior, Nashville.
Kwan, M., and Johnston, R. E., 1980, The role of vaginal secretion in hamster sexual behavior: males' responses to normal and vaginectomized females and their odors, J. Comp. Physiol. Psychol., 94:905.
Landauer, M. R., and Banks, E. M., 1973, Olfactory preferences of male and female golden hamsters, Bull. Ecolog. Soc. Amer., 54:44.
Landauer, M. R., Wiese, R. E., Jr., and Carr, W. J., 1977, Responses of sexually experienced and naive male rats to cues from receptive and nonreceptive females, Anim. Learn. Behav., 5:398.
Landauer, M. R., Banks, E. M., and Carter, C. S., 1978, Sexual and olfactory preferences of naive and experienced male hamsters, Anim. Behav., 26:611.
LeMagnen, J., 1952, Les phéromones olfacto-sexuals chez le rat blanc, Archives des Sciences Physiologiques, 6:295.
Lydell, K., and Doty, R. L., 1972, Male rat odor preferences for female urine as a function of sexual experience, urine age, and urine source, Horm. Behav., 3:205.

Macrides, F., Bartke, A., Fernandez, F., and D'Angelo, W., 1974, Effects of exposure to vaginal odor and receptive females on plasma testosterone in the male hamster. Neuroendocrinol., 15:355.

Macrides, F., Bartke, A., and Dalterio, S., 1975, Strange females increase plasma testosterone levels in male mice, Science, 189:1104.

Macrides, F., Johnson, P. A., Schneider, S. P., 1977, Responses of the male golden hamster to vaginal secretion and dimethyl disulfide: attraction versus sexual behavior, Behav. Biol., 20:377.

Maruniak, J. A., and Bronson, F. H., 1976, Gonadotropic responses of male mice to female urine, Endocrinol., 99:963.

Mugford, R. A., and Nowell, N. W., 1971a, Endocrine control over production and activity of the anti-aggression pheromone from female mice, J. Endocrinol., 49:225.

Mugford, R. A., and Nowell, N. W., 1971b, The relationship between endocrine status of female opponents and aggressive behaviour of male mice, Anim. Behav., 19:153.

Nyby, J., Wysocki, C. J., Whitney, G., and Dizinno, G., 1977, Pheromonal regulation of male mouse ultrasonic courtship (Mus musculus), Anim. Behav., 25:333.

Nyby, J., and Whitney, G., 1980, Experience affects behavioral responses to sex odors, in "Chemical Signals in Vertebrates and Aquatic Animals", D. Müller-Schwarze and R. M. Silverstein, ed., Plenum Press, New York.

Nyby, J., Wysocki, C. J., Whitney, G., Dizinno, G., Schneider, J. and Nunez, A. A., 1981, Stimuli for male mouse (Mus musculus) ultrasonic courtship vocalizations: presence of female chemosignals and/or absence of male chemosignals, J. Comp. Physiol. Psychol., 95:623.

O'Connell, R. J., Singer, A. G., Stern, F. Lee, Jesmajian, S., and Agosta, W. C., 1981, Cyclic variations in the concentration of sex attractant pheromone in hamster vaginal discharge, Behav. Neur. Biol., 31:457.

O'Connell, R. J., Singer, A. G., Macrides, F., Pfaffmann, C., and Agosta, W. C., 1978, Response of the male golden hamster to mixture of odorants identified from vaginal discharge, Behav. Biol., 24:244.

Pfaff, D. W., and Pfaffman, C., 1969, Behavioral and electrophysiological responses of male rats to female urine odors, in Olfaction and Taste, Vol. III, C. Pfaffman, ed., Rockefeller Univ. Press, New York.

Ruddy, L. L., 1980, Discrimination among colony mates' anogenital odors by guinea pigs (Cavia porcellus), J. Comp. Physiol. Psychol. 94:767.

Stern, J. J., 1970, Responses of male rats to sex odors, Physiol. Behav., 5:519.

Singer, A. G., Macrides, F., and Agosta, W. C., 1980, Chemical studies of hamster reproductive pheromones, _in_ D. Müller-Schwarze, and Silverstein R. M., (eds.), Chemical Signals, Plenum Press, New York.
Singer, A. G., Agosta, W. C., O'Connell, R. J., Pfaffman, C., Bowen, D. V., and Field, F. H., 1976, Dimethyl disulfide: an attractant pheromone in hamster vaginal secretion, _Science_, 191:948.
Taylor, G. T., 1975, Male aggression in the presence of an estrous female, _J. Comp. Physiol. Psychol._, 89:246.
Taylor, G. T., 1980, Sex pheromones and aggressive behavior in male rats, _Anim. Learn. Behav._, 8:485.
Wysocki, C. J., 1979, Neurobehavioral evidence for the involvement of the vomeronasal system in mammalian reproduction, _Neurosci. Biobehav. Rev._, 3:301.

COMMUNICATION DISPARITIES BETWEEN GENETICALLY-DIVERGING POPULATIONS OF DEERMICE

Glenn Perrigo and F. H. Bronson

The Institute of Reproductive Biology
Department of Zoology, University of Texas
Austin, Texas 78712

In nature there is considerable genetic variation between populations belonging to the same species. Members of a population share a common gene pool. Gene pools, however, are uniquely dynamic, and since no two populations experience identical selection pressures, one expects that even closely related populations will diverge genetically with time. Developing reproductive isolation is a normal consequence of such divergence and, by classical definition, the process of speciation is complete when the populations can no longer interbreed. Speciation can be rapid if an abrupt chromosomal change occurs, or it can be a slow process involving the steady accumulation of subtle behavioral and physiological incompatibilities that progressively reduce the probability of successful reproduction between members of two populations.

Behavioral isolating mechanisms that operate through incompatibilities in social communication are the most efficient mechanisms for preventing intermating between divergent populations. In insects, for example, closely related species often are effectively isolated reproductively by minor structural changes in chemical compounds or even by the use of different geometric isomers of a molecule serving as a sex attractant (e.g., Roelofs and Cardé, 1974). Admittedly, communication among mammals is a more complicated affair than it is in insects. Communication between mammals usually encompasses a simultaneous and synergistic involvement of most of the individual's sensory modalities. Furthermore, the cues transmitted by these modalities typically are quite complex, and the responses they elicit often are graded and usually dependent upon context. Nevertheless, since chemical cues often provide the dominant mode of communication among mammals, one should expect developing disparities in chemical communication to be relatively common among diverging populations. This possibility has received only enough attention to verify its general veracity.

At the level of an interaction between two species of mammals, for example, it has been demonstrated that house mice easily can discriminate their own odors from those of deermice (Bowers and Alexander, 1967), and that chemical recognition apparently provides the primary means by which two closely related species of Peromyscus are able to discriminate between each other (Moore, 1965). In a similar study involving two other species of Peromyscus, however, Doty (1972) found that only estrous females showed a reliable preference for homospecific male odors. At the within-species level, the discrimination of subspecific odors has been documented for bank voles (Godfrey, 1958), mole rats (Nevo, et al., 1976) and deer (Müller-Schwarze, 1974), and gas chromatography has been used to establish subspecific differences in the urine and glandular extracts of some rodents (Stoddart, 1977), and tamarins (Epple, et al., 1979).

Given all of the above, it seems highly probable that incompatibilities in chemical communication indeed often do provide the substrate for reproductive isolation in mammals, but it is evident also that we probably have little appreciation of the importance of this phenomenon at this time. In an attempt to more systematically study the relationship between genetic divergence and chemical communication, we have begun a search for communication incompatibilities among spatially-separated populations of each of two species of wild rodents -- the house mouse (Mus musculus) and the deermouse (Peromyscus maniculatus). This chapter constitutes a progress report of these efforts. In general, we will show that communication incompatibilities may or may not exist between two populations of the same species, depending upon their degree of genetic divergence. Where present, some of the incompatibilities we have isolated result in dramatic behavioral maladjustments, and some operate in the dimension of endocrine priming. At this time, however, our evidence for a chemical basis for these incompatibilities is only circumstantial.

The remainder of this chapter will be organized as follows. First, we will describe our experimental populations and our rationale for choosing them. Second, we will note briefly the generalities of our experimental paradigms. Third, we will summarize our experimental results to date, concentrating first on behavioral disparities and then on priming disparities. Finally, we will suggest a working model -- actually a set of potential relationships -- that may prove advantageous in guiding future research of this type.

Experimental Populations and Test Paradigms

The two species chosen for our research, the house mouse and the deermouse, differ markedly both in their general reproductive strategies and in the degree to which they have been subjected to local selection on this continent. Endemic to North America, the genus *Peromyscus* exhibits patent evidence of genetic divergence. This genus includes some 56 recognized species, and of these, the deermouse is the most widely distributed, now including a score of well defined subspecies, each of which has undergone intense local adaptation for survival and reproductive efficiency (Dice, 1968; Baker, 1968). In contrast, the house mouse is an Asian-derived, opportunistic colonizer that became established on this continent only with the arrival of European settlers. House mice exhibit little evidence of genetic divergence in North America and, reproductively, they remain remarkably flexible and well-adapted to a wide variety of environmental conditions, both when living commensally with man and in feral situations (Bronson, 1979).

The four experimental stocks chosen for our studies include a northern and a southern population of each species, both separated by roughly 3000 km. The two deermouse populations represent *P. m. borealis*, a robust northern subspecies, from Fairview, Alberta, and *P. m. pallescens*, a smaller prairie-dwelling subspecies, from near Austin, Texas. The two house mouse populations were obtained from granaries near Calgary, Alberta, and from barns near Austin. The two deermouse populations have been geographically isolated for at least 10,000 years. The two populations of house mice may have been separated for only a few hundred years or, alternatively, they may differ in their genetic founders (French *vs.* Spanish origins) and, hence, they may have been separated for a somewhat longer period of time (Schwarz and Schwarz, 1943). Progenitors of these four populations were trapped in 1979 to establish our breeding colonies which are now in their third generation. The details of our management of these colonies may be found in Perrigo and Bronson (1982).

The dispersal of weanling animals in the wild has a profound impact on population demography, thus, the social paradigm within which we searched for incompatibilities of communication usually involved an interaction between an adult male and a prepubertal female. Two general types of experiments were conducted, one concentrating on each of the two major dimensions of intersexual communication: (a) the intersexual signaling necessary to organize sexual behavior itself and (b) the endocrine priming of one sex by another which frequently must precede sexual behavior. The signaling dimension was examined simply by observing an adult male's behavioral responses when he encountered a young female. Of the many measures quantified during such encounters, the most meaningful for our purposes seemed to be the amount of aggression shown by the male and the tendency of the two animals to occupy a single nest (Perrigo and Bronson, 1982). Priming compatibility was assessed by measuring uterine growth in a young female after a period of cohabitation with an adult male. In both types of experiments we compared the interactions between a male and a female of the same population with those of a male and a female of the two different populations of the same species. All such interactions took place in the male's home cage.

Signaling Compatibility

Signaling cues always form the basis for mate recognition. In this first study, a prepubertal female of the same or the different population as the test male was placed in the male's cage, left overnight, and then checked quietly the following morning for shared *vs.* separate nesting. As expected, there was no suggestion of any signaling incompatibility between our two house mouse stocks. While separate nesting was common in this species, and while two young females actually were killed by their test males, there was no indication that cross population interactions were different from same population interactions in either regard (Figure 1).

All combinations of pairs of deermice resulted in shared nesting except the male Alberta/female Texas combination. Apparently the majority of the male Alberta deermice did not recognize the Texas females as reproductively viable members of their own species and profound aggression resulted (Figure 1). Over half of the female Texas deermice caged with Alberta males were killed and cannibalized and some others were severely bitten and near death when found the next morning. This is dramatic evidence of a gross signaling disparity of some kind between these two populations. Further observation confirmed that this was not an unusual result of housing this particular combination of animals together. Indeed, 50% of our male Alberta deermice routinely exhibit a killing response when presented with Texas females of their own species. When placed in the home cage of these males, Texas females usually are attacked and dispatched immediately, whether anesthetized or awake, and cannibalization typically follows killing (Perrigo and Bronson, 1982). In contrast, Alberta females always are received amicably by males of their own population, with frequent naso-nasal and naso-genital contact, and mutual grooming that is typical of normal male/female interactions in this species (Eisenberg, 1968).

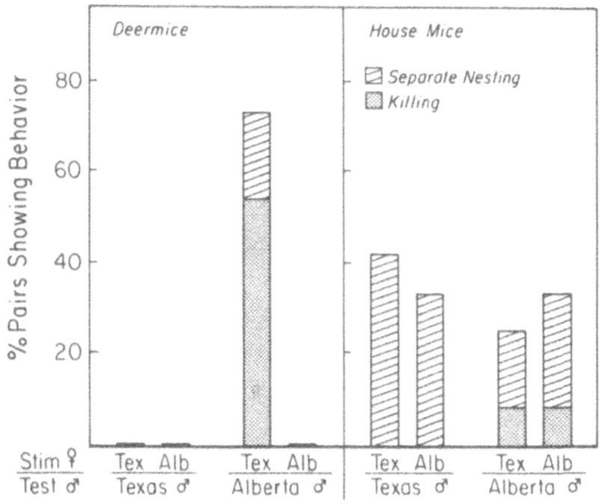

Fig. 1. Frequency of killing and separate nesting among pairs of adult males and young females of the same vs. different populations of the same species. N = 11-12 for each comparison.

The killing behavior described above prompted us to ask whether or not such killing also would result when Alberta and Texas deermouse pairings involved other combinations of sex and age. In this unpublished experiment, 48 animals of each sex of Alberta deermice were isolated at weaning and, when adult, half the animals of each sex were presented either with prepubertal females or with prepubertal males of both the Alberta and Texas stocks. Thus each adult received two randomized presentations of young animals of the same sex of either subspecies, separated by a 7-10 day interval between presentations.

As shown in Figure 2, adult female Alberta deermice generally displayed less aggression than did Alberta males when presented with young Texas deermice, regardless of the sex of the latter. Again, however, almost half of the Alberta males killed and cannibalized young Texas females, but only a quarter of them killed young Texas males. Thus, for some reason, killing is much more common in the adult male Alberta/prepubertal female Texas combination than in any other. As an interesting aside, despite their

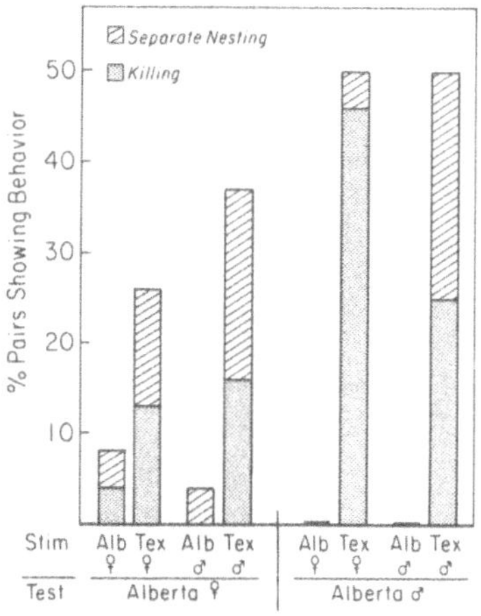

Fig. 2. Frequency of killing and separate nesting when adult male or female deermice from Alberta were cohabited with young males or females of the same vs. the Texas population. N = 24 for each comparison.

age, some of our male Alberta deermice had juvenile testes and these particular males were just as likely to kill prepubertal females as were males with fully developed testes. The killing behavior of our Alberta males therefore appears to be independent of androgenic influence.

We have conducted two types of experiments in an effort to estimate the degree to which the killing behavior represents a disparity in chemical communication. Several times we tried placing urine from an Alberta female on the back of a female Texas deermouse to see if this would eliminate killing. The results suggested an ameliorating effect, but by no means did this procedure yield a complete cessation of killing. Thus we conducted another type of test, wherein anesthetized females were presented to males in complete darkness, thereby eliminating all modalities except the chemical and the tactile/textural. Two groups of Alberta males were so tested. One group consisted of selected killers, verified as such by previous tests. The other group consisted of males that had no previous contact with other animals since weaning; this was the same type of male used in the experiments reported previously. Similar to the preceding experiment, each male received two randomized presentations, 7-10 days apart, of prepubertal females of the Alberta and Texas subspecies.

The results of this experiment, shown in Table 1, can be summarized as follows. The selected killer Alberta males killed and cannibalized their anesthetized Texas female test animals 78% of the time, while the comparable figure for the naive males was 50%. Under the conditions of these tests, however, 36% of the selected killers also killed anesthetized females of their own Alberta population and 21% of the naive males did likewise. Thus, removing all sensory input except that involving passive chemical and tactile/textural cues did not alter the expected 50% rate at which naive Alberta males kill Texas females (c.f. Figs. 1 and 2), but it did for the first time yield killing of females from the Alberta male's own population. Tentatively, then, we must conclude that the killing behavior shown by our Alberta male deermice is mediated by more than one type of sensory input, but that chemical cues must be a major factor in eliciting this behavior.

The adaptive function of the killing behavior exhibited by male Alberta deermice is of obvious interest. Two possibilities suggest themselves immediately. First, Alberta deermice might be somewhat carnivorous and, thus, a failure to recognize a smaller animal as belonging to the same species might simply elicit predation. Alternatively, the killing could represent a form of

Table 1. Number of cannibalizations/number of pairings when male Alberta deermice were presented with anesthetized females in complete darkness for 1 hour.

Selected Killers		Naive Males	
Alberta Females	Texas Females	Alberta Females	Texas Females
9/26	20/26	6/28	14/28

genetic exclusion, i.e., a direct and violent elimination of potential genetic competitors when they are encountered. To resolve this question we exposed naive, adult Alberta males to prepubertal females of a spectrum of deermouse stocks differing in varying degrees of genetic divergence from the Alberta population. The geographic origin of these stocks is shown in Figure 3. More specifically, we presented adult *P. m. borealis* males from Alberta with prepubertal females of the following stocks: *P. m. borealis*, *P. m. nebrascensis*, *P. m. bairdii*, *P. m. pallescens*, *P. leucopus* (the white footed mouse) and *Mus musculus*. The *P. m. nebrascensis* stock was trapped in South Dakota whereas the *P. m. bairdii* and *P. leucopus* stocks originated in Michigan. The house mice were from our Alberta colony.

The results of an overnight cohabitation with these diverse stocks, shown in Figure 4, demonstrate that killing of young females by Alberta male deermice is limited largely to within the species *P. maniculatus* itself. Furthermore, the proportion of *P. maniculatus* females that are killed by Alberta males increases as a function of decreasing genetic relatedness and increasing geographic distance relative to the Alberta population. Apparently, the compatibility of intersexual signaling degenerates rapidly along these two gradients. In regard to the purpose of this experiment, our results certainly suggest a form of *genetic exclusion*, as opposed to a confused predation basis for the

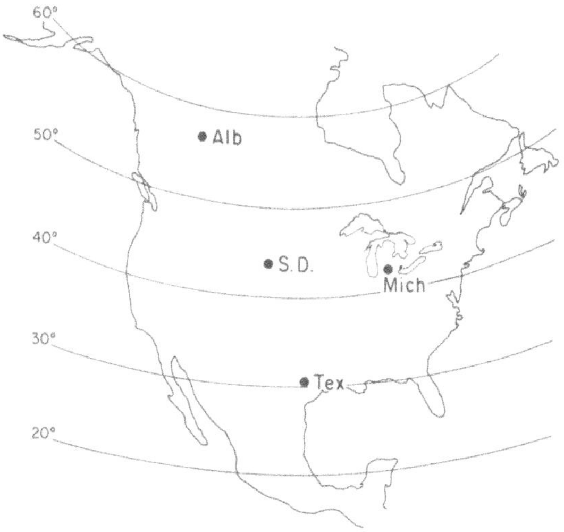

Fig. 3. Populations of deermice tested for their ability to to elicit killing behavior by Alberta male deermice.

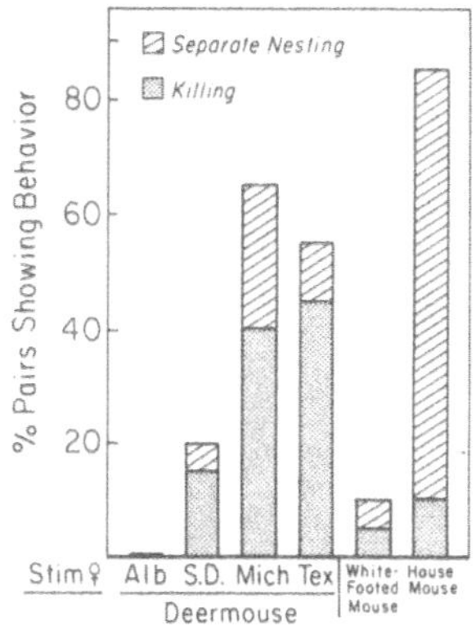

Fig. 4. Frequency of killing and separate nesting when adult male Alberta deermice were cohabited with young females of a variety of stocks. N = 20 for each comparison.

killing behavior. Almost all pairings of Alberta males with female white-footed mice, a closely related species of the same genus, resulted in shared nesting; only one female was killed but was not eaten (Fig. 4). Two house mice were killed and the majority of these pairings resulted in separate nesting. The interpretation of these cross-species observations will be presented later.

Priming Compatibility

It is well established in both house mice and deermice that the female's ovulatory cycle is markedly dependent upon urinary cues emanating from the male (reviewed by Bronson, 1982). It is also well established that these chemical cues act with the support of tactile cues, at least in house mice (Drickamer, 1974; Bronson and Maruniak, 1975). The effects of such primer modulation are particularly dramatic in the prepubertal female where assessment of uterine growth has been used routinely as a bioassay of primer activity (Vandenbergh, 1969; Bronson and Desjardins, 1974; Vandenbergh, et al., 1975). In an effort to determine if priming incompatibilities exist between our various test populations we conducted the following experiment. Prepubertal females of the Texas and Alberta stocks of both deermice and house mice were paired for several days with adult males of the same or the different populations of their own species, or they were housed in isolation in male-free rooms to serve as controls for unstimulated uterine growth. Considerable pilot work yielded the procedures used in this experiment (Perrigo and Bronson, 1982).

As shown in Figure 5, female house mice displayed a three-fold increase in uterine weight in response to male exposure, but the genetic stock of the male was not important. Female Alberta deermice reacted to males of their own population with a two-fold increase in uterine weight, but they displayed no response to

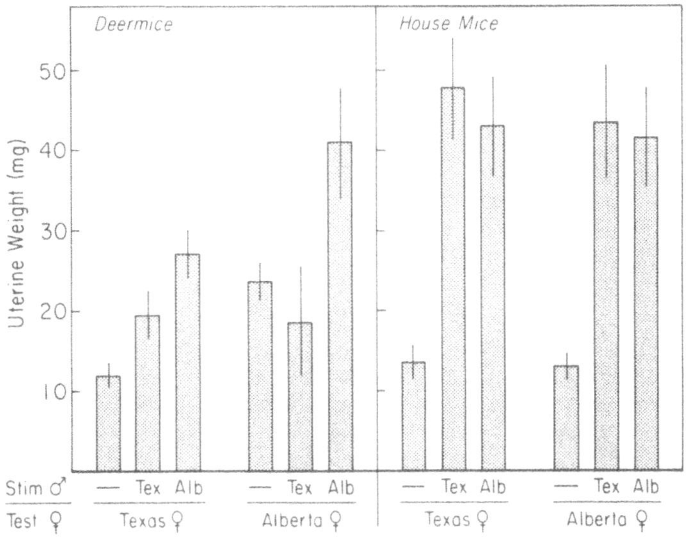

Fig. 5. Mean (±SE) uterine weights of females held in isolation (—) or exposed to males of either the same or different population of the same species. N = 10-11 for each comparison, except Texas females where N = 17-18.

Texas males. In contrast, female deermice of the Texas subspecies showed significant uterine growth in response to both Texas and Alberta males, with no significant difference between the abilities of the two stocks of males to induce this response. It should be noted here that the results shown in Figure 5 for the male Alberta/female Texas deermouse combinations represent surviving pairs only; once again, many Texas females were killed and eaten by Alberta males.

Conclusions and Related Concerns

Intuitively one should expect genetic divergence to act upon the efficiency of communication generally in the manner shown in Figure 6. That is, as the gene pools of two populations become more and more divergent, the individuals of the two populations should experience progressively increasing difficulty in communicating with each other. As noted earlier, the communication necessary for successful mating in mammals typically is complex and multi-sensorial. Thus one can view the information flow between two potential mates as consisting of many separate bits of information, or cues, that in their totality either promote or inhibit mating. Barring immediate speciation via chromosomal change, then, one can view diverging populations as steadily replacing compatible cues with incompatible cues until reproductive isolation occurs. Importantly, such a situation may create a zone of confusion, arising at a stage of divergence when the individuals of the two populations are providing each other with an ambiguous mixture of compatible and incompatible cues. In classical terms, this zone should center at a level of genetic divergence denoted by the taxonomist as that of incipient speciation, i.e., at the level of subspecies.

The studies summarized herein constitute an attempt to test the validity of some of the relationships suggested in Figure 6. Thus we chose populations of rodents that would yield contrasting levels of divergence and, hence, contrasting potentials for communication incompatibility. We expected no disparity in the capacity of our two house mouse populations to communicate with each other for the many reasons stated earlier and, indeed, in neither the priming nor in the signaling dimension of intersexual communication did we observe any incompatibilities. Furthermore, mating tests revealed absolutely no differences in litter production among same vs. different population pairs of our two house mouse stocks (Perrigo and Bronson, 1982).

In contrast, our deermouse stocks were chosen to yield a much greater degree of genetic divergence and, hence, a greater probability of detecting communication incompatibilities. The Texas and Alberta populations of deermice differ taxonomically at

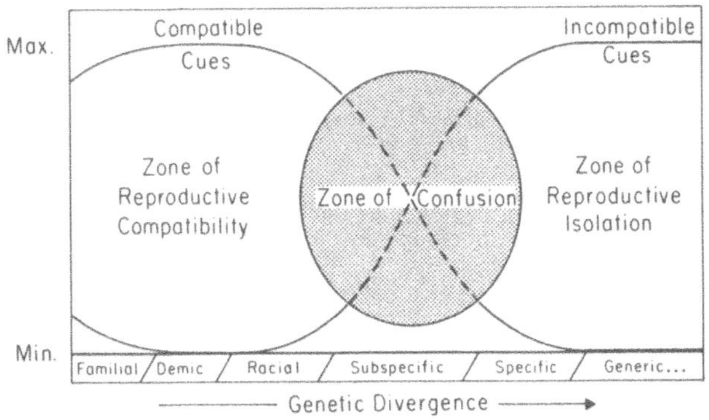

Fig. 6. The probable relationship between genetic divergence and the capacity to communicate.

the subspecific level and, apparently, they are partially but not completely isolated reproductively at this time. When potential mates from these two populations encounter each other they display both signaling and priming disparities and, importantly, these two classes of disparities have evolved independently. The killing behavior exhibited by many Alberta male deermice when they encounter a Texas female of the same species is evidence of a profound signaling incompatibility promoting sexual isolation (also see Dice, 1933). However, when Alberta males do not kill Texas females, these males have no trouble priming and impregnating Texas females and reproduction is not further depressed (Perrigo and Bronson, 1982). On the other hand, while we observed no signaling incompatibilities between adult male Texas deermice and young Alberta females (Figure 1), the former could not prime the latter and, as a consequence, pregnancies are extremely rare in this particular combination (Perrigo and Bronson, 1982). Thus, our two most distant deermouse populations seem to be on the verge of complete reproductive isolation, having independently developed incompatibilities of both the signaling and the priming type, but still sharing enough compatible cues to accomplish reproduction in some combinations of individuals. In terms of Figure 6, then, it seems obvious that the degree of divergence attained by our Texas and Alberta deermice falls in the zone of confusion. Our two house mouse populations, on the other hand, interact in the zone of compatibility.

Although somewhat speculative and not tested herein, one further concept has been incorporated in the model shown in Figure 6; primarily that the zone of compatibility probably is not a homogeneous zone at all but only a background against which differential mating preferences can act. As one illustration of this possibility, outbreeding preferences have been established in mammals. Strictly in regard to chemical communication, for example, Yamazaki, et al. (1980) have established that the loci of the ubiquitous mammalian MHC (major histocompatibility) gene complex are the source of individual and strain specific odors in mice. Highly inbred house mouse strains differing only in the composition of their MHC genes often show mating preferences for animals with dissimilar MHC types (e.g., Yamazaki, et al., 1976). Thus, the slight inward inflection of the zone of compatibility at the level of familial divergence is meant to represent such outbreeding preferences as well as the possibility of inbreeding avoidance (e.g. Hoogland, 1982).

Still in regard to the model depicted in Figure 6, it should be noted that the relationship between the various taxa listed on the ordinate of this figure and the capacity to communicate holds only for cues that support reproductive isolation. Given the multi-sensory nature of mammalian communication, and the many functions it must serve, it certainly is probable that some gestures, whether chemical or not, are conserved and used for other adaptive purposes over a broad range of taxa. Indeed, some of the results of our behavioral studies with deermice may provide an example of such a situation. Peaceful, communal nesting occurred in 90% of the cases where young female white-footed mice were presented to male Alberta deermice (Figure 4). Communal nesting between naturally-occurring individuals of these two species has been noted repeatedly during northern winters (e.g., Howard, 1951). Thus the observation of their communal nesting in the laboratory may reflect a conserved set of gestures whose compatibility is necessary for interspecific cooperation in thermoregulation. Our house mice/deermice pairs, on the other hand, did not share nests to any great degree, but house mice often did not engage in communal nesting even among themselves (Figure 2). Furthermore, deermice and house mice probably are strong ecological competitors throughout their sympatric ranges and they frequently interact aggressively with each other (Grant, 1972).

Finally, the relevance of the findings reported in this paper to chemical communication *per se* is not entirely resolved. As noted earlier, we still do not know the precise degree to which specific disparities in chemical messages form the basis for the developing reproductive isolation that we have observed in our two deermouse stocks. Certainly the chemical modality dominates the communication of these strictly nocturnal rodents (Bronson, 1982), and the results we have obtained so far certainly suggest that these disparities are primarily but not exclusively chemical. Even on this tenuous basis, however, we can still comment upon the meaning of our observations for research in chemical communication in general. The study of mammalian chemical communication has developed largely as a laboratory science conducted by physiologists, behaviorists, and chemists. As such, most of our concepts rely upon the static and outmoded view of the species as a relatively homogeneous and unchanging entity which encompasses mainly individual and sexual variation. The population biologist, however, views the natural world much differently -- and certainly more realistically -- namely as a hierarchy of degrees of genetic relatedness including familial, demic, racial, subspecific and specific variation as well as individual variation. Furthermore, gene frequencies change constantly at each of these levels and, hence, there is little that is static in the real world. Communication incompatibilities related to reproductive isolation, such as those described herein, probably are relatively common throughout Class Mammalia, particularly in mammals with distinct populations or in highly social species. Thus our past reliance on a narrow view of the species undoubtedly has obscured a wealth of interesting variation concerning both the stimulus and the response side of communication which, if understood, could generate a much more dynamic view of chemical communication in general. Indeed, simple putative species signals and simple responses that are universal among all populations of a species, may be the exception and not the rule in mammalian chemical communication.

REFERENCES

Baker, R. H. 1968. Habitats and distribution. In Biology of Peromyscus (J. A. King, ed.) Spec. Pub. No. 3, The American Society of Mammalogists.

Bowers, J. M. and Alexander, B. K. 1967. Mice: Individual Recognition by olfactory cues. Science 158:1208-1210.

Bronson, F. H. 1979. The reproductive ecology of the house mouse. Quart. Rev. Biol. 54:265-299.

Bronson, F. H. 1982. Chemical communication in house mice and deermice: functional roles in the reproduction of natural populations. In Recent Advances in the Study of Mammalian Behavior (J. F. Eisenberg and D. Kleiman, eds.) Spec. Pub. No. 7, The American Society of Mammalogists, in press.

Bronson, F. H. and Desjardins, C. 1974. Circulating concentrations of FSH, LH, estradiol and progesterone associated with acute, male-induced puberty in female mice. Endocrinology 94:1658-1668.
Bronson, F. H. and Maruniak, J. 1975. Male-induced puberty in female mice: evidence for a synergistic action of social cues. Biol. Reprod. 13:94.
Dice, L. R. 1933. Fertility relationships between some of the species and subspecies of mice in the genus Peromyscus. J. Mammal. 14:298-305.
Dice, L. R. 1968. Speciation. In Biology of Peromyscus (J. A. King, ed.) Spec. Pub. No. 3, The American Society of Mammalogists.
Doty, R. L. 1972. Odor preferences of female Peromyscus maniculatus bairdii for male mouse odors of P. m. bairdii and P. leucopus noveboracensis as a function of the estrous state. J. Comp. Physiol. Psychol. 81:191-197.
Drickamer, L. C. 1974. Contact stimulation, androgenized females and accelerated sexual maturation in female mice. Behav. Biol. 12:101-110.
Eisenberg, J. F. 1968. Behavior Patterns. In Biology of Peromyscus (J. A. King, ed.) Spec. Pub. No. 3, The American Society of Mammalogists.
Epple, G., Golob, N. F. and Smith, A. B. III. 1979. Odor communication in the tamarin Saquinus fuscicallis (Callitrichidae): Behavioral and chemical studies. In Chemical Ecology: Odour Communication in Animals (F. J. Ritter, ed.) Elsevier/North Holland, pp. 117-130.
Godfrey, J. 1958. The origin of sexual isolation between bank voles. Proc. Roy. Phys. Soc. Edinburgh 27:47-55.
Grant, P. R. 1972. Interspecific competition among rodents. Ann. Rev. Ecol. Syst. 3:79-106.
Hoogland, J. L. 1982. Prairies dogs avoid extreme inbreeding. Science 215.1639-1641.
Howard, W. E. 1951. Relation between low temperatures and available food to survival of small rodents. J. Mammal. 32:300-312.
Moore, R. E. 1965. Olfactory discrimination as an isolation mechanism between Peromyscus maniculatus and Peromyscus polionotus. Am. Midl. Nat. 73:85-100.
Müller -Schwarze, D. 1974. Olfactory recognition of species, groups, individuals and physiological states among mammals. In Pheromones (M. C. Birch, ed.) Elsevier/North Holland, pp. 316-326.
Nevo, E., Bodmer, M. and Heth, G. 1976. Olfactory discrimination as an isolating mechanism in speciating mole rats. Experientia 32:1511-1512.

Perrigo, G. and Bronson, F. H. 1982. Signaling and priming communication: Independent roles in the reproductive isolation of spatially-separated populations of rodents. Behav. Ecol. Sociobiol., in press.

Roelofs, W. L. and Cardé, R. T. 1974. Sex pheromones in the reproductive isolation of lepidopterous species. In Pheromones (M. C. Birch, ed.) Elsevier/North Holland, pp. 96-114.

Schwarz, E. and Schwarz, H. K. 1943. The wild and commensal stocks of the house mouse, Mus musculus L. J. Mammal. 24: 59-72.

Stoddart, D. M. 1977. Two hypotheses supporting the social function of odorous secretions of some Old World rodents. In Chemical Signals in Vertebrates (D. Müller-Schwarze and M. Mozell, eds.) Plenum Press, New York, pp. 333-355.

Vandenbergh, J. G. 1969. Male odor accelerates female sexual maturation in mice. Endocrinology 84:658-660.

Vandenbergh, J. G., Finlayson, J. S., Dobrogosz, W. J., Dills, S. S., and Kost, T. A. 1976. Chromatographic separation of puberty accelerating pheromone from mouse urine. Biol. Reprod. 15:260-265.

Yamazaki, K., Boyse, E. A., Mike, V., Thaler, H. T., Mathieson, B. J., Abbot, J., Boyse, J., Zayas, Z. A. and Thomas, L. 1976. Control of mating preferences in mice by genes in the major histocompatibility complex. J. Exp. Med. 144: 1324-1335.

Yamazaki, K., Yamaguchi, Y., Beauchamp, G. K., Bard, J., Boyse, E. A. and Thomas, L. 1981. Chemosensation: An aspect of the uniqueness of the individual. In Biochemistry of Taste and Olfaction (R. H. Cagan and M. R. Kare, eds.) Academic Press, New York, pp.85-91.

THE ECOLOGICAL IMPORTANCE OF THE ANAL GLAND SECRETION OF YELLOW VOLES (*LAGURUS LUTEUS*)

Fan Zhiqin

Institute of Zoology
Academia Sinica
Beijing, China

INTRODUCTION

Yellow voles, found in the northern part of the Xinjiang Autonomous Region, live as family groups in elaborate burrows. They have three or four litters per year. Yellow vole populations fluctuate considerably, reaching high numbers some years, causing great damages to grassland.

Olfactory communication appears to be an important aspect of the biology of yellow voles as social animals. They have three different scent glands: anal glands, preputial glands and Harderian glands, all of which have been shown to be important sources of pheromones. The present study was designed to determine the ecological role of anal secretion glands. The potential for using anal gland in an integrated program for the control of voles is also discussed here.

MATERIALS AND METHODS

1. All the voles tested were obtained from their natural habitats. Some animals were killed for their anal glands and adrenal glands, which were then measured while fresh. The remainder were raised in the laboratory, to be used later for ethological study and experiments involving stimulation with pheromones. Great care was taken to maintain and test the animals in such a way that no contamination from other sources might confound the results.

2. Lipid extracts were made from the anal gland specimens. Cotton sponges soaked with the extracts were applied to the noses of

the yellow voles once a day for 15-30 days. Voles in the control group, on the other hand, were exposed to cotton sponges soaked with the solvent only. The variables of daily observation during the experimental period include: body weights, vaginal smears, and behavioural responses.

3. As a measure of pheromonal release, urinary marks deposited on a clean filter paper during the testing were scored under UV light (3650A). To minimize bias caused by overlapping marks, each filter paper was replaced at 20 minute intervals. Maps with the discrete marks outlined with pencil were then compared to evaluate the experimental effects.

4. All animals were killed and autopsied at the end of the experiment. The data obtained from the relevant organs were taken to obtain further information for comparison.

5. Effects of anal secretions on the behaviour of voles were also studied in their natural habitats. Four observation plots and four control plots, at least half a mile apart, were established in selected areas of relatively high vole density. The study lasted from the end of April to August 1978 and 1979. The voles in two plots were live-trapped, marked, and then released. In the other two plots, the burrow openings were stopped up periodically, and the number of those reopened was recorded as a measure of the effect of the pheromones applied.

RESULTS

Sexual and Seasonal Difference in the Weight of Anal Glands

In the yellow voles, both male and females, a pair of conspicuous spindle-shaped glands are found lateral to the rectum. The glandular cavity is filled with a light brown, thick oily fluid which gives a strong odor.

The mean weight of the anal glands of adult males was found to be 49.4 ± 3.93 mg., making up about 0.7% of the total body weight, while that of adult females was 35.2 ± 2.56 mg., comprising 0.74% of the body weight. The difference in gland weights between adult males and females was 14.2 mg. This was significantly different at the 0.01 level.

A marked seasonal variation of the gland weights was also noted, as shown in Figure 1.

In both male and female yellow voles, the glands diminished in size in winter. Histological examination of the glands showed atrophy of the secretory part; in addition the glands secreted very

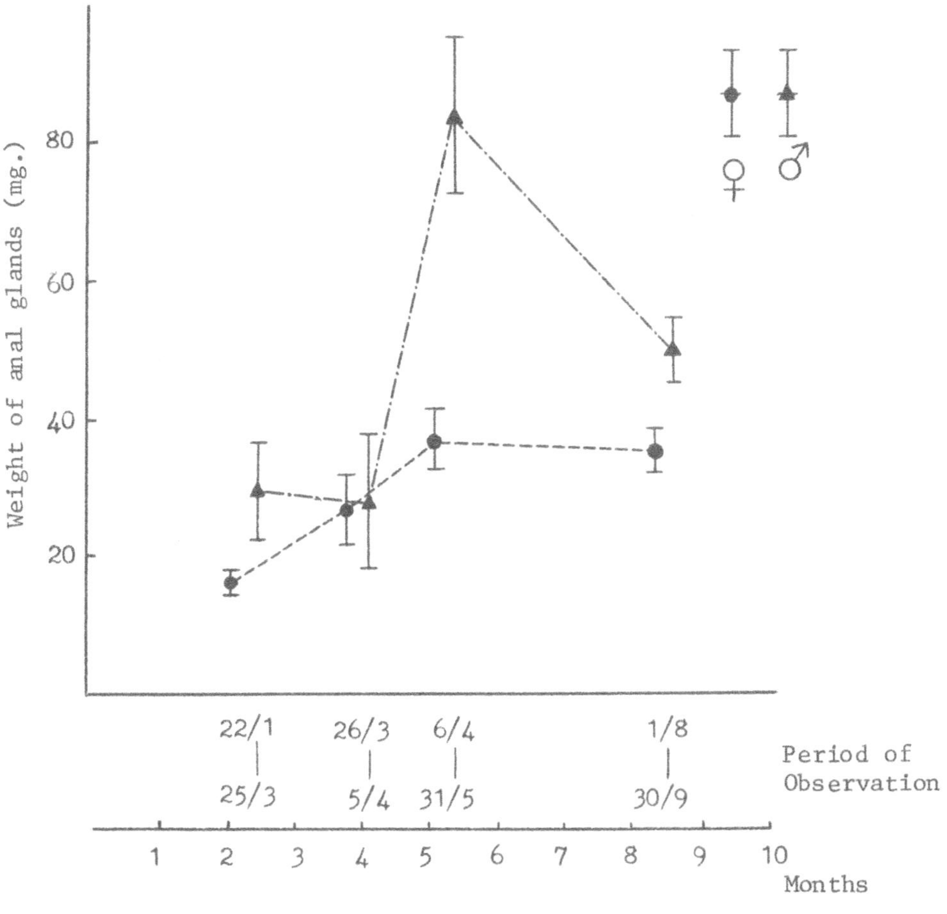

Figure 1. Seasonal variation in weight of anal glands of Lagurus luteus (mean ± S.D.).

little. The gland weights of adult males and females were found to be only 25.35 ± 3.62 mg. and 17.03 ± 3.77 mg., respectively. In April, however, the glands showed rapid growth with enhanced secretory activity, reaching values of 75.16 ± 8.54 mg. and 31.18 ± 5.92 mg., respectively. A markedly significant difference was confirmed by the t-test ($p<0.01$). The glands diminished in weight in autumn, with an average of 49.4 ± 4.46 mg. and 35.24 ± 2.56 mg., respectively. But the difference was still found to be significant at the 0.05 level.

The anal glands were found to be very small during early life. The anal glands of a 15 g. young male vole, for instance, weigh only 0.7 mg. while that of a 16 g. young female vole, only 2.6 mg. The glands, however, grew rapidly afterwards, showing a nearly 70 fold increase as compared with an increase from 15.0 g. to only 70.0 g. in body weight, that is, no more than five-fold. The weight of adult anal glands varied widely from 15.0 mg. to 150 mg.

Spatial and Temporal Distribution of Fecal Pellets

The secretions of the anal glands are easily carried by feces to the outside. We have observed that voles also spread their secretions by directly rubbing their anal region on objects in their environment. As the feces passed by the voles appear to be the normal vehicle for the spreading of the secretion as chemical information, we made quantitative observations of the distribution of the fecal pellets passed by the yellow voles under experimental conditions.

The fecal pellets were not distributed randomly. They appeared to be denser near their nests and along the boundaries with adjacent colonies. On average, 64 pellets were found there for each vole during 24 hours, making up 60% of the total pellets. Many fewer were found in other places, though they covered 3/4 of the whole area. The male produced a larger number than the female did, in approximate proportion to the size of their glands.

The production of fecal pellets was found to be highest in daytime from 12 a.m. to 5 p.m., giving about half of the total pellets passed in a day. Equal amounts were found in the afternoon and after sunset. The smallest number was produced at night (Table 1).

It is of great interest to note the obvious temporal parallelism between the number of fecal pellets passed and the activity of the animals; that is, the more active they became, the more pellets they produced.

Functions of the Anal Secretions

The Behavioural responses to anal secretions. The yellow voles tended to gnaw at the places where a piece of cotton sponge soaked with lipid extract of anal secretion had been placed. With such stimulation, the voles usually showed increased activities, moving about cautiously, exploring the surroundings by nose, climbing the wire of the experimental cage, and depositing fecal pellets alongside of the odour sources. Generally both male and female voles responded strongly to anal secretions of the same sex, while much weaker responses were observed with secretions of the opposite sex.

Table 1. Number of fecal pellets of yellow voles at different times of day.

	Time			
Sex	9.00-12.00	12.00-17.00	17.00-20.00	20.00-9.00
Males	23	42	34	6
Females	24	42	30	9

They reacted less to their own secretions (Table 2, Fan Zhiqin et al., 1981).

The repellent effect of anal secretion observed in field conditions.

(Plot 1). A single dose was administered to each burrow opening that had been in frequent use. As a result, all the voles traversed openings that had been abandoned for a long time, or openings that were newly dug. On the second day, only 23% of the contaminated openings were used by the voles again (Figure 2).

(Plot 2). Anal secretions were administered to only half of the burrow openings that had been in frequent use. As a result, the burrows blocked after administration of anal secretion were found to be reopened in only 25% as opposed to one of 55% in the control burrows. This difference is statistically significant ($p<0.05$). On the second day, the experimental group gave a value of 31%, while the control one 75% ($p<0.05$). However, no significant difference was found on the third and fourth day.

(Plot 3). All the voles were live-trapped, marked, and then released every day. Over five doses of secretions were administered to each of two burrows (1 & 2), while the other two burrows (3 & 4) were used as controls. Yellow voles of all the experimental burrows 1 & 2 as well as the control burrow 3 disappeared after the treat-

Table 2. Behavioural response to different anal secretions of yellow voles.

Sex	No. of yellow voles	Odour sources	Frequency of behaviours in 20 minutes per yellow voles					
			Gnawing	Defecation	Climbing	Grooming	Traveling	Feeding
Males	21	Anal gland from yellow voles of same sex	14	20	23	7	16	5
Females	22		12	14	35	5	17	7
Males	22	Anal gland from yellow voles of opposite sex	9	6	11	6	21	2
Females	20		7	7	5	3	9	2
Males	5	Own anal gland	1	3	10	5	7	6
Females	5		0	2	6	5	9	7

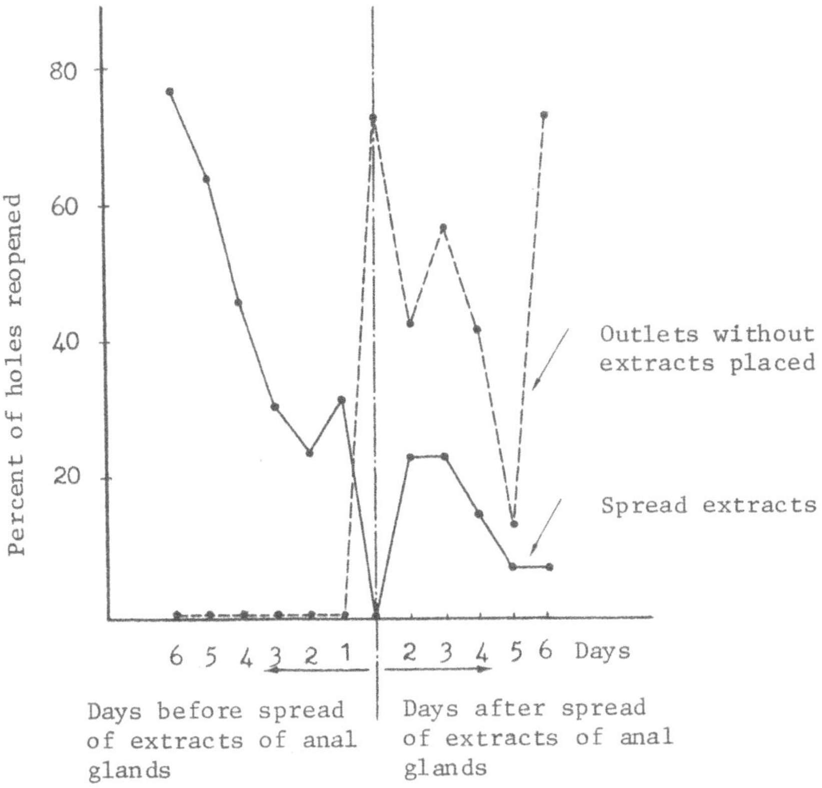

Figure 2. Influence of anal gland secretion on frequency of entering the holes.

ment. The yellow voles, on the other hand, remained in the control burrow 4. The observation lasted about a month. At the end of the observation period, four unmarked young voles were captured together with those that had been marked.

Effects of anal odour on some endocrine glands (gonads, adrenal glands) and exocrine glands (preputial and anal glands). Lipid extracts of anal glands were rubbed on the noses of experimental voles, while odourless cottons were applied to voles with same body weights as control.

Twenty-four voles were exposed to anal secretions for 15 days. At the end of the 15 days the body weights of the experimental voles were only slightly lower than those of the control. The mean weights of testicles in the experimental group, however, were 38.1 $\pm$ 3.45 mg., as compared with 44.36 $\pm$ 4.13 mg. ($p<0.05$) in the control. The ovarian weights of the test group were found to be lower also, though no significant difference was confirmed by statistical calculation.

Twenty-eight yellow voles were exposed to male anal odours for 30 days. The weights of testicles, preputial glands, anal glands, and the whole body were found to be lower than those in the controls. But the weight of adrenal glands was higher, increasing progressively during the 30 days of exposure. All the indices in female yellow voles, however, were slightly higher than in the control, though the differences were not as marked as in the case of male voles (Table 3).

Effects of anal secretions of Lagurus luteus on Lagurus lagurus. Lipid extract of anal gland of *Lagurus luteus* was placed in one corner of the cage holding *Lagurus lagurus* voles. They showed ceaseless running, decreased frequency of feeding, reduced food intake and wanted to escape after sniffing the extract.

Avoidance behaviour was noted in *Lagurus lagurus* when extracts were placed at the burrows opening in natural habitats. The rate of entering the burrows decreased from the 100% before the treatment to 36.1% after.

Periodical rubbing of the noses of *Lagurus lagurus* with lipid extracts of anal glands of *Lagurus luteus* for two weeks led to decrease in body weight by 1.1 g., as compared with 0.2 g. in control voles. Since many male voles died during the experiment, comparison of testicular weights could not be made. It is uncertain if inhibition on female gonads occurs. The ovarian weight was 0.8 mg. in the test group as compared to 1.5 mg. in the control, but the difference was not statistically significant. In summary the anal odours of male *Lagurus luteus* exerted negative influences on feeding, reproduction and activities of *Lagurus lagurus*, though not so marked as on conspecifics.

Table 3. Body and gland weights after exposure to anal odour of strange males in *Lagurus luteus*.

Group	Sex	Mean weight					
		Body (g)	Testicle & Ovary (mg)	Adrenal gland (mg)	Preputial gland (mg)	Anal gland (mg)	Harderian gland (mg)
Treated	Males	80.17±6.61	74.42±6.28	56.19±6.23	59.12± 8.37	75.57±5.28	84.9±14.34
	Females	72.38±1.83	6.26±0.61	16.81±2.08	38.78± 5.01	35.16±3.37	94.83± 7.19
Control	Males	91.14±3.97	83.72±5.57	43.69±4.02	94.97±15.29	77.84±7.78	75.69± 8.61
	Females	68.38±4.72	5.77±0.73	15.58±1.76	37.89±6.75	30.81± 5.30	81.13± 9.20

The effects of anal extract of Lagurus lagurus on Lagurus luteus were found to be very small.

DISCUSSION

Yellow voles are social animals that live in large groups. The social behaviour increases the demand for chemical communication (Thiessen, 1971). Our experiment has confirmed the important role played by chemical communication in yellow voles, of which anal odour is an important aspect.

A territory-marking role of the anal odour is suggested by the way the anal secretions are distributed and the responses elicited. Territory is chosen primarily by the yellow voles to insure proper breeding of offspring. The animal usually marks its territory by some chemical markers, odourous substances of course, such as the anal secretion in our experiment. Immersed in an atmosphere of familiar odour, the animal seems to feel at ease, moving about and feeding confidently. It makes attacks on intruders and quite often emerges as the winner. But when the circumstances are reversed, i.e., when an animal is put in a territory of another conspecific animal, it shows expressions of alarm and restlessness, and cannot carry out the normal activities such as feeding and breeding.

A clear correlation can be seen between seasonal changes of anal glands and territory-marking activity. At the end of autumn, the voles move to suitable locations where they live through the winter. There exists no strict territories in these temporary and crowded areas. The anal glands of both sexes shrink rapidly at that time. In spring, the animals become active again. Males and females establishe their territories. The anal glands size and secretory activity increase. Upon entering a new environment, an area is marked with anal secretions. Males seem to take the chief responsibility in territory-marking behaviour; they have larger glands to produce more secretion and deposit it well distributed over the area. The territorial marks serve to keep off conspecific animals, especially of the same sex, i.e., the potential sexual rivals. The scent deposited in the space becomes a symbol of the animal itself and continues to be an effective message even in the absence of the signalling animal.

Animals may respond to foreign secretions placed within their own territories as if they themselves had entered foreign territories. An important component of the responses is the frequent ejection of fecal pellets. Apparently, when the animal finds it difficult to escape from an area enveloped by hostile scents, it ejects more fecal pellets, thereby producing more scent of its own to counteract the foreign scents.

In field studies anal secretions have been shown to have a repellent effect, which, however, is only temporary, being marked on the first day but decreasing gradually later on.

Within the space in which odours of its own prevail, an animal acquires a physiological state which allows it to complete various biologically important functions and notably to participate in various stages of reproduction. Exposed to foreign scent given by one male, the conspecific males showed decreased body weights and reduced preputial glands and gonads, that indicate inhibition of sexual activity. Yet the weights of adrenal glands increased, an indication of enhanced stress functions in response to unfavourable conditions. In striking contrast to the males, the females responded only slightly to male secretion, which support the assumption that the male voles take the chief responsibility of territory-marking behaviour.

The anal secretion of Lagurus luteus exerts a similar, though much less, effect on Lagurus lagurus. In the northern part of Xinjiang, these two species occur together. Generally speaking, there is competition between two ecologically close species, and usually species of larger size gain dominance over the smaller and less agressive ones. The anal scent of Lagurus luteus repels Lagurus lagurus, and inhibits the growth of body weight and gonads. Conversely, the anal secretions of Lagurus lagurus exert no obvious effect on Lagurus luteus. This fact bears out the dominance of the larger Lagurus luteus over the smaller Lagurus lagurus.

In a sense, there exists an "odorous space" within which the mutually excluding breeding niches occupied by animal couples are defined by the specific "odorous clouds" they generate. Or by analogy to physics, we may speak of an "odorous field" within which there is mutual repulsion among conspecific animals of the same sex but unidirectional repulsion among different species along the order of relative dominance.

Apart from the theoretical implication of our present results, we are particularly interested in the practical aspects. The experiments described here have clearly shown the repellent and inhibitive effects of anal pheromones on conspecific animals. We might be able to use the anal pheromone as a rodent control. This pheromonal approach is biological, and, there is no pollution of the environment by foreign substances.

REFERENCES

Fan Zhiqin, Hao Shoushen, Liang Yingnan, Wang Mingyue, and Zhao Yong, 1981, Chemical signals and their role in social groups of two voles (*Lagurus luteus* and *Lagurus lagurus*), *Acta Ecologica Sinica*, 1:66.

Thiessen, D. D., Owen, K., and Lindzey, G., 1971, Mechanisms of territorial marking in the male and female Mongolian gerbil (*Meriones unguiculatus*), *J. Comp. Physiol. Psychol.*, 77:38.

ODOR AS A COMPONENT OF TRAP ENTRY BEHAVIOR IN SMALL RODENTS

D. Michael Stoddart

Department of Zoology
University of London King's College
Strand
London WC2R 2LS
U.K.

INTRODUCTION

For many years students of small rodent ecology have utilized live-traps and capture-mark-release techniques for examining basic population parameters. In a great majority of studies, estimates of population size, and related parameters such as density, sex and age distribution, are based on one of a number of derivatives of Petersen's index, all of which demand traps to have an equal sampling efficiency. Since a trap remains passive, and effects capture only after a rodent has purposely entered it, it cannot be used to sample a population of rodents in the same way that, for example, a drift net will sample a population of pelagic fish. Rodents do not blunder into live traps (although a few studies have employed pitfall traps) - they will only be caught if they can overcome any inhibition which they show towards the trap (Shillito 1963). Although phenomena such as "trap-shyness" and "trap-proneness" have long been recognized, (in which certain individuals are caught less or more frequently than would be expected by chance if the traps were sampling the population randomly (Chitty and Kempson 1949)), it is surprising that so little attention has been paid to whether the ability of a given trap to catch a given individual is influenced by residues left behind by a previous occupant. If residues do influence the subsequent pattern of trap entry, estimates and analyses will be inaccurate and distorted by an unknown amount.

The objectives of this paper are threefold. Firstly the results of an intensive study of the role of trap odor on the trappability of a mixed community of rodents consisting mainly of Microtus agrestis together with much smaller numbers of Apodemus sylvaticus, A. flavicollis, Clethrionomys glareolus and three species of shrews are presented in order to establish whether part of the heterogeneity of the trapping results can be attributed to odorous residues remaining in the traps from a previous occupant, and whether there are any clearly demonstrable trends in the pattern of trap entry. Secondly, the effects on Microtus capture of the residues of other species, and vice versa, are examined to determine whether a community effect exists, and thirdly the usefulness of the mousetrap as a self-contained olfactory laboratory, in which careful behavioral bioassay can be conducted, is discussed. The requirement for sound bioassay is recognized by all olfactory biologists, and the lack of suitable techniques is impeding the development of this branch of behavioral ecology. The paper investigates whether the mousetrap has a useful role to play in this important area of olfactory biology.

MATERIALS AND METHODS

All the fieldwork herein reported was conducted at King's College Field Station at Rogate, in Sussex, England. A population of short-tailed voles, M. agrestis, inhabits an overgrown and ungrazed meadow at the station and has been the subject of a number of investigations into the effects of interspecific odors on rodent trappability (Stoddart 1976; 1980). A standard grid of 98 Longworth live traps was used, being dispersed in pairs at each of the 49 intersections of a 7 x 7 arrangement of rows. The distance between rows of traps was 10 m. At each trap site (i.e. grid intersection) the pair of traps was positioned as close together as possible, and with their entrances never more than 5 cm apart. In the present study, the traps at each site were labelled a) and b), and the relative position of one to the other (i.e. which one lay to the left of the other when viewed by the operator following an identical route around the grid) was determined anew each time the traps were set by a toss of a coin. Even though the trap entrances were very close together, this procedure was designed to overcome any spatial bias. The pattern of entry into each trap at each site was recorded. Traps were set each evening between 1600 h G.M.T. and 1700 h G.M.T., and checked from 0730 h G.M.T. They remained closed during the day in order to reduce, as far as possible, immigration by non-resident rodents. The data upon which the following analyses are made were collected over an intensive trapping program, from July 20 through September 1 1980, and when all rodents were in reproductive condition. Traps were meticulously cleaned before trapping commenced, and there was no period of free-baiting. Altogether a total of 1048 captures of small mammals (824 M. agrestis; 142 Apodemus spp.; 24 C. glareolus and 58 shrews) were obtained in 2450 trap-nights. This program was part of an investigation into the influence of trap odor in the estimation of population size of *M. agrestis* (Stoddart, *in press*), and is examined here not so much as a statement on *Microtus* behavioral ecology (though much of it is concerned with this species), but as a demonstration of the effect of one behavioral influence which operates on small rodent trappability under normal field conditions at one time of the year.

RESULTS

The data matrix upon which most of the subsequent analyses are performed is displayed in Table 1. It shows for each category of capture on trapping occasion 't' (where 't' is any occasion from 1 through 12), the category of capture obtained on 't + 1'. The very high value of χ^2 indicates that the data are far from homogeneous and that subsequent captures may be influenced by previous ones. Close inspection of the data reveals that over half the total χ^2 is contributed by two cells only - the much higher than expected numbers of male and female *M. agrestis* which were caught in traps which had immediately before held a member of the same sex.

Table 1. Analysis of sequential captures of *Microtus agrestis* and other species of small mammal during an intensive trapping program. 't' = any day from 1 through 12. Expected values in round brackets; contribution to total χ^2 in square brackets.

captured on day 't' \ captured on day 't + 1'	0*	Male *M. agrestis*	Female *M. agrestis*	Other†
0*	828 [35.3] (673.6)	98 [21.9] (156.7)	139 [32.3] (224.1)	85 [1.17] (95.6)
Male *M. agrestis*	109 [14.4] (156.4)	116 [174.1] (36.4)	36 [4.92] (52.0)	6 [11.82] (22.2)
Female *M. agrestis*	137 [32.5] (222.0)	33 [6.7] (51.6)	200 [215.2] (73.9)	9 [26.1] (31.5)
Other†	74 [5.1] (96.1)	17 [1.3] (22.3)	10 [15.1] (32.0)	63 [179.4] (13.6)

Total χ^2 = 767.3

Notes * = empty trap
† = other species *Apodemus sylvaticus*, *A. flavicollis*, *Sorex* spp., *Clethrionomys glareolus*, *Neomys fodiens*.

It stands to reason that if the rodents are not randomly dispersed, or if one sex has a much larger individual range than the other, bias will occur. The problem with the measurement of dispersion is that data have to be gathered by trapping; an independent means of measurement must be used if the effect of this population attribute on trappability is to be assessed. Although traps in some parts of the grid caught voles more frequently than traps in some others, there was no evidence of extreme patchiness in dispersion. It is acknowledged that a degree of bias stems from the non-random dispersion of voles. On the question of individual range, previous workers have reported male voles to occupy ranges twice as large as those of females (Myllymäki 1977). If this is so, they would be expected to re-enter the same trap on 't + 1' in which they themselves had been held in on 't' less frequently than would females with their smaller ranges. The data were examined for any bias of this nature and it was found there was no significant difference between the proportion of males and females being retrapped in the same trap as that in which they had been caught on 't' (χ^2 = 0.67; 1.d.f.; n.s.). Does this mean that male voles have similar sized individual ranges to females, or does it mean that males may have larger ranges but make a greater effort to seek out a particular trap? The mean number of captures per utilized trap site was calculated for males and females independently and subjected to analysis by Student's 't' test. As is shown in Table 2 there is no significant difference between the values, and so it would appear that there is no evidence to indicate that males have larger individual ranges than females. Additionally, the almost identical values expressed in the column in Table 2 headed "Mean number of captures per utilized trap site per vole" indicate that neither sex is more readily trapped than the other.

Table 2. Utilization of the trapping grid by male and female Microtus agrestis. Data analysed by 'Student's 't' test.

	Number of voles caught	Number of captures	Number of trap sites utilized	Mean number of captures per utlized trap site	Mean number of captures per utilized trap site per vole
Males	35	181	36	5.028*	0.143
Females	41	210	37	5.676*	0.138

* 't' = 0.75 1.d.f.; n.s.

As is clear from Table 1, traps remained empty on a great number of occasions. Analysis of the pattern of entry of voles into traps which had previously held either nothing, or a vole, indicated that voles were as likely to enter one as the other ($\chi^2 = 0.47$; 1.d.f.; n.s.). Further analysis on the likelihood of an empty trap following upon a capture of a vole, or another vole following, indicated that there was no significant difference between the sexes ($\chi^2 = 1.44$; 1.d.f.; n.s.), suggesting that neither sex inhibits the entry of M. agrestis more than the other. This analysis uses males and females together, as opposed to an empty trap, but when only those traps which had held a Microtus are examined with respect to the sex of the subsequent visitor, it is seen that males are significantly more likely to be retrapped in traps which had previously held a female, than vice versa (Table 3 $\chi^2 = 5.68$; 1.d.f.; p = < 0.02; > 0.01). This result points to the existence of a perceptible trace of the previous occupant emerging from the trap, and to which males and females react differently. Soiled traps do not appear to attract more initial captures than clean traps, and there is no significant difference between the proportions of males (42.4%) or females (47.0%) making their first appearances in dirty traps. Although the data are too few for a detailed analysis, the most recent occupants of each of the 26 soiled traps which first trapped previously unmarked M. agrestis were adult male voles (15 traps), adult female (6 traps), juveniles (4 traps) and another species of mammal (1 trap).

Table 3. Analysis of the relationship between sex of previous occupant and subsequent captures in Microtus agrestis. Expected values in round brackets, contribution to total χ^2 in square brackets.

	Number of captures on 't + 1' in traps which had held, on 't'.	
	a member of the opposite sex	a member of the same sex*
Males	[2.8] 36 *(27.2)*	[0.6] 116 *(124.7)*
Females	[1.8] 33 *(41.7)*	[0.4] 200 *(191.2)*

* including self Total $\chi^2 = 5.68$; 1.d.f.; p = < 0.02; > 0.01

Consideration of Table 1 shows that both male and female voles are significantly drawn to traps which had, before, held members of the same sex. As the mean number of trap sites utilized by males was 2.7 ($\pm$ 1.14 S.D. n = 20) and by females was 1.95 ($\pm$ 0.87 S.D. n = 37) (data for voles caught at least five times), this attraction cannot be explained by the voles having very limited movement. It seems that voles seek out a trap which appears familiar in some way, even if the exact position of such a trap is not the same as it was previously. The apparently high number of "other" species following themselves is due to a patchy distribution of *Apodemus flavicollis* along one edge of the study area (for further explanation, see Stoddart *in press*).

The most abundant species of rodent on the grid, after *M. agrestis*, was the woodmouse, *Apodemus sylvaticus*. Analysis of the interaction of captures of this species on those of the short-tailed vole indicates that there is a strong association between the species caught on 't' and that caught in the same trap on 't + 1' (Table 4). It appears that *M. agrestis* follows *A. sylvaticus* into traps far more readily than *vice versa*, and that captures of *M. agrestis* have a depressive effect on the captures of *A. sylvaticus*. No such analysis is possible with *A. flavicollis* because this species was caught almost exclusively in the peripheral rows where *Microtus* was virtually absent.

Table 4. The interaction of *Apodemus sylvaticus* and *Microtus agrestis* captures during an intensive trapping program. Fisher's exact probability test.

	't + 1' Number of occasions upon which captures of *A. sylvaticus* and *M. agrestis* were followed by:	
't'	the same species or empty trap	the other species
A. sylvaticus	27	5
M. agrestis	370	2

p = 0.000042

DISCUSSION

The data presented reveal two fundamental points about the efficacy of live traps for studying small rodent communities. First, it is obvious from Table 1 that the data they provide are not homogenous. Although this paper is concerned with the existence of recognizable traces left behind by a previous occupant, marked heterogeneity can result from violent fluctuations in species density or sex ratio from one part of the grid to another, as much as from selective attraction. Examination of the raw data for voles reveals that there is no evidence of extreme patchiness in dispersion, although they are by no means randomly dispersed. Bearing in mind that the precise location of each trap was determined anew and by a random process each time they were set, it would appear that the major cause of the perturbation was something other than patchy dispersion of the rodents. Second, examination of the trapping record indicates that the species and sex of the last occupant of a trap are important factors in determining the species and sex of the next (Table 1 and 3). The only possible traces which could effect such a determination are those left behind by the last occupant, and they are likely to be associated with urine, feces, or sebaceous, salivary and sex gland secretions.

It is clear that *M. agrestis* exhibits a substantial degree of discrimination between different odor signals emanating from previously soiled traps (Stoddart 1982), and it naturally follows that the state of cleanliness of the traps is a feature of the trap-oriented behavior of the voles which cannot be ignored. A number of ecologists have recently addressed this problem and examined whether dirty traps are equally effective as clean ones in trapping rodents. Boonstra and Krebs (1976) reported that the use of previously soiled traps considerably decreases the time necessary for trapping "a substantial portion of the population" of *Microtus townsendii*. They indicate that previously unmarked voles are more likely to enter soiled traps than clean, and they suggest that the residual odor helps to reduce the 'new object reaction' (Shillito 1963). No such preference for soiled traps by unmarked voles was observed during the present study. Mazdzer and his co-workers (1976) showed that *Peromyscus leucopus* was more readily trapped in soiled traps, and Daly *et al* (1980) showed the same to be true for *P. maniculatus*, *P. eremicus*, *Dipodomys agilis* and *Perognathus fallax*, at least during the breeding season. In an earlier study Daly *et al* (1978) reported that all species avoided traps smelling of deer mice when they were sexually quiescent, but this was not so in the spring when they became sexually active. Summerlin and Wolfe (1973) noted that adult female cotton rats, *Sigmodon hispidus*, were markedly attracted to soiled traps - adult males and juveniles were not. And working with a captive colony of house mice, Rowe (1970) observed that adults of each sex were attracted to traps smelling of the opposite sex. Montgomery (1979) investigated interspecific, sexual and individual biases affecting the trappability of a small rodent community inhabiting a mixed deciduous woodland and concluded that the previous occupant of a trap was an important determinant of future trapping success, but his trapping regime did not correct for spatial bias at the trap site and so must be interpreted with caution.

How useful is the mousetrap as an olfactory laboratory? Perhaps the greatest advantage to be gained from conducting odor discrimination and responsiveness tests in the field is that the results are far freer from experimental artefact than those obtained during laboratory studies. They are not totally without bias, however, and two obvious problems can be enumerated.

1. The density and dispersion pattern of the population, or community, is not known. Spatial segregation of species, together with their different activity patterns may alter the availability of traps to the species under study. To illustrate this point, Fig. 1 shows the proportion of the total number of captures of both _A. sylvaticus_ and _M. agrestis_ taken on each day of a thirteen day trapping program. It is clear that an approximately equal proportion of the total number of _M. agrestis_ captures occurs on each day, though there are some minor fluctuations. _A. sylvaticus_ on the other hand exhibits a slow build-up from a low proportion at the start of trapping to a high proportion later on. Factors other than increasing familiarity with the traps may cause this pattern, however, but correlation with mean night-time temperature or rainfall revealed nothing in this instance. The data in Fig. 1 suggest that _A. sylvaticus_ is more trap-shy than is _M. agrestis_, and that a considerable period of acclimatization is required before they will enter readily.

2. The sample sizes cannot be determined in advance. The number of _potentially_ responding individuals is not known, so the actual number which do respond to a given trap-borne odor is an unknown proportion of the whole. Not only does this demand a critical appraisal of the data but it necessitates care in the choice of statistical analyses.

Consideration of these problems suggests that mousetraps have a real and effective part to play in rodent odor detection and behavior studies, but that the simplest of questions only must be asked of them. For example Stoddart has demonstrated on three occasions (1976, 1980; _in press_) that three different predator odors bring about a reduction in the number of individuals caught. The technique used was to introduce into the trap tunnel a small piece of filter paper soaked with urine or anal gland secretion,

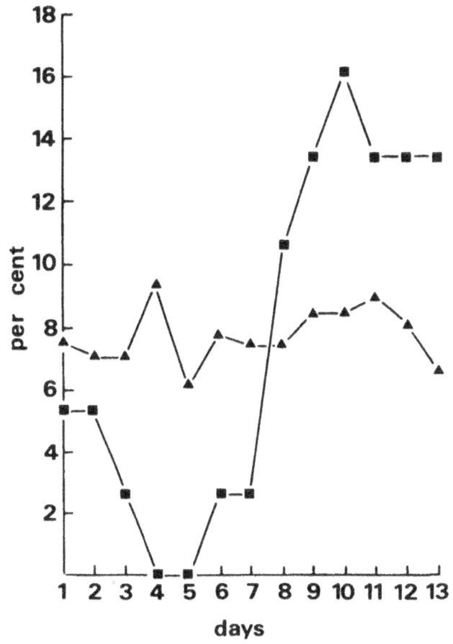

Fig.1 Graph to show the percentage of the total captures of M. agrestis and A. sylvaticus taken on each day of a thirteen day trapping program.
▲——▲ M. agrestis ■——■ A. sylvaticus

and the resulting population estimates - before and after the experiment as well as during it - compared. In the most recent study (Stoddart, in press) the mean trappability of a group of M. agrestis which avoided capture when tiger urine was introduced into the traps, but which were caught in scent-free controls both before and after, was significantly lower than that of those voles which were not repelled by the odor ('t' = 3.46; 34 d.f., p = < 0.001). These data indicate a degree of behavioral heterogeneity within the vole population and the technique by which they are derived may be a quick method for distinguishing between the "trap-prone" and the "trap-shy" members of the population.

The use of tainted traps in the field allows broad species differences in response to odor to be made. Table 5 shows the number of males and females of the three most abundant species of small rodent caught during three trapping periods each of three nights' duration. An experimental period, in which tiger urine was introduced into the traps, was flanked immediately before and after by scent-free controls. The most apparent species effect is that no female A. flavicollis were captured during the experimental session; female A. sylvaticus and M. agrestis were not so severely inhibited though the total number of M. agrestis captures was considerably reduced. Although the data are too sparse to justify firm conclusions, it would appear that these three species respond in slightly different ways to this particular trap-born odor.

Mousetraps in the field have a significant part to play in the examination of odor-influenced behavior of rodents, allowing behavioral heterogeneity within a natural population to be studied. Until more is known of the ecological significance of this heterogeneity, it is best to keep the experiments simple and for the controls to be carefully executed and scrupulously monitored. Mousetraps provide the means whereby field experimentation may be performed under as natural conditions as possible, and they should be seen as complementary to olfactometers in the laboratory.

Table 5. The numbers of captures of both sexes of M. agrestis, A. sylvaticus and A. flavicollis obtained during an experimental trapping system, in which tiger urine was introduced into the traps, and in preceding and following scent-free controls.

		Control	Experimental	Control
M. agrestis	Male	28	14	16
	Female	43	24	29
	Total	71	38	45
A. sylvaticus	Male	9	12	11
	Female	7	6	6
	Total	16	18	17
A. flavicollis	Male	6	5	9
	Female	5	0	6
	Total	11	5	15

REFERENCES

Boonstra, R. and C.J. Krebs, 1976, The effect of odour on trap responses in Microtus townsendii. J. Zool. Lond. 180: 467-476.

Chitty, D. and Kempson, D.A. 1949, Prebaiting small mammals and a new design of live trap. Ecology 30: 536-542.

Daly, M., Wilson, M.I. and Faux, S.F. 1978, Seasonally variable effects of conspecific odors upon capture of deer mice (Peromyscus maniculatus gambelii). Behav. Biol. 23: 254-259.

Daly, M., Wilson, M.I. and Behrends, P. 1980, Factors affecting rodents' responses to odours of strangers encountered in the field: experiments with odour-baited traps. Behav. Ecol. and Sociobiol. 6: 323-329.

Mazdzer, E., Capone, M.R. and Drickamer, L.R. 1976, Conspecific odors and trappability of deer mice (Peromyscus leucopus noveboracensis). J. Mammal. 57: 607-609.

Montgomery, W.I. 1979, An examination of interspecific, sexual and individual biases affecting rodent captures in Longworth traps. Acta Theriologica 24: 35-45.

Myllymäki, A. 1977, Intraspecific competition and home range dynamics in the field vole, Microtus agrestis. Oikos 29: 553-569.

Rowe, F.P. 1970, The response of wild mice (Mus musculus) to live traps marked by their own and foreign mouse odour. J. Zool. Lond. 162: 517-520.

Shillito, E.E. 1963, Exploratory behaviour in the short-tailed vole, Microtus agrestis. Behaviour 21: 145-154.

Summerlin, C.T. and Wolfe, J.L. 1973, Social influences on trap response of the cotton rat, Sigmodon hispidus. Ecology 54: 1156-1159.

Stoddart, D.M. 1976, Effect of the odour of weasels (Mustela nivalis L) on trapped samples of their prey. Oecologia 22: 439-441.

Stoddart, D.M. 1980, Some responses of a free living community of rodents to the odors of predators in "Chemical Signals in Vertebrates", Müller-Schwarze, D., and Silverstein, R.M., eds. Plenum, NY, p.1-10.

Stoddart, D.M. 1982, Demonstration of olfactory discrimination by the short-tailed vole, Microtus agrestis L. Anim. Behav. 30: 293-294.

Stoddart, D.M. (in press) Does trap odour influence estimation of population size of the short-tailed vole, Microtus agrestis L? J. Anim. Ecol. 51:

EXPERIMENTAL MODULATION OF BEHAVIOR OF FREE-RANGING MAMMALS BY SEMIOCHEMICALS

Dietland Müller-Schwarze

College of Environmental Science and Forestry
State University of New York
Syracuse, New York 13210 USA

INTRODUCTION

This paper reviews the mammalian species and behavioral contexts for which we know the role of behavior-modulating conspecific chemicals (=pheromones) well enough to be able to change the behavior of freely moving single individuals consistently and predictably in their natural physical and social environment.

Since time immemorial hunters, trappers, and animal breeders have been aware of the role that "scents" play in the behavior of many wild mammals and have applied their knowledge to facilitate behaviors that are in man's interest.

Do modern chemical ecologists understand chemical communication well enough to be able to set a chemical stimulus and to elicit selectively a predictable response, even though many other responses are possible? I.e., are we able to demonstrate a strong or even obligatory stimulus-response sequence? Laboratory findings have been suspected of being artifacts, possibly due to spatial restrictions, unnatural social units, an impoverished structure of the environment, forced choices imposed on the animals, and non-biological stimuli or concentrations used. In addition, mammalian behavior is generally considered to be highly complex and flexible. Have we analyzed the often intricate behavior well enough to pinpoint precisely behavioral effects of social odors? What steps are necessary next to understand better the place of odor signals under natural conditions?

1. Rodents

What have we learned about the responses by free-ranging rodents to conspecific scents? Are general rules emerging from the studies that have been conducted? The two experimental approaches have been to observe behavior (or other effects) at experimental scent marks, and to capture small rodents with scented traps.

a. Behavioral responses to experimental scent marks. Since most rodents are either nocturnal or active under cover, or both, Observations of behavior at scent marks in the wild are rare. An example is the response of free-ranging beaver (*Castor canadensis*) to experimental scent marks (Hogdon, 1978). A recent experiment (Müller-Schwarze et al., in prep.) follows.

At each of two ponds two artificial scent mounds (ASM) were constructed from mud and topped with 1 gram castoreum. They were placed near the inhabited lodge of an established beaver colony. In addition, the beaver had their own scent mounds ("beaver scent mounds" BSM) which were left intact. The responses of the adult beaver were observed after their emerging from the lodge in the evening. The behavior on four such subsequent "scent nights" was compared with that on the preceding four "control nights." Blocks of four control nights and four scent nights alternated during the period from June 22 to August 30.

The adult beaver, after emerging between 19.00 and 19.30 hrs, floated in the water, presumably defecating, then patrolled the bank areas near lodge and dam, by swimming with head lifted and turning right and left, apparently sniffing, looking and listening. On control nights, with only BSMs present, the beaver, after cruising near dam and lodge, swam to their feeding grounds upstream and remained out of sight until nightfall.

On "scent nights" however, the behavior of the beaver was quite different. After their initial quiet floating they swam up and down the pond's edge, hissed (an aggressive vocalization), sniffed toward the ASMs and went ashore. They headed for an ASM, smelled it, walked on top of it, and marked it with rubbing movements of their cloaca. Often they obliterated the ASM with pawing movements of their front feet, and even carried the mud from the ASM to their own BSMs, walking bipedally.

The behaviors that increased significantly from control nights to scent nights were: number of tail slaps in water, number of land visits, time spent on land, percent of scent mounds visited, number of visits per scent mound, frequency of pawing scent mounds, depositing mud, straddling scent mound (marking), and number of activity bouts near the scent marks.

The beaver kept responding during the 70 days of the experiment, even though there occurred habituation of most of the responses over the summer, especially at one of the two sites.

In summary, there are behavioral changes, the direction of which is predictable. There is no single, stereotyped response that is "released" by strange castoreum, but instead a whole spate of motor patterns is activated or intensified. This syndrome has as its common denominator the examination and elimination of a strange conspecific scent mark, and the re-establishing of the exclusive colony odor at the scent marks.

Furthermore, experimental scent mounds with castoreum decreased the probability of vacant beaver lodges becoming occupied by dispersing beaver (Müller-Schwarze and Heckman, 1980). This indicates a deterrent function.

Yellow voles (Lagurus luteus) will open up and re-enter burrows that have been plugged up experimentally. But if the burrow in addition has been treated with anal gland secretion, the voles are inhibited from re-opening the hole and will move on (Fan, 1981; Fan, this volume).

b. Responses to scented traps. Several field studies have demonstrated differential entering of traps that carry different conspecific odors. The limitation of these studies is that no other behavior such as investigation, approach, withdrawal or scent marking can be recorded. Free-ranging rodents entering scented traps may "prefer" (i.e., are attracted to) conspecific to heterospecific odor (Clethrionomys and Apodemus: Hansson, 1967), opposite over same sex odor (Mus musculus: Rowe, 1970; Peromyscus leucopus: Mazdzer et al., 1976; Microtus brandtii: Fan, 1978), but also same over opposite sex odor (Mus musculus: Rowe, 1970; Ondatra zibethicus: Ritter et al., unpubl.; Microtus brandtii: Fan, 1978; Peromyscus maniculatus: Daly et al., 1978). Especially males are attracted to the odor of the same sex. This is true for Mus musculus, and Ondatra zibethicus (possibly). Reproductive condition often determines response: non-reproductive adults of both sexes enter unscented traps, while individuals in reproductive condition are attracted to conspecific odor (Perognathus and Dipodomys: Daly et al., 1980; Peromyscus maniculatus: Daly et al., 1978). But young individuals or those new to an area (or the traps) may also prefer conspecific odor over blanks (Microtus townsendii: Boonstra and Krebs, 1976; Peromyscus maniculatus: Daly et al., 1978). Finally, social experiences may determine trap response: dominant individuals may prefer conspecific scent, while low-ranking ones enter unscented traps, as in Sigmodon hispidus (Summerlin and Wolfe, 1973). Reproductive (and dominant) condition appears to heighten sensitivity to conspecific odors and to lead to more intraspecific contact and confrontation. Similarly, male Mus musculus entered preferentially traps that were scented with soiled wood chips from dominant conspecific males. Females, on the other hand, entered clean traps and those scented with soiled bedding from subordinate males (Wuensch, 1982). The odor was even effective across species: Peromyscus maniculatus males entered traps with scent from dominant and subordinate male house mice, while females entered almost only clean traps (Wuensch, 1982).

Are there any rules discernible in these results? At the level of the species, activation of behavior by conspecific rather than heterospecific or control odors is to be expected. The kind of response would depend on sex, age and condition of the signaler, as well as that of the receiver, resulting in numerous possible behaviors. In addition, the condition of an animal will determine the behavioral and spatial context in which it will experience a scent: it can be dispersing, immigrating, or territorial.

The muskrat (Ondatra zibethicus) has been trapped in North America since centuries, and in Europe since its introduction there in 1905. Innumerable scent concoctions have been used on traps and sworn by. However, with the exception of a preliminary study by Williams (1951) who used food bait and muskrat musk on traps consecutively, it is only now that the responses of muskrats to conspecific musk are critically examined.

In the Netherlands in a long trapping experiment, lasting from September 15, 1980 to December 31, 1982, 78 muskrats were caught. Of these 30 (0.085 animals/trap night) were caught with gland extract from Dutch muskrats, 24 (0.065 animals/trap night) with musk from American muskrats, 15 (0.043 animals/trap night) with a plant extract (sweet flag) and 9 (0.027 animals/trap night) in blanks. The difference between 0.085 and 0.027 animals/trap night is significant at the 5% level (Ritter et al. 1982, Brüggemann, pers. comm.).

c. Interspecific effects. Short-tailed voles (Microtus agrestis) enter traps with predator urine (weasel, or jaguar) less often than clean traps or those that are scented with rabbit urine as a control odor. Female voles avoid the predator-scented traps more than males (Stoddart, 1980).

2. Ungulates

a. Alert odor. The black-tailed deer, Odocoileus hemionus columbianus, has a metatarsal gland that measures about 6 x 2 cm (Quay and Müller-Schwarze, 1970). The gland discharges an odor when a deer is stressed, startled, handled, cornered or chased (Müller-Schwarze, 1971).

To examine the responses of conspecifics to the metatarsal odor, a new bioassay was developed. It allows metatarsal odor that had been collected from a sedated deer to be presented by remote control to a group of freely moving deer that are at ease, and are feeding (Müller-Schwarze, 1980).

The metatarsal odor increased the level of alertness of females only. Control odors, such as black-tailed deer urine, did not have this effect. Male metatarsal odor is slightly more effective than that from females. This may be due to a large amount of secretion from the larger gland of the male. The effect is independent of ambient temperature, but increases with relative humidity. Responses at dusk and during daylight did not differ. There are seasonal differences: females are more responsive to metatarsal odor in May and June, which is the fawning season, and least in September and October, the pre-rut period (Müller-Schwarze and Altieri, in prep.).

b. Interspecific odors. The Block Island experiment: In spring and summer 1980 we conducted a field experiment on Block Island, in the Atlantic Ocean, off Rhode Island. White-tailed deer (*Odocoileus virginianus*) foraged in gardens, and predator odors that were expected to act as deer repellents were placed around the gardens.

In May 1980, 12 gardens, ranging in size from about 10 x 10 to 50 x 50 meters were surveyed for deer activity and damage. On May 23 and 30 ten were treated with odors, and on August 4 and 25 eight of these were treated again. The odor samples were rotated between different gardens. Thus we have a spring and a summer sample. Owners were given data sheets on which they recorded deer seen, deer tracks and deer damage for seven days after treatment.

Odor samples were alcohol extracts of wolf (*Canis lupus*) dung (spring: 450 gr, summer: 250 gr dung in 100 ml EtoH); coyote (*Canis latrans*) dung (spring: 450 gr, summer: 115 gr in 100 ml EtoH), coyote urine, male dog urine, black-tailed deer (*Odocoileus hemionus columbianus*), metatarsal secretion (collection from 8 glands in 30 ml polyethylene glycol, and onion extract (100 grams onion in 50 ml EtoH). Polyethylene glycol and ethyl alcohol served as blanks, and muskrat (*Ondatra zibethicus*) musk (4 glands extracted in 16 ml EtOH) and beaver (*Castor canadensis*) castor extracts (1 gram gland tissue in 25 ml EtOH) as control odors.

The odor samples (0.5 ml each) were applied on 12 rocks (or 16 at large gardens) spaced 1 m apart, and placed on the side(s) where the deer approached the gardens. The samples were code-labelled and their nature was not revealed to the owners.

The initial survey of the gardens, supplemented by interviews with the owners, established that a total of 64 plant species had been taken by the deer. Of these, 27 were vegetables, 19 trees and shrubs, and 18 ornamental flowers.

Most preferred vegetables (damaged in three or more gardens) were tomato, beans, asparagus, and beets, while rhubarb, mint and lettuce were never taken. Preferred trees and shrubs were ivy, juniper, and rhododendron; those never touched by deer included blackberry, blue spruce, cotoneaster, holly, and Scotch and Japanese black pines. Among the flowers, tulips, yellow day lilies, and sunflowers were most often eaten, while the deer consistently ignored bleeding heart, daffodil, marigold, and peony. Intermediate, i.e. taken only in one or two gardens, were 22 kinds of vegetable, 16 species of trees and shrubs, and 14 flower species. The rejection of certain plant species may indicate chemical and mechanical defenses against mammalian herbivores. Marigold, for instance, has long been known to be resistant against plant insects. In feeding experiments with black- and white-tailed deer, bleeding heart, iris and calendula were always rejected after sniffing, while marigold was rejected by all of 8 black-tailed deer, and peony was tasted by seven of ten deer (Müller-Schwarze, unpubl.).

The number of days with deer visits during the pre-treatment period was impossible to ascertain accurately. It was merely established that deer had been present within the last several days as evidenced by tracks and feeding signs.

The effectiveness of a sample was measured as the number of days that deer were absent as a percentage of the total number of days that the sample was used in the field. All treated areas had experienced some deer activity, either on the day immediately prior to the treatment or within the preceding week.

For five samples at least 25 days of reliable data were available. Gardens treated with wolf scats were visited least often, and those with onion extract most often. Metatarsal odor of black-tailed deer, beaver castor, and coyote scats were intermediate. The distribution of these percentages is significantly different from chance ($x^2 = 11.03$; $p<0.05$). Onion differs from wolf scats and metatarsal odor (Table 1).

There was more overall deer activity in August than in May ($p< 0.005$). Furthermore, the relative ranking of the samples remained the same from spring to summer (Table 1, last two columns). The difference between deer activity on the first day after treatment (19.3%) and on the seventh (26%) is not significant.

Table 1. Number of Days with Deer Visits to Gardens after Application of Odors.

Odor	No. of Days Used	Deer Activity: Both Seasons No. of Days	Deer Activity: Pooled Percent	Deer Activity: May %	Deer Activity: August %
Wolf Scats	42	5	12[a]	0	18
Black-tailed Deer MT Secretion	27	4	15[b]	7	23
Beaver Castor	25	4	16	0	31
Coyote Scats	32	11	34	17	57
Onion	25	11	44	33	54
		Average:	24.2	16.7[c]	39.8

[a]Different from onion ($x^2 = 6.6$; $p = 0.01$). [b]Different from onion ($x^2 = 4.3$; $p < 0.05$). [c]Difference between May and August: $x^2 = 10.4$; $p < 0.005$.

3. Carnivores

Responses to scent marks. The behavior of canids, such as coyotes (*Canis latrans*) and wolves (*Canis lupus*) at experimental scent marks has been reviewed several times (e.g., Müller-Schwarze, 1974; Shumake, 1977).

Are we in a position to predict canid behavior vis-a-vis specified stimuli? The most successful attempt at eliciting predictable responses to specific stimuli has been the presentation of synthetic fox urine to free-ranging red foxes, *Vulpes vulpes* (Whitten et al., 1980). The synthetic urine consisted of eight components, two of them containing sulphur. Wild foxes marked the artificial marks more often than odorless control solutions or a control odor (citronellal). Most of the marking was done by males as evidenced by sign of raised-leg urination.

River otters (*Lutra canadensis*) and Everglades mink (*Mustela vison evergladensis* are attracted to their respective species' own anal gland secretion, and more so at breeding time during the late wet season (fall) than at any other time of the year (Humphrey and Zinn, 1982).

DISCUSSION

During the past five years a promising start has been made in manipulating free-ranging mammals with conspecific and predator odors. It is now necessary to standardize methods. Especially the scented trap studies are difficult to compare because of the different methods of collecting soiled bedding, and the different amounts used in traps, if it was measured at all. Beyond that step of precise dosage of complex stimuli, such as urine or "soiled bedding," fractionation and isolation of single compounds, as in the study of the red fox (Whitten et al., 1980) will be necessary. Single components or mixtures of urine constituents that influence behavior can then be singled out. In addition to this chemical sophistication, we need information in three more areas if we are to understand the signalling system: stimulus context, i.e., where does it have to be applied, and when (season, time of day); condition of the responding animal; and environmental factors, such as temperature, humidity, or air currents.

The muskrat provides a rare opportunity to compare the chemical stimuli it uses, as well as its behavior, in its original range with those in Europe which it has invaded. There may be a pronounced divergence of chemical signals.

The white-tailed deer's responses to odors at gardens give some clues for future work: First, there appears to be an effect on deer activity by certain odors. Second, odor samples differ in their effect. Third, there is a greater effect in spring than in summer. Fourth, the relative ranking of the odor samples remained the same in spring and in summer. Fifth, the plant species that were never touched should be investigated further. Extracts from such plants as rhubarb, mint, daffodil, marigold, peony, and bleeding heart should be tested for repellency. If effective, isolation and identification of plant defense substances and their application against herbivores could be a worthwhile effort.

The difference in response between spring and summer is probably due to a combination of factors: more forage is available in the gardens, and the grass on the outside dries out. This makes the gardens more attractive. Furthermore, in the spring the deer just start to visit the gardens, while in summer they have established a pattern of visiting, and their experience may override any deterrent effect. Also, higher ambient temperatures probably result in faster evaporation of the odor.

Experimentation with free-ranging wild mammals may appear the ultimate challenge. Detailed knowledge of their behavior, however, will reveal their "weak spots" for manipulation. This will make possible management and control by chemical signals, combined with other methods.

ACKNOWLEDGMENTS

The beaver work was supported by a grant from the Edmund Niles Huyck Preserve, Rensselaerville, New York; the muskrat study by a NATO grant; and the deer work by several grants from the National Science Foundation.

REFERENCES

Boonstra, R. and Krebs, C. J., 1976, The effect of odour on trap response in Microtus townsendii, J. Zool. (Lond.), 180:467.

Daly, M., Wilson, M. I., and Faux, S. F., 1978, Seasonally variable effects of conspecific odors upon capture of deer mice (Peromyscus maniculatus gambelii), Behav. Biol., 23:254.

Daly, M., Wilson, M.I., and Behrends, P., 1980, Factors affecting rodents' responses to odours of strangers encountered in the field: experiments with odour-baited traps, Behav. Ecol. Sociobiol., 6:323.

Fan, Z., 1978, The use of sexual attractant pheromones in controlling Brandt's voles, Acta Zool. Sin., 24:366.

Fan, Z., Hao, S., Liang, Y., Wang, M., and Zhao, Y., 1981, Chemical signals and their role in social groups of two voles (Lagurus luteus and Lagurus lagurus), Acta Ecol. Sin., 1:66.

Hansson, L., 1967, Index line catches as a basis for population studies on small mammals, Oikos, 18:261.

Hodgdon, H. E., 1978, Social dynamics and behavior within an unexploited beaver (Castor canadensis) population. Ph.D. thesis, Univ. Massachusetts, Amherst, 292 pp.

Humphrey, S. R., and Zinn, T. L., 1982, Seasonal habitat use by river otters and Everglades mink in Florida, J. Wildl. Manage., 46:375.

Mazdzer, E., Capone, M. R., and Drickamer, L. C., 1976, Conspecific odors and trappability of deer mice (Peromyscus leucopus noveboracensis), J. Mammal., 57:607.

Müller-Schwarze, D., 1971, Pheromones in black-tailed deer (Odocoileus hemionus columbianus), Anim. Behav., 19:141.

Müller-Schwarze, D., 1974, Application of pheromones in mammals, in: "Pheromones," M. C. Birch, ed., North Holland/American Elsevier, Amsterdam, New York.

Müller-Schwarze, D., 1980, Chemical signals in alarm behavior of deer, in: "Chemical Signals," D. Müller-Schwarze and R.M.

Silverstein, eds., Plenum, New York.
Müller-Schwarze, D., and Heckman, S., 1980, The social role of scent marks in beaver, Castor canadensis, J. Chem. Ecol., 6:81.
Müller-Schwarze, D., Heckman, S., and Stagge, B., 1982, Responses of free-living beaver to experimental scent marks, Proc. First Worldwide Beaver Symp. Helsinki, Aug. 1982, in press.
Müller-Schwarze, D., and Altieri, R., in prep., An alert odor in black-tailed deer, Odocoileus hemionus columbianus.
Ritter, F., Bruggemann, I., Gut, J., Persoon, S. C. J., and Verweil, P., 1982, Chemical stimuli of the muskrat, in: "The Determination of Behaviour by Chemical Stimuli," Proc. 5th ECRO Minisymp. Jerusalem, Israel, Nov. 1981.
Rowe, F. P., 1970, The response of wild mice (Mus musculus) to live traps marked by their own and foreign mouse, Odor J. Zool. (Lond.), 162:517.
Shumake, S. A., 1977, The search for applications of chemical signals in wildlife management, in: "Chemical Signals in Vertebrates," D. Muller-Schwarze and M. M. Mozell, eds., Plenum, New York.
Stoddart, D. M., 1980, Some responses of a free living community of rodents to the odors of predators, in: "Chemical Signals," D. Müller-Schwarze and R.M. Silverstein, eds., Plenum, New York.
Summerlin, C. T., and Wolfe, J. L., 1973, Social influences on trap response of the cottonrat, Sigmodon hispidus, Ecology, 54:1156.
Whitten, W. K., Wilson, M. C., Wilson, S. R., Jorgenson, J. W., Novotny, M., and Carmack, M., 1980, Induction of marking behavior in wild red foxes (Vulpes vulpes L.) by synthetic urinary constituents, J. Chem. Ecol., 6:49.
Williams, R. M., 1951, The use of scent in live-trapping muskrats, J. Wildl. Mgt., 15:117.
Wuensch, K. L., 1982, Effect of scented traps on captures of Mus musculus and Peromyscus maniculatus, J. Mammal., 63:312.

MECHANISMS OF INDIVIDUAL DISCRIMINATION IN HAMSTERS

Robert E. Johnston

Department of Psychology
Cornell University
Ithaca, NY 14853

The capacity to discriminate between individuals has been demonstrated in a variety of animal species (see Halpin, 1980, for a recent review). Among mammals, the cues that have been studied most frequently as sources of individually distinct information have been olfactory; for example, mice, rats, gerbils, dogs, wolves and mongooses all have been shown to be capable of discriminating between the odors of different individuals of the same species (Bowers and Alexander, 1967; Carr et al., 1970; Halpin, 1976; Brown and Johnston, this volume; Rasa, 1973). In most of these studies it is the capacity to discriminate between individuals that has been demonstrated, either in a discrimination learning task or in an habituation-dishabituation paradigm. Neither of these tasks demonstrate that animals actually use individual discrimination in their normal social interactions. The habituation paradigm is preferable in this regard, since the animals tested in this type of task do spontaneously notice and pay attention to differences in the odors of individuals (e.g., see Brown and Johnston, this volume). One situation in which functional use of individual discrimination has been repeatedly demonstrated among mammals is the recognition of young in a variety of ungulates, in which both odors and vocalizations have been shown to be important (Klopfer and Gamble, 1967; Espmark, 1971).

Another situation in which individual discrimination is likely to have direct functional significance is in the recognition of mates. Hamsters live solitarily, one to a burrow, but they may mate promiscuously; a male, for example, could conceivably mate with more than one female within the course of a day, and he would benefit by being able to recognize females so he would know which ones he had already mated with. In nature there would of course be other information to go on, such as the location of the different females, but nonetheless the ability to recognize individuals would be valuable. In the laboratory a similar situation has been extensively studied in which an animal is tested with a series of sexual partners to determine the effects of a new partner on copulatory performance. The basic paradigm is that an animal, most often a male, is allowed to mate with one female until he is exhausted and not interested in mating any longer. The first female is removed and a new female is introduced, and the male then becomes rejuvenated and attempts to mate with the new female. This effect is known as the Coolidge effect and has been observed in a variety of mammalian species, such as hamsters, rats, bulls, montane voles and meadow voles (Bunnell, et al., 1977; see Dewsbury, 1981, for a review). Although it has primarily been males that have been tested in this paradigm females of some species show similar effects (Dewsbury, 1981). It has been suggested that the ability of satiated males to be sexually aroused by a new partner is an adaptive feature of some species and that promiscuous species should have this capacity whereas monogamous species should not, since monogamous individuals would presumably not have the opportunity to benefit from it (Wilson, Kuehn and Beach, 1963; Thomas and Birney, 1979). Although the magnitude of the effect is not always great, the majority of the data that are available fit this interpretation (Dewsbury, 1981).

The presence of the Coolidge effect suggests the importance of individual discrimination but does not prove it. There are several alternative mechanisms by which a male could discriminate between a female he had mated with and a female that was novel to him and that had not recently copulated. New females could arouse the male more by courting him more vigorously than females that had already mated with him for some time. It is also possible that males are aroused not by discrimination of a different female but by cues that indicate that the female has not recently been engaged in copulatory activity.

Our first experiment was designed to determine if odors of females were the likely cues that were responsible for the Coolidge effect in hamsters and if it was the individuality of a female's odors that was the critical feature. We tested males with anesthetized females, which effectively eliminates the influence of the female's behavior and vocalizations on the performance of males.

The basic paradigm was that males were allowed to mate with a female in an observation box until they showed no mounting attempts for 10 minutes, our operational definition of satiety. Males generally interacted with females for about 50-60 minutes before reaching this criterion. Males were then returned to their home cage for 8 minutes, after which they were introduced into the 2 x 3 foot testing arena, in which they encountered two anesthetized females, placed about 1 foot apart. They were observed for 10 minutes and the amount of time they spent sniffing the body, sniffing the genital region, licking the genital region and attempting to mount each of the females was recorded on an event recorder. For the purposes of this paper these measures have been combined to yield a single measure of total interest in the two females.

Three groups of 10-12 males were tested in three conditions. In the first condition the stimulus females were, (1) the same one that the male had just mated with (SAME), and (2) a female that the male was not familiar with and that had not recently copulated (FRESH). This condition is a replication of the basic Coolidge effect paradigm, only using anesthetized females. We expected males to show more interest in the FRESH females and attempt to mount them more often than the SAME females. In the second condition a male encountered, (1) the SAME female and (2) a female that was new to him but that had just copulated with another male (MATED). If males are reinvigorated by the odors of an individual new to them, they ought to prefer the MATED female over the SAME female, but if males are reinvigorated by females that have not recently mated they should show no preference and be relatively uninterested in both stimulus females. In the third condition males encountered (1) a MATED female, and (2) a FRESH female. In this case both females are new to the male, so if he is reinvigorated by encountering a new individual he should be interested in both and show no preference. On the other hand, if he is avoiding a recently mated female he should prefer the FRESH female.

The results are shown in Fig. 1. It can be seen that males did prefer to investigate and attempt to mount the FRESH females more than the SAME females ($p<.01$, Wilcoxon signed ranks test). These results indicate that the testing paradigm using anesthetized females yields results similar to those obtained with awake females (Bunnell et al., 1977), and that the Coolidge effect in hamsters is largely based on odor differences between females. It is worth noting that all the behavioral measures that were obtained showed the same pattern and were significantly different. For example, nine of 12 males attempted to mount anesthetized females, but only five of them attempted to mount the SAME female; all nine that did attempt to mount did so more often with the FRESH females ($p<.01$).

Males also demonstrated a strong preference for novel, MATED females over the SAME females ($p<.01$), suggesting that males were choosing on the basis of odors of individuals, not on the basis of odors developed or picked up during copulation. Again all of the measures analyzed individually were significantly different. The interpretation that males were primarily responding to the differences between the odors of individual females is further strengthened by the results from the last group of males, that were tested with

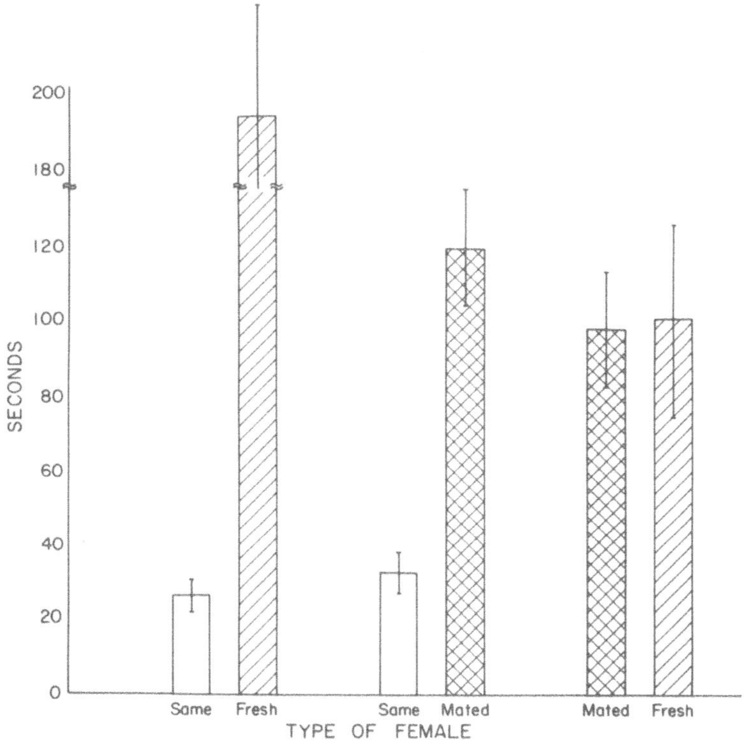

Fig. 1. Mean (± SE) amount of time that sexually satiated males spent sniffing, licking and attempting to mount anesthetized stimulus females (see text for details).

MATED and FRESH females, both of which were novel. Males were attracted to and sexually aroused by both females, and showed no preference between them.

There is, however, another plausible explanation for the results of this experiment. It could be that a male recognizes his own odors on the female he has copulated with and therefore responds less strongly to her. We evaluated this hypothesis by testing a group of males in the same manner as described above, but this time the males encountered, (1) a novel, unmated female (FRESH), and (2) a novel, unmated female that had been scented with the odors of the test male (FRESH SCENTED). Two methods were used to transfer scent from males to females. First, odors from the flank gland and head region (Harderian glands and ear glands) of the male were rubbed onto glass plates and these glass plates were rubbed against the stimulus female. Second, the male himself was held firmly and rubbed against the female, including his head, flank and genital regions. If satiated males are not aroused by females on which they have deposited their own odors, in this experiment test males should prefer the FRESH females to the FRESH SCENTED females. In fact, males showed no preference: they spent a mean of 143.2 ± 13.8 seconds sniffing, licking and attempting to mount the FRESH females and 146.9 ± 9.8 seconds with the FRESH SCENTED females. Thus males seem to be primarily responding to the individuality of the odors of females and not to odors that might have been transferred to the females from the males during copulation.

Next we investigated the sources of odors that might be involved in discrimination of individual females. Might one scent gland be specialized for the function of providing individually variable odors, thus facilitating individual discrimination and recognition? On the face of it this possibility seems unlikely, since we know that genetic differences between individuals result in slightly different chemical compositions in nearly every organ of the body (Williams, 1956). By extension, all scent glands or sources of odorous materials ought to provide sufficient cues for individual discrimination. On the other hand, why shouldn't some scent glands be especially adapted for this function? Most species of mammals have several different scent glands, presumably because different glands are specialized for different functions. In the dwarf mongoose there is suggestive evidence that individuals can recognize others on the basis of anal gland secretions but not on the basis of cheek gland secretions (Rasa, 1973). An analogy with individual recognition in humans is useful here, since the human face is a particularly salient source of cues for individual recognition and may have been specialized to some extent for this function, but it is not the only source of individually discriminable features.

Hamsters have four sources of scent that are known to be involved in communication. The Harderian glands are sexually dimorphic glands situated behind the eyeball. The glands of males are larger than those of females and they contain two types of cells, whereas the glands of females contain only one type of cell but have accumulations of porphyrin pigments, substances lacking in the glands of males (Clabough and Norvell, 1973). These sexual differences appear to be androgen dependent (Payne et al., 1977). The secretion from the Harderian gland is transported to the nares and is distributed around the head region when an animal grooms (Thiessen, 1977). The ear glands and the flank glands are both modified sebaceous glands that are larger in males than females, the differences being due to differences in androgen titers (Plewig and Luderschmidt, 1977; Hamilton and Montagna, 1950). Finally, vaginal secretions are known to be a source of odors that influence males in a number of ways, most strikingly by being highly attractive, stimulating copulatory behavior, and eliciting hormonal responses (see Johnston, 1977b for review). There is evidence linking all of these scent glands to sexual recognition and/or attraction (Payne, 1979; Landauer et al., 1980; Johnston, 1975, 1977a, 1977b).

For the purposes of a first experiment we divided these scent sources on the basis of regions of the body and tested males with odors from the vagina, the flank region, and the head region (ear and Harderian glands). Three groups of 12 males were tested in a manner similar to that described above. Males were mated to satiety, removed to their home cage for 8 minutes, and then placed in a testing arena. In the arena they encountered two 3 x 5 inch glass plates, placed about 1 foot apart, on which were distributed odors from one of the body regions of one of two females. The odors were collected from females (all of which were in estrus) before the beginning of the mating encounter by rubbing the head or flank gland against the glass plate or by collection of vaginal secretion with a spatula and spreading it onto the glass plate. Males were observed for 5 minutes and the time they spent sniffing and licking the glass plates was recorded.

The results, shown in Fig. 2, were quite surprising. Males did not discriminate between the odors from the head region or from vaginal secretions of the two females, but they spent significantly more time sniffing secretions from the flank region of the FRESH female than those of the SAME female ($t=3.4$, $df=11$, $p<.01$). Males spent less time investigating the flank gland scent of SAME females than they did investigating all of the other scents that were tested. These results thus suggest that the flank gland may be a particularly salient source of cues involved in individual recognition. Further experiments in which we eliminated the flank glands of all stimulus females have shown that the flank glands are not necessary for the Coolidge effect to be observed toward anesthetized females. Thus the flank gland may be much like the face for humans: a source of the most salient information for individual recognition, but not necessary for such recognition.

Finally, we were interested in whether the capacity to discriminate individuals was a function of the main olfactory system or the vomeronasal system. In the recent explosion of research on the functions of the vomeronasal system there has been a relative lack of theory about the kinds of functions the two systems might have. Several lines of evidence have suggested that the two systems

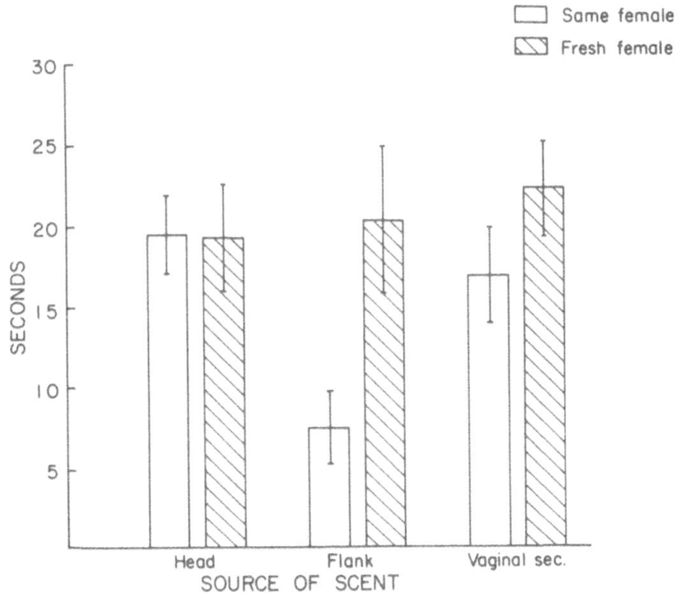

Fig. 2. Mean (± SE) amount of time that sexually satiated males spent sniffing and licking odors from familiar (SAME) and novel (FRESH) females.

might be responsible for responses to two different classes of compounds, with the vomeronasal system primarily responsive to large, nonvolatile molecules and the olfactory system primarily responsive to smaller, volatile compounds (see, for example, Wysocki et al., 1980). I would like to suggest a distinction between the two systems at a more functional level of analysis: the vomeronasal system may be primarily involved in mediation of relatively specific responses with direct biological significance, such as copulatory behavior or hormonal responses to odors, whereas the main olfactory system may be more of an all-purpose system involved in making discriminations between odorants, exploration of the environment, and recognition of objects or other animals. Correlated with this difference in function might be a tendency for the vomeronasal system to be specialized for the reception of specific compounds with relatively specific functions (something like the classic insect pheromones) whereas the main olfactory system might be specialized for making discriminations between complex mixtures of chemicals.

These speculations are based on two relatively slender lines of evidence. First, the two systems differ dramatically in the nature of their projections to the central nervous system. Receptors in the vomeronasal organ project to the accessory olfactory bulb, which sends its efferent projections to regions of the medial and posterior amygdala, which in turn sends fibers to the hypothalamus and bed nucleus of the stria terminalis. In contrast the main olfactory bulb receives its afferents from the olfactory mucosa and projects to a much broader set of areas, including the olfactory tubercle, anterior olfactory nucleus, primary olfactory cortex, periamygdaloid region, anterior cortical nucleus of the amygdala and the anterior hippocampus. Wysocki (1979) has characterized the differences in these projections as indicating that the olfactory system is primarily related to telencephalic cortex and the associated thalamic nuclei, whereas the vomeronasal system is more closely linked to the diencephalon, particularly the hypothalamus and bed nucleus of the stria terminalis, via the amygdala. He suggests that these neuroanatomical differences implicate the vomeronasal system in both the behavioral and neuroendocrine aspects of reproduction. At a very general level there seem to be two points of significance: (1) the greater diversity of projections of the olfactory system suggests a greater diversity of functions than the vomeronasal system, and (2) the first order connections of the olfactory system to cortex and the second order projections to the thalamus suggest the importance of this system in complex odor discriminations and memories, whereas the more limited projections of the vomeronasal system to the diencephelon suggest more basic biological functions. It is known, for example, that lesions of the pathway from the area of the thalamus (the mediodorsal nucleus) that receives olfactory system projections disrupt odor discriminations and social odor preferences (Eichenbaum et al., 1980; Sapolsky and Eichenbaum, 1980).

The second line of evidence supporting my hypothesis is that most of the effects so far described following vomeronasal system lesions relate to fundamental biological processes. The vomeronasal system has been implicated in copulatory behavior of hamsters, maternal behavior of rats, and several odor-mediated hormone effects such as pregnancy block and puberty acceleration (Winans and Powers, 1977; Fleming et al., 1979; Bellringer et al., 1980, Keneko et al., 1980). In contrast the vomeronasal system is not necessary for species-specific responses to odors in hamsters (Murphy, 1980). This evidence, although more direct, is less compelling, since the range of situations in which the involvement of the two systems have been tested is still quite small.

In hamsters both the vomeronasal and olfactory systems have been implicated in the control of male sexual behavior; if both systems are eliminated male hamsters don't mate (Winans and Powers, 1977; see also Winans, this volume). The vomeronasal system seems to play a more important role, however, since lesions of the vomeronasal nerves result in reductions of copulatory performance but lesions of the olfactory system alone have no effect (Winans and Powers, 1977; Meredith et al., 1980). Despite this evidence of a predominant role of the vomeronasal system, the ideas developed above predict that in the Coolidge effect situation, in which a male's copulatory behavior with the second female is dependent on his discrimination of her as a different individual, the main olfactory system should be important but the vomeronasal system should not.

We tested this prediction by lesioning either the vomeronasal organ or the olfactory mucosa and examining the behavior exhibited by lesioned and sham lesioned males in our standard test paradigm. Vomeronasal lesions were produced in 14 males by inserting a flat electrode, bare on the lower edge, into the nose and down into the groove above the vomeronasal organ; radio frequency cauterization of the organ was produced by applying 5 volts for 10 sec. with a Grass LM-3 lesion maker. Sham lesions were produced in 9 males by turning the electrode upside down and cauterizing the roof of the non-olfactory nasal mucosa directly above the vomeronasal organ.

The amount of time that lesioned and sham lesioned males spent investigating and attempting to mount SAME and FRESH anesthetized females is shown in Fig. 3. It can be seen that the vomeronasal lesions did not alter the males' ability to discriminate the odors of the two females nor their preference for the FRESH female; both the lesioned males and the sham lesioned males demonstrated the normal preference for the FRESH females ($p<.05$ for shams, Wilcoxon signed ranks test). Lesioned males demonstrated a significant preference for the fresh female in all the measures of their behavior that were obtained (sniff body, sniff genital region, lick genital region and attempted mounts). It is also interesting that the magnitude of the scores was similar for the lesioned and the sham lesioned groups.

Histological examination of the snouts of the lesioned males showed considerable variability in the extent of vomeronasal organ damage, ranging from complete destruction to about 30% destruction. There was no correlation between the extent of the damage and the measures of behavior that we obtained, however. The anterior portion of the organ was severely damaged in all animals, and since the opening of the organ into the nasal cavity is at this end it is probable that access of chemicals to the organ was eliminated in all of them.

The variability in the extent of the lesions was disturbing, however, so we repeated this experiment using a different method of vomeronasal destruction. This time the vomeronasal organ was

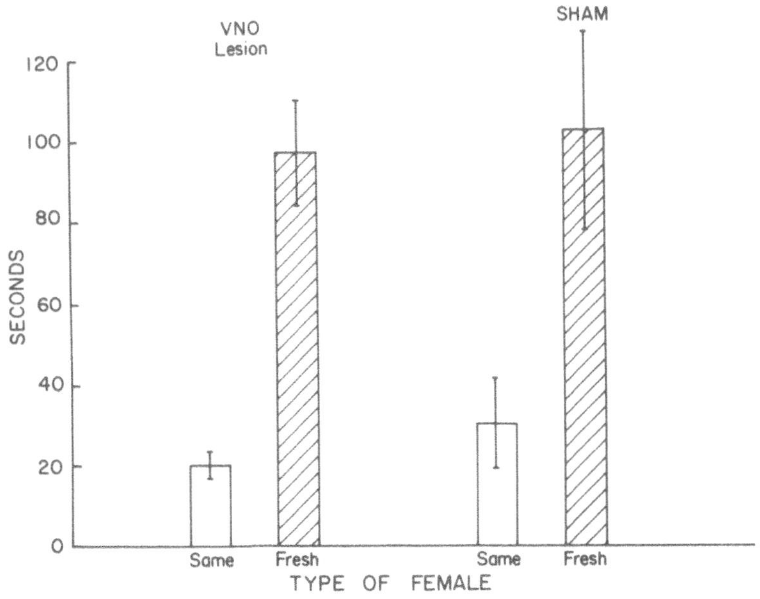

Fig. 3. Mean (± SE) amount of time that sexually satiated males with vomeronasal organ lesions or control lesions spent investigating and attempting to mount familiar and novel anesthetized females.

surgically removed via an incision in the roof of the mouth, using methods developed by Wysocki (1982). Histological verification of these procedures is not yet complete, but since the surgery is done under visual control it is unlikely that any significant portion of the vomeronasal organ remained. Even if some did remain, the entire vomeronasal capsule is destroyed, so it is unlikely that any function remained. Males (N=6) with their vomeronasal organs removed were similar to those with radio frequency lesions. They spent a mean of 13.2 ± 2.5 seconds investigating and attempting to mount the SAME females versus 76.5 ± 21.6 seconds with the FRESH females ($p<.05$, t test), thus indicating again that males with vomeronasal lesions but with functional main olfactory systems do discriminate between a familiar and a novel female and become more sexually aroused by the novel female.

Several other aspects of the behavior of males with vomeronasal lesions or removals are worth noting. All of these males did mate reasonably vigorously with the first female, and even after mating to satiety some, but not all, attempted to mount anesthetized females. Thus the copulatory performance of these males was considerably more robust than that obtained by Winans and Powers (1977) with vomeronasal nerve cuts. The reason for this difference is not known, but it is possible that the nerve cut procedure causes more general trauma to the central nervous system or causes damage to the main olfactory bulb which contributes to the deficits seen using that method.

Destruction of the olfactory mucosa was accomplished by irrigation of the nasal cavities with $ZnSO_4$, following the procedure of Winans and Powers (1977) which spares the vomeronasal mucosa. Ten males received treatment with $ZnSO_4$ and 7 received a sham treatment, irrigation with physiological saline. Two days after these treatments males were tested in the standard testing paradigm, first mating to satiety and then encountering anesthetized SAME and FRESH females for 10 minutes.

Males with their olfactory mucosa destroyed did not discriminate between SAME and FRESH females, as shown by the lack of preference for the FRESH females (Fig. 4). Presumably because males

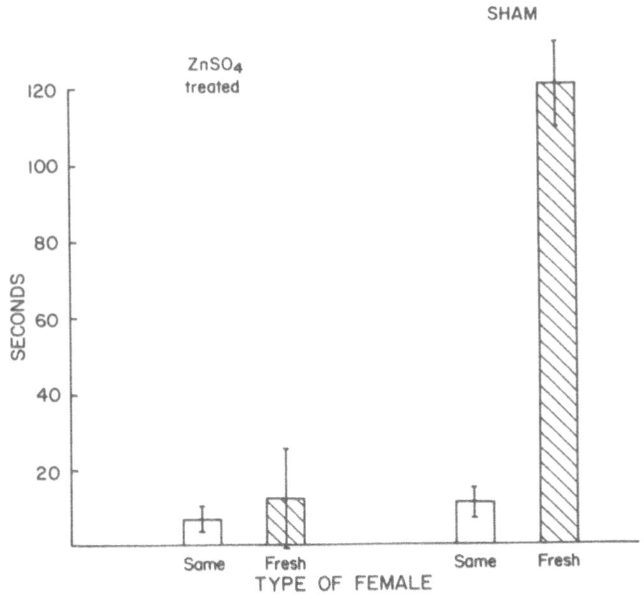

Fig. 4. Mean (± SE) amount of time that sexually satiated males with olfactory system lesions or a control treatment spent investigating and attempting to mount familiar and novel anesthetized females.

lacked the ability to notice the presence of a new female they were not attracted to her and only three of the 10 males attempted to mount. All three did, however, mount the FRESH female more than the SAME female. In contrast the males treated with saline behaved just as normal males behaved in the first experiment, showing a strong preference for the FRESH female ($p<.02$). Five of the seven control males mounted the FRESH female. It should be noted that both the experimental and the control males did mate vigorously with the females during the satiety phase of the experiment, so it is clear that lesions of the olfactory mucosa did not eliminate sexual arousal in general or the ability to copulate.

This experiment suggests that the ability of males to discriminate individual females is mediated by the main olfactory system but not by the vomeronasal system. This is the first report I am aware of which has shown these two systems to be differentially involved in this kind of social discrimination, although the olfactory system is implicated in the ability to make species discriminations as well (Murphy, 1980). In sexually satiated males sexual arousal is dependent on the recognition of a novel female, and thus the arousal caused by a novel female is also dependent on the olfactory system. Finally, the results of this last experiment also indicate that discrimination of the two anesthetized females cannot be accounted for on the basis of visual cues.

In summary, this set of experiments demonstrate that: (1) male hamsters can discriminate individual females on the basis of their odors and that in some situations such discrimination influences sexual arousal, attraction and copulatory performance, (2) one scent source, the flank gland, may be particularly important as a source of chemicals which allow individual recognition, (3) individual discrimination is mediated by the main olfactory system but not by the vomeronasal system.

REFERENCES

Bellringer, J. F., Pratt, H. P. M., and Keverne, E. B., 1980, Involvement of the vomeronasal organ and prolactin in pheromonal induction of delayed implantation in mice, J. Reprod. Fert., 59:223.

Bowers, J. M. and Alexander, B. K., 1967, Mice: individual recognition by olfactory cues, Science, 158:1208.

Bunnell, B., Boland, B. D., and Dewsbury, D. A., 1977, Copulatory behavior of golden hamsters, Mesocricetus auratus, Behaviour, 61:180.

Carr, W. J., Krames, L. and Costanzo, D. J., 1970, Previous sexual experience and olfactory preference for novel versus original sex partners in rats, J. Comp. Physiol. Psychol., 71:216.

Clabough, J. W. and Norvell, J. E., 1973, Effects of castration,

blinding, and the pineal gland on the Harderian glands of the male golden hamster, Neuroendocrinology, 12:344.
Dewsbury, D. A., 1981, Effects of novelty on copulatory behavior: The Coolidge effect and related phenomena, Psych. Bull., 89: 464.
Eichenbaum, H., Shedlack, K. J. and Eckmann, K. W., 1980, Thalamo-cortical mechanisms in odor-guided behavior. I. Effects of lesions of the mediodorsal thalamic nucleus and frontal cortex on olfactory discrimination in the rat, Brain Behav. Evol., 17:255.
Espmark, Y., 1971, Individual recognition by voice in reindeer mother-young relationship. Field observations and playback experiments, Behaviour, 40:295.
Fleming, A., Vaccarino, F., Tambosso, L., and Chee, P., 1979, Vomeronasal and olfactory modulation of maternal behavior in the rat, Science, 203:372.
Halpin, Z. T., 1976, The role of individual recognition by odors in social interactions of the mongolian gerbil (Meriones unguiculatus), Behaviour, 48:117.
Halpin, Z. T., 1980, Individual odors and individual recognition: review and commentary, Biol. Behav., 5:233.
Hamilton, J. B., and Montagna, W., 1950, The sebaceous glands of the hamster. I. Morphological effects of androgens on integumentary structures, Amer. J. Anat., 86:191.
Johnston, R. E., 1977, Scent marking by male hamsters. I. Effects of odors and social encounters, Z. Tierpsychol., 37:75.
Johnston, R. E., 1977a, The causation of two scent marking behaviors in female golden hamsters, Anim. Behav., 25:317.
Johnston, R. E., 1977b, Sex pheromones in golden hamsters, in: "Chemical Signals in Vertebrates," D. Müller-Schwarze and M. M. Mozell, eds., Plenum, NY.
Kaneko, N., Debski, E. A., Wilson, M. C., Whitten, W. K., 1980, Puberty acceleration in mice. II. Evidence that the vomeronasal organ is a receptor for the primer pheromone in male mouse urine, Biol. Reprod., 22:873.
Klopfer, P. H., and Gamble, J., 1967, Maternal "imprinting" in goats: The role of chemical senses, Z. Tierpsychol., 23:588.
Landauer, M. R., Liu, S., and Goldberg, N., 1980, Responses of male hamsters to the ear gland secretions of conspecifics , Physiol. Behav., 24:1023.
Meredith, M., Marques, D. M., O'Connell, R. J., and Stern, F. L., 1980, Vomeronasal pump: significance for male hamster sexual behavior, Science, 207:1224.
Murphy, M. R., 1980, Sexual preferences of male hamsters: importance of preweaning and adult experience, vaginal secretion and olfactory or vomeronasal sensation, Behav. Neural Biol., 30: 232.
Payne, A. P., McGadey, J., Moore, M. R., and Thompson, G., 1977, Androgenic control of the Harderian gland in the male golden hamster, J. Endocr., 75:73.
Payne, A. P.,

Rasa, O. A. E., 1973, Marking behaviour and its social significance in the African Dwarf Mongoose, Helogale undulata rufula, Z. Tierpsychol., 32:293.
Sapolsky, R. M. and Eichenbaum, H., 1980, Thalamocortical mechanisms in odor-guided behavior. II. Effects of lesions of the mediodorsal thalamic nucleus and frontal cortex on odor preferences and sexual behavior in the hamster, Brain Behav. Evol., 17:276.
Thiessen, D. D., 1977, Thermoenergetics and the evolution of pheromone communication, Progress in Psychobiology and Physiological Psychology, 7:91.
Thomas, J. A. and Birney, E. C., 1979, Parental care and mating system of the prairie vole, Microtus ochrogaster, Behav. Ecol. Sociobiol., 5:171.
Williams, R. J., 1956, "Biochemical Individuality.", Wiley, NY.
Wilson, J. R., Kuehn, R. E. and Beach, F. A., 1963, Modification in the sexual behavior of male rats produced by changing the stimulus female, J. Comp. Physiol. Psychol., 56:636.
Winans, S. S. and Powers, J. B., 1977, Olfactory and vomeronasal deafferentation of male hamsters: histological and behavioral analyses, Brain Research, 126:325.
Wysocki, C. J., 1979, Neurobehavioral evidence for the involvement of the vomeronasal system in mammalian reproduction, Neuroscience and Biobehavioral Reviews, 3:301.
Wysocki, C. J., 1982, The vomeronasal organ: its influence upon reproductive behavior and underlying endocrine systems, In: "Olfaction and Endocrine Processes," W. Breipohl, Ed., Information Retrieval Limited, London.
Wysocki, C. J., Wellington, J. L. and Beauchamp, G. K., 1980, Access of urinary nonvolatiles to the mammalian vomeronasal organ, Science, 207:781.

HUMAN OLFACTORY COMMUNICATIONS

Michael J. Russell

University of California
Brain-Behavior Research Center
Eldridge, California 95431

INTRODUCTION

It has been known for some time that most mammals use odors as a means of intraspecific communication (see Cheal and Sprott, 1971; Bronson, 1971; Doty, 1976; Shorey, 1976; Müller-Schwarze and Mozell, 1977, Mykytowycz, 1977; for reviews). However, until recently little attention has been paid to the possibility that humans might also use smell as a means of communication. This review examines the areas of olfactory communication that are known to be of significance in animals: individual recognition, scent marking, sexual identification, sexual attraction, and reproductive synchrony, and discusses the evidence for human olfactory communication in each of these areas. Suggestions are made for further research in this field.

INDIVIDUAL RECOGNITION

As early as 1877, Darwin observed that an infant would turn its head towards its mother when its eyes were closed and her breast was brought near. Darwin suggested that the infant might be attracted to either the mother's odor or her body heat. Pratt et al. (1930) reported that Pryer, Canestrini and Peterson Rainey did experiments to see if infants were attracted by the odor of their mothers' milk, but these early experiments yielded conflicting results and were not followed up. Then in the mid-seventies two laboratories working independently (MacFarlane, 1975; Russell, 1976) reported work that was nearly identical in procedure and strikingly similar in result. Both of these studies examined the head-turning and routing reflex of hungry sleeping infants to the presence of odors from breast pads that had been worn by either their own mother or by an unfamiliar mother. Both of these studies found that the infants showed no measurable responses to either of the odors at two days post-partum, but that some infants were able to discriminate their own mother's odor from an unfamiliar mother's odor by six days (MacFarlane, 1975) and virtually all of the infants were able to make the discrimination by six weeks (Russell, 1976). A later study by Schaal et al. (1980) also examined the response of infants to breast pads. By analyzing film recordings of infants, they determined that babies reduced their body movements more when they were exposed to the odors of their own mothers than when they were exposed to odor from unfamiliar mothers.

The results of the above studies indicate that infants are not able to identify their mothers at two days of age, but can discriminate their own mother from an unfamiliar mother by the second week after birth. Two hypotheses can be drawn from these results. One hypothesis is that the babies are capable of making the discrimination at birth, but that we are not able to demonstrate the infants' ability by either of these methods of measurement. This "in utero" hypothesis suggests that the baby is exposed to some mixture of the mother's body chemistry while living in an amniotic fluid and that the infant is simply responding to similar airborne constituents that are perceived as odors after the baby is born. None of the experiments done to date either proves or disproves this hypothesis. The present evidence suggests that a second hypothesis based on "extra-uterine" learning is more plausible, however. The olfactory threshold of infants is quite high at birth and drops steadily for the first week (Lipsitt et al., 1963), suggesting that the infant may not be capable of making the relatively subtle olfactory discriminations necessary for identification of its mother until after the first week of life. Also, the evidence from the other sensory modalities (Carpenter, 1974) indicates that an infant is not able to identify its primary caretaker until the second week of life. MacFarlane (1976, p. 112) has pointed out that historically humans have had a high maternal death rate, and that it is probably not advantageous for human infants to form the immediate attachments found in some other species (Hess, 1973). It is to the infant's advantage to have an initial period of attachment flexibility before forming a bond with a primary caretaker. Delaying the ability to identify this caretaker would be a simple means of postponing this attachment bonding and encouraging "Allomothering," or adoption by alternate parents.

Data on specific types and sources of odor that human infants use for maternal identification are not available, but work with primates suggests that infants are both able to recognize odors that are indigenous to the infant's primary caretaker (Kaplan, 1977), and to identify their own odor which they use to mark their primary caretaker (Kaplan and Russell, 1974). In these studies infant squirrel monkeys (Saimiri sciureus) were raised on cloth covered surrogates to determine how infants identify their mothers during the early stages of life, and to assess the effects that perceptual qualities have on early attachment. Infants were raised in a variety of conditions to test their preference for visual, olfactory and tactile cues. Beginning when the infants were four weeks of age and continuing at four-week intervals until they were six months of age, the infant squirrel monkeys were tested to determine their preferences for the color and odor of the surrogates that they had been raised on. A series of two-choice and four-choice preference tests showed that the infants recognized and greatly preferred surrogates upon which they had deposited their own odors, or surrogates that had been perfumed by the experimenter during early rearing (Kaplan, 1977) to either familiar visual cues or unfamiliar odors during early infancy. Later in life the animals showed a preference for visual cues. It is probable that human infants also use odors which they have deposited on their mothers, and their mothers' indigenous odors, as cues for olfactory identification.

If an infant is able to identify its mother by olfactory cues, can a mother also identify her infant? Schaal et al. (1980) have shown that at two days post-partum, mothers are able to discriminate the "T" shirts worn by their own infants from those worn by other infants. In a series of experiments at our laboratory (Russell et al., 1982), we have found that mothers are able to identify their own infants immediately after delivery. In this test, the mothers were blindfolded and presented with their own infant and two other infants, and asked to identify their own after sniffing the infants' heads. The mothers correctly identified their own infant in 61% of the trials. In an identical test, the fathers were not able to make this discrimination when tested from 24 to 48 hours after the birth of their children. The fathers achieved a correct response in only 37% of the trials in a three choice test. This may be because of the differing experiences of the mother and father with the child or differences in olfactory thresholds. Whatever the reason, in this experiment mothers were able to recognize their infants and fathers were not. Although it has not yet been demonstrated that fathers can identify their infants later in life, it seems likely that they would be able to do so with sufficient exposure.

Several studies have now been done to demonstrate that humans can also recognize other individuals by smell. Porter and Moore (1981) have shown that infants are able to identify odors of siblings and that parents can identify the odors of their children's "T" shirts. Unfortunately, this study did not consider the specific responses of fathers, so it is impossible to determine from the report if men are able to identify their children. Both men and women are able to identify both their own body odor as well as that of their mate (Schleidt, 1980; Hold and Schleidt, 1977; Wallace, 1977; Russell, 1976).

SCENT MARKING IN HUMANS

One of the advantages of olfactory cues over signals from other sensory modalities is that they persist over time, and some odors deposited on an object can be used later by the same individual that deposited it, or by different individuals when the depositor is no longer present. Scent marking is a common behavior in animals and is used for a variety of purposes ranging from marking territory to the expression of social status (Stoddart, 1980; Johnston, 1975; Theissen, 1968; Mykytowycz, 1965; Whitten and Bronson, 1970; Epple, 1974; Ralls, 1971; Barrette, 1977 and many others). Among the more common examples of scent marking studied in animals is the identification of home territories and nest sites. Scent marking often begins immediately after birth with the mutual exchange of odors by mothers and infants as a means of enhancing the ability of each to identify the other. Infant animals, including primates (Kaplan, 1977), kittens (Rosenblatt, 1972), sheep (Alexander, 1978) and rats (Leon and Moltz, 1972) recognize, and are strongly influenced by odors which are associated with their mothers, litter-mates and home area. Most often these infant animals show reduced signs of stress when their own odors or their mothers' odors are present in the environment.

There is evidence that recognition of one's own odor or the odor of kin are a significant means of reducing stress in human infants. Children will often scent mark a blanket or cloth toy with their own odor and then carry this marked object around with them. These "security blankets" are then used by the infant as an object of attachment when it is tired or away from its parents (Passman and Weisberg, 1975). Recently, I observed a particularly inventive three-year-old boy hold a cloth toy in a position which allowed the child to simultaneously suck its thumb, touch a soft portion of the toy, and have a select portion of the toy near his nose on several occasions when he was tired. When the child was asked why it held the toy in that position he responded "to get the smell right." In this manner, he was able to get contact comfort by using his thumb as a pacifier, touching the soft fur for tactile stimulation, and positioning the cloth for inhaling a familiar odor; all of which seemed to calm him and make him more relaxed when stressed or tired. The observation of this form of human scent marking is common, and has even been the subject of a popular cartoon character (Schultz), but those authors that have examined it have done so in the context of contact comfort rather than scent marking (Hong and Townes, 1976; Passman and Weisberg, 1975).

The objects are usually sucked and rubbed about the face by the child, so the mouth or the face may be the sources of the odor, although any familiar odor might have the same function and the source may be irrelevant. There is evidence which suggests why some infants attach to these objects and others do not. Hong and Townes (1976) have looked at attachment cross-culturally and found that "an infant's attachment to inanimate objects is lower in a culture or social group in which infants receive a greater amount of physical contact, including a higher rate of breast feeding, and in which the mother is more physically involved and available when the infants go to sleep."

With the possible exception of the sucking and rubbing motions of nursing infants, humans do not display behaviors that are readily identifiable as scent marking. Human body odors are deposited, however, in our environment on cloths, bedding, furniture, etc. Shaal et al. (1980) reported that infants spent more time in contact with "T" shirts that had been worn by their own mothers than with identical "T" shirts worn by other mothers. It may be that the familiarity of the odor is more important to the infant than whether the odor came from the infant itself or from its mother.

ODORS AND HUMAN SEXUAL BEHAVIOR

It is well established that olfactory cues play a significant role in the sexual behavior of a wide variety of animals, but the role olfactory cues play in human sexual behavior is poorly understood and often controversial (Hopson, 1979; Rogel, 1978; Doty, 1976; Comfort, 1974). In humans the three areas most frequently discussed are: 1) sexual identification, 2) sexual attraction, and 3) hormonal synchronization.

Sexual Identification

It is generally believed that human beings can detect differences between the sexes on the basis of odors given off by sexually specific sources such as vaginal secretions, smegma, and seminal fluid (Doty, 1977). Although no studies have been done to verify this commonly held belief, it seems likely that it is a correct assumption, at least in sexually experienced individuals.

The ability to discriminate genders by olfactory cues from sexually shared sources, such as underarm and hand odors, have been examined by several authors. Russell (1976), and Hold and Schleidt (1977) have tested the ability of adults to discriminate between sexes on the basis of axillary odors, and reported that both sexes were able to make gender identifications by axillary odors. Wallace (1977) found that sexual identifications could also be made on the basis of hand odors. McBurney et al. (1977) and Doty (1977) found that when odors were equated for intensity by a magnitude estimation procedure, the discriminations were much more difficult, and that strong odors are usually classified as male and less intense odors classified as female. This reflects the fact that women have, on the average, smaller apocrine glands and produce less intense body odor than men (Hurley and Shelly, 1960; Shehadah and Kligman, 1963; Doty, 1977). Thus, for shared structures, gender identification does occur, but is based more on odor intensity than on the presence of some specific volatile compound.

Sex Attractants

The idea that olfactory cues can act as sexual attractants in humans goes back several centuries (see Walton, 1958 for a review) and is supported by a variety of clinical observations and psychoanalytic theories (Brill, 1932; Ellis, 1905; and Wiener, 1966). There are two categories of possible attractants that may be similar in their effects, but are different in their origins. The first is associative learning; e.g., an individual is exposed to specific odors each time he comes into contact with members of the opposite sex and soon learns to associate those odors with sexual identity or sexual experience. The second type of attractant is a pheromone; e.g., one or more specific volatile compounds given off by one sex which causes sexual excitement in the other without behavioral conditioning. Both types of olfactory attractant have been suggested to exist in humans (Wilson, 1963; Wiener, 1967; Comfort, 1974), but it is frequently difficult to distinguish between the two in controlled animal studies (Goldfoot, 1981; Keverne, 1980; Michael et al., 1976), and may be impossible to completely separate them in human research, although the current evidence supports associative learning as the primary source of sexual attraction in humans.

The arguments for the existence of human sexual attractant pheromones are based on three observations. The first is that in many other species of mammals the female advertizes her sexual receptivity by emitting an attractant pheromone when she is ovulating (dogs are the most frequently observed example, Beach and Gilmore, 1949); the second is the large number of anecdotal reports suggesting some perfumes are powerfully attracting to some individuals; and the third is that humans have a set of apocrine and sebaceous glands that become more active at puberty and are responsive to emotional and/or sexual stimulation (Kuno, 1934; Shelley and Hurley, 1953).

Many mammalian females give off an attractant pheromone during ovulation, usually along with other visual and behavioral cues that are used to advertize their sexual receptivity (Beach, 1976; Müller-Schwarze, 1974). One of the most exceptional aspects of human sexuality is, however, that evolution has made us unaware of human ovulatory periods. Women are unique in that they may be sexually receptive at any time of their cycle and do not exhibit a phase of physiological or behavioral cues that signal ovulation (Butler, 1974; Alexander and Noonan, 1979; Benshoof and Thornhill, 1979; Strassmann, 1981). The fact that other primate species do have pronounced periods of estrus signaling (Hall, 1962; Goodall, 1965; Hausfater, 1975) suggests that our ancestors may have also had this trait, but that the evolutionary trend has been for women not to provide any cues of their period of ovulation for males or themselves. Any odor that was given off as an attractant during ovulation would not be consistent with this evolutionary development, because it would identify when ovulation was occurring. Further, studies that have examined the incidence of sexual activity for increases in coital behavior around the ovulatory period do not show the significant increase that would be expected if such an attractant were present (James, 1971; Spitz, et al. 1975).

One possible hypothesis is that women have adapted to continuous sexual receptivity by constantly giving off an attractant odor. Cowley et al. (1977) have studied the effects of two odorous compounds, androstenol (5α-16-androsten-3α-ol, the boar attractant), and "Copulin" (a mixture of aliphatic acids found in vaginal secretions) on the responses of male and female students in an Assessment-Of-People test. In this study, subjects were asked to evaluate the qualities of other individuals while wearing masks that had been treated with one of the odorous compounds. The subjects did not know the purpose of the study or that they were being exposed to the odors. This study found that the women tended to judge the males more highly when androstenol was present, but that there were no changes in the responses of the males. Kirk-Smith et al. (1978) also used treated masks to examing the influence of the odor of androstenol on the subjects' judgments of photographs. In contrast to the results of Cowley et al., both male and female subjects rated photographs of women as "more attractive" and "better" in the presence of androstenol. In a third study of the effects of androstenol, McCollough et al. (1981) had subjects read an erotic passage while being exposed to either androstenol or Rose Water and found no change in emotional responsiveness in either men or women as measured by the Differential-Emotions-Scale questionnaire. While these findings cannot be compared directly because of the differences in techniques, it appears that there are enough differences in the results to make any general conclusions about the attractiveness of androstenol tentative at best. However, even if the results of Cowley et al. are confirmed, it will not be sufficient evidence to demonstrate a pheromonal response, as it is quite possible that any preference for this compound which is present in male secretions (Brooksbank et al., 1974; Sastry et al., 1980) is a learned preference due to associative learning. To demonstrate a pheromonal effect the chemical must stimulate particular aspects of behavior, in this case sexual attraction, and associative learning should be specifically excluded (Goldfoot, 1981; Beauchamp et al., 1976). Goldfoot (1981) has suggested some specific criteria for determining the existence of a pheromonal effect in primates and these same criteria should be used when considering human pheromones. These are: 1) compound specificity - a specific chemical or mixture of chemicals must be shown to have behavioral potency in conspecifics; 2) behavior specificity - a chemical must stimulate particular aspects of behavior rather than general responses such as arousal; 3) species specificity - a chemical must be active only in the same or related species; and 4) innate response - the behavior must be genetically based. Goldfoot (1981) has included "imprinting" in this last criterion, but Beauchamp et al. (1976) would exclude it.

Sexual Attraction

Another hypothesis is that odors may not be used for sexual attraction at all, but rather for negative sexual selection. In this hypothesis, the partner's odor is considered to be undesirable unless the aversion to the odor is overridden by some other factor, such as sexual arousal or habituation to the aversive odor. Stoddart (1980, p. 103) has suggested that in some animal species females are able "to discriminate between less desirable and more desirable males by their noses. When deprived of this ability, they accept all comers." This view is supported by the observation that sexual receptivity is enhanced by the removal of the olfactory bulbs of female rats (Satli and Aron, 1976).

The bulk of the evidence in humans shows that odors given off by males are found to be aversive rather than attractive. Much of the personal hygiene practices found in our society are activities which either reduce the intensity of our natural smell, or attempt to eliminate personal odors altogether. Studies of odor preferences in our culture demonstrate that male body odors collected on "T" shirts are viewed by both males and females as unpleasant (Hold and Schleidt, 1976; Schleidt, et al., 1981). Since the current evidence suggests that the major difference between male and female odors is the greater intensity of male odors, and these odors can be identified as belonging to specific individuals, it may be that males are using odors as a means of signaling their physical presence rather than as a sexual attractant. If this were combined with associative learning on the part of females, it would lead to a situation in which some male odors were attractive while others were aversive, depending on the personal experiences of the individuals involved. Lawless and Cain (1975) have shown that learned recognition of odors can be very long-lasting, and it is possible that early experiences may play a role in any associative learning which may occur in humans.

It is too early to rule out the possibility that human attractant pheromones exist, but the strong interest by the popular press and perfume industry makes it important that any claims for such a discovery be viewed cautiously. It is also important that research into this question be continued. Any improvement in our understanding of the behavioral and physiological functions of olfactory cues in sexual attraction or sexual development is likely to have significant social consequences.

Hormonal Synchronization

Two types of hormonal synchronization have been suggested in humans that may be related to olfactory cues; in one the continued presence of a male may increase the frequency of a woman's menstrual cycle; and in the other, women who are living in close proximity have concurrent menstrual cycles. Both of these effects have been demonstrated in a variety of animals including primates (Rowell and Dixson, 1975; Rosenblum, 1968; Harrington, 1975; Conaway and Sade, 1965; Vandenbergh and Vessey, 1968). The first demonstration that similar phenomena were occurring in humans was presented in a landmark paper by McClintock (1971). McClintock noted that menstrual synchrony was often reported by women in all-female living groups. She examined the timing of the onset of menstrual cycles for roommates and close friends on a college campus by asking dormitory residents to report the timing of their cycles. She questioned 135 women aged 17-22, and found significant correlations in the timing of cycles of women who spent time in close proximity (e.g., roommates). Further, McClintock found that women who reported that they were in the company of males more than three times a week tended to have shorter menstrual cycles than those who spent less time with males. McClintock concluded that "there is some interpersonal physiological process which affects the menstrual cycle" and suggested that it could be pheromonal in nature. Subsequent to this report, Graham and McGrew (1980), Russell et al. (1980), and Quadagno et al. (1981) have also reported menstrual synchronization in women. Two of these studies (Graham and McGrew, and Quadagno et al.) also looked for a shortening of the menstrual cycle for women with close contacts with men, but found none.

It is possible that this failure to find a shortening of the cycles with exposure to men is due to the fact that in the later studies male isolation time varied. Additional studies are needed to establish that increases in the frequency of menstrual cycles occur due to exposure to males.

Menstrual synchrony in humans is of particular interest because it has been well demonstrated that estrus cycles in animals are influenced by olfactory cues (Grau, 1976; Whitten, 1969; Vandenbergh et al., 1975; Müller-Schwarze, 1974; Bronson, 1971). McClintock demonstrated that the human phenomenon is not due to changes in diet, awareness of menstrual timing, or lunar cycles, and suggested that the only significant factor seemed to be the amount of time that the women spent together and the relative length of their cycles. That is, they are controlled by changes in unknown volatile compounds that are not caused by the individuals learning to respond. Rogel (1978), in a critical evaluation of the possible existence of higher primate pheromones suggested that "the most promising area in which to search for pheromonal control of higher primate sexual and reproductive behavior is menstrual synchrony, a phenomenon that suggests the action of a primer pheromone." (p. 862).

Russell et al. (1980) examined the menstrual cycles of college women after exposing them to the axillary secretions of another donor female. A single donor was selected who had a regular menstrual cycle of 28 days, and who claimed a previous history of "driving" another woman's cycle. That is, a friend had become synchronous with her when they roomed together in the summer and dissynchronous when they moved apart in the fall. The subjects were divided into two groups, one which received the odor on an alcohol treated pad and the other which received only the alcohol treated pad as a control odor. The odorant was applied to the upper lip three times a week. The number of participants was five in the experimental group and six in the control group. The study included a one-month pre-treatment period and a four-month treatment period. The subjects were informed of the nature of the experiment, but did not know which group they were in. The dates of the onset of the subjects' menstrual cycles were determined by questioning. The mean difference between the onset of the cycle of the subjects and the onset of the donor cycle was 9.3 days in the pre-treatment month and 3.4 days post-treatment for the experimental group; and 8.0 days for the pre-treatment month and 9.2 days post treatment in the control group. The results were statistically significant and support the view that odor is the communicative element in human menstrual synchrony.

Menstrual synchrony does suggest that some type of pheromone exists in humans and that the mode of action is similar to that found in other mammals, but it also raises a number of questions. First, this was a pilot study done with a single donor and these results should not be generalized to other women. Does this particular donor have some unique characteristics that make her able to modify the cycles of other women or would any woman with a regular cycle have a similar effect? Is the phenomenon really olfactory in the traditional sense? While the experiment was conducted in the context of an olfactory stimulus, none of the women involved in the study reported an awareness of changes in the odor of the sample that they received, suggesting that conscious awareness of an odor may not be necessary for modification of the cycle. We do not know from these studies which phase of the cycle is changing or how the phenomenon of synchrony is related to changes in ovulation. These questions and many others, will require further research in this area.

CONCLUSIONS

Humans do use their natural body odors as sources of information for non-verbal communication in a number of contexts. The best established of these is the recognition of self and individual family members by infants and adults. In addition to recognition of family members by odors that are indigenous to those individuals, it has been suggested that infants may be depositing scent marks on their mothers in a manner similar to other primates. These scent marks seem to be used to reduce anxiety in infants who place them on cloth toys or blankets, and may have similar functions at other times. Humans are also capable of gender identification through the use of olfactory cues from both sexually unique sources, such as vaginal secretions and seminal fluid, and sexually shared sources such as hand and axillary regions. The primary basis for the discrimination of the shared sources is a sexual dimorphism of greater odor intensity for males. Several studies have been done which suggest that odors are a significant factor in sexual attraction, but currently there is not sufficient evidence to support claims for an attractant pheromone. The hypothesis is presented that associative learning better explains the existing observations of sexual attraction in humans, but further research should be done which examines the role that odorants have on psychosexual function. Studies are also cited which report that body odors, particularly male ones, are aversive to many people; and that they may function to signal an individual's presence rather than as sexual attractants. There is evidence that a pheromone may exist for menstrual synchrony, however, and it is suggested that more research is needed to replicate and expand the work that has already been done in this area. Particularly, work should be done to determine what the relationship is between menstrual synchrony and ovulation. The identification of olfactory cues that could alter hormonal levels, change time of ovulation or effect implantation of the ovum would be a significant achievement, but there is little or no research to determine if any of these effects are possible.

REFERENCES

Alexander, R. D., and Noonan, K. M., 1979, Concealment of ovulation, parental care, and human social evolution, in: N. Chagnon, and W. Irons (eds.),"Evolutionary Biology and Human Social Behavior," North Scituate, Mass., Duxbury.

Alexander, G., 1978, Odour and the recognition of lambs by Merino ewes, Appl. Anim. Ethol.,4:153.

Barrette, C., 1977, Scent marking in captive muntjacs, *Muntiacus reevesi*, Anim. Behav., 35:536.

Beach, F. A., 1976, Sexual attractivity, proceptivity, and receptivity in female mammals, Horm. Behav., 7:105.

Beach, F. A., and Gilmore, R.W., 1949, Response of male dogs to

urine from females in heat, F. Mammal., 30:391.
Beauchamp, G. K., Doty, R. L., Moulton, D. G., and Mugford, R. A., 1976, The pheromone concept in mammalian chemical communication: a critique, in: R.L. Doty (Ed.), "Mammalian Olfaction, Reproductive Processes and Behavior," New York, Academic Press.
Benshoof, L., and Thornhill, R., 1979, The evolution of monogamy and concealed ovulation in humans, Int. J. Soc. Biol. Struct., 2:94.
Brill, A. A., 1932, The sense of smell in neuroses and psychoses, Psychoanalytic Qu., 1:7.
Brooksbank, B. W. L., Brown, R., and Gustafsson, J. A., 1979, The detection of 5α-androst-16-en-3α-ol in human male axillary sweat, Experientia, 30:864.
Bronson, F. H., 1970, Rodent pheromones, Biol. Reprod., 4:344.
Butler, H., 1974, Evolutionary trends in primate cycles, Contributions Primatol., 3:2.
Carpenter, G., 1974, Mother's face and the newborn, New Sci., 21 March, 742.
Cheal, M. L., and Sprott, R. L., 1971, Social olfaction: a review of the role of olfaction in a variety of animal behaviors, Psychol. Rep., 29:195.
Comfort, A., 1974, The likelihood of human pheromones, in: M. Birch (Ed.), "Pheromones. Frontiers of Biology," 32:386.
Conaway, C. H., and Sade, D. S., 1965, The seasonal spermatogenic cycle in free ranging rhesus monkeys, Folia Primat., 3:1.
Cowley, J. J., Johnson, A. L. and Brooksbank, B. W. L., 1977, The effect of two odorous compounds on performance in an assessment-of-people test, Psychoneuroendocrinol., 2:159.
Darwin, C., 1877, Mind, 7:282.
Doty, R. L., (Ed.), 1976a, "Mammalian Olfaction, Reproductive Processes and Behavior," New York, Academic Press.
Epple, G., 1974, Olfactory communication in South American primates, in: W. S. Cain (Ed.), "Odors: Evaluation, Utilization, and Control," Annals New York Acad. Sci., 237:261.
Ellis, H., 1905, "Sexual Selection in Man," v. 4, New York, Davis.
Graham, C. A., and McGrew, W. C., 1980, Menstrual synchrony in female undergraduates living on a coeducational campus, Psychoneuroendocrinol., 5:245.
Grau, G., 1976, Olfaction and reproduction in ungulates, in: R. L. Doty (Ed.), "Mammalian Olfaction, Reproductive Processes and Behavior," New York, Academic Press.
Goldfoot, D. A., 1981 , Olfaction, sexual behavior, and the pheromone hypothesis in rhesus monkeys: a critique, Amer. Zool., 21:153.
Hall, K. R. L., 1962, The sexual, agonistic and derived social behavior patterns of the wild chacma baboon, Papio ursinus, Proc. Zool., Society London, 139:238.
Harrington, J. P., 1975, Olfactory communication in Lemur fulvus, in: R. D. Martin, G. A. Doyle, and A. C. Walker (Eds.), "Prosimian Behavior," Duckworth, Cloucester Crescent.
Hausfater, G., 1975, Dominance and reproduction in baboons (Papio

cynocephalus): A quantitative analysis, "Contributions to Primatology," 12 S. Darger, Basel.
Hess, E. H., 1973,"Imprinting: Early Experience and the Developmental Psychobiology of Attachment,"New York, Van Nostrand Reinhold.
Hong, K. M. and Townes, B. D., 1976, Infants' attachment to inanimate objects: a cross cultural study, J. Am. Acad. Child Psychiat., 15:49.
Hurley, H. J., and Shelley, W. B., 1960, "The Human Apocrine Sweat Gland in Health and Disease,"Springfield, Thomas.
Hold, B., and Schleidt, M., 1977, The importance of human odour in non-verbal communication, Z. Tierpsychol., 43:225.
Hopson, J., 1979,"Scent Signals: The Silent Language of Sex," New York, William Morrow.
James, W. H., 1971, The fecundability of U. S. women, Popul. Stud., 27:493.
Johnston, R. E., 1975, Scent marking by male hamsters, Z. Tierpsychol., 37:75.
Kaplan, J., 1977, Perceptual properties of attachment in surrogate-reared and mother-reared squirrel monkeys, in: S. Chevalier-Skolnikoff and F. E. Poirier (Eds.), "Primate Bio-Social Development: Biological, Social and Ecological Determinants," New York: Garland.
Kaplan, J., and Russell, M. J., 1974, Olfactory recognition in the infant squirrel monkey, Develop. Psychobiol., 7:15.
Keverne, E. B., 1980, Olfaction in the behaviour of non-human primates, Symp. Zool. Soc. Lond., 45:313.
Kirk-Smith, M., Booth, D. A., Carroll, D., and Davies, P., 1978, Human social attitudes affected by androstenol, Res. Commun. Psychol. Psychiat. Behav., 3:379.
Kuno, Y., 1934, "The Physiology of Human Perspiration," London, Churchill.
Lawless, H. T., and Cain, W. S., 1975, Recognition memory for odours, Chem. Senses Flavor., 1:331.
Leon, M., and Moltz, H., 1972, The development of the pheromonal bond in the albino rat, Physiol. Behav., 8:683.
Lipsitt, L. P., Engen, T., and Kaye, H., 1963, Developmental changes in the olfactory threshold of the neonate, Child Dev., 34:371.
McBurney, D. H., Levine, J. M., and Cavanaugh, P. H., 1977, Psychophysical and social ratings of human body odor, Pers. Soc. Psychol. Bull., 3:135.
McClintock, M. K., 1971, Menstrual synchrony and suppression, Nature, 229:244.
McCollough, P. A., Owen, J. W. and Pollak, E., 1981, Does androstenol affect emotion,? Ethol. Sociobiol., 2:85.
MacFarlane, A., 1975, Olfaction in the development of social preferences in the neonate, in: "The Human Neonate in Parent-Infant Interaction," Ciba Found. Symp., Amsterdam, 33:103.
Michael, R. P., Bonsall, R. W., and Zumpe, D., 1976, "Lack of effects of vaginal fatty acids, etc.": A reply to Goldfoot et al., Hormones and Behav., 7:365.

Müller-Schwarze, D., 1974, Social functions of various scent glands in certain ungulates and the problems encountered in experimental studies of scent communication, in: V. Geist and F. Walther (Eds.), "The Behaviour of Ungulates and Its Relation to Management," New Series No. 24, Morges, Switz.: IUCN Publications.
Müller-Schwarze, D., 1974, Olfactory recognition of species, groups, individuals and physiological states among mammals, in: M. C. Birch (Ed.) "Pheromones," Amsterdam: North Holland.
Müller-Schwarze, D., and Mozell, M. J. (Eds.), 1977, "Chemical Signals In Vertebrates," New York, Plenum Press.
Mykytowycz, R., 1977, Olfaction in relation to reproduction in domestic animals, in: Müller-Schwarze, D. and Mozell, M. M. (Eds.), "Chemical Signals In Vertebrates," New York, Plenum Press.
Mykytowycz, R., 1968, Territorial marking by rabbits, Sci. Amer., 218:116.
Passman, R. H., and Weisberg, P., 1975, Mothers and blankets as agents for promoting play and exploration by children in a novel environment: the effects of social and nonsocial attachment objects, Develop. Psychol., 11:170.
Porter, R. H., and Moore, J. D., 1981, Human kin recognition by olfactory cues, Physiol. and Behav., 27:493.
Pratt, K. C., Nelson, A. K., and Sun, K. S., 1930, "The Behavior of the Newborn Infant," Columbus, Ohio State University Press.
Quadagno, D. M., Shuveita, H. E., Deck, J., and Francoeur, D., 1981, Influence of male social contacts, exercise and all-female living conditions on the menstrual cycle, Psychoneuroendocrinol., 6:339.
Ralls, K., 1971, Mammalian scent marking, Science, 171:443.
Rogel, M. J., 1978, A critical evaluation of the possibility of higher primate reproductive and sexual pheromones, Psychol. Bull., 85:810.
Rosenblatt, J. S., 1972, Learning in newborn kittens, Sci. Amer., 227:18.
Rosenblum, L., 1968, Some aspects of female reproductive physiology in the squirrel monkey, in: L. A. Rosenblum and R.W. Cooper (Eds.) "The Squirrel Monkey," New York, Academic Press.
Rowell, T. E., and Dixson, A. F., 1975, Changes in social organization during the breeding season of wild talapoin monkeys, J. Reprod. Fert., 43:419.
Russell, M. J., Mendelson, T., and Peeke, H. V. S., 1982, Mothers' identification of their infants' odors, In preparation.
Russell, M. J., 1976, Human olfactory communication, Nature, 260: 520.
Russell, M. J., Switz, G. M., and Thompson, K., 1980, Olfactory influences on the human menstrual cycle, Pharmac. Biochem. Behav., 13:737.
Sastry, S. D., Buck, K. T., Janak, J., Dressler, M., and Preti, G., 1980, Volatiles emitted by humans, in: G. R. Waller and O. C.

Dermer (Eds.), "Biochemical Applications of Mass Spectrometry, First Supplementary Volume," New York, John Wiley and Sons.

Satli, M. A., and Aron, C., 1976, New data on olfactory control of estral receptivity of female rats, Compte rendu hebdomadaire des séances de l"Academie des Sciences, Paris, 282:875.

Schleidt, M., Hold, M., and Attili, G., 1981, A cross-cultural study on the attitude towards personal odors, J. Chem. Ecol., 7:19.

Schleidt, M., 1980, Personal odor and nonverbal communication, Ethol. Sociobiol., 1:225.

Schaal, B., Montagner, H., Hertling, E., Bolzoni, S., Moyse, A., and Quichon, R., 1980, Les stimulations olfactives dans les relations entre l'enfant et la mère, Reprod. Nutr. Develop., 20: 843.

Schulz, C., 1982, Peanuts, United Feature Syndicate, in: San Francisco Chronicle.

Shelley, W. B., and H. J., 1953, The physiology of the human axillary apocrine sweat gland, J. Invest. Dermatol., 20:285.

Shorey, H. H., 1976, "Animal Communication by Pheromones," New York Academic Press.

Spitz, C. J., Gold, A. R., and Adams, D. B., 1975, Cognitive and hormonal factors affecting coital frequency, Arch. Sex. Behav., 4:249.

Strassmann, B. I., 1981, Sexual selection, parental care, and concealed ovulation in humans, Ethol. Sociobiol., 2:31.

Stoddart, D. M., 1980, "The Ecology of Vertebrate Olfaction," New York, Chapman and Hall.

Vandenbergh, J. C., 1975, Acceleration and inhibition of puberty in female mice by pheromones, J. Reprod. Fert. Suppl., 19:411.

Vandenbergh, J. G., and Vessey, S., 1968, Seasonal breeding of free ranging rhesus monkeys and related ecological factors, J. Reprod. Fert., 15:71.

Wallace, P., 1977, Individual discrimination of humans by odor, Physiol. and Behav., 19:577.

Walton, A.H., 1958, "Aphrodisiacs: From Legend to Prescription," Westport, Connecticut, Associated Booksellers.

Whitten, W. K., 1969, Mammalian pheromones, in C. Pfaffmann (ed.), "Olfaction and Taste: Proceedings of the Third International Symposium," New York.

Whitten, W. K., and Bronson, F. H., 1970, The role of pheromones in mammalian reproduction, in: J. W. Johnson et al. (eds.), "Advances in Chemoreception I: Communication by Chemical Signals," New York, Appleton-Century-Crofts.

Wiener, H., 1966, External chemical messengers: I. Emission and reception in man, N. Y. State J. Med., 66:3153.

Wilson, E. O., 1963, Pheromones, Sci. Am., 208:100.

Goodall, J., 1971, "In the Shadow of Man," Boston, Houghton-Mifflin.

STUDIES OF THE CHEMICAL COMPOSITION OF SECRETIONS FROM SKIN GLANDS OF THE RABBIT *ORYCTOLAGUS CUNICULUS*

B.S. Goodrich

Division of Wildlife Research, CSIRO
P.O. Box 52, North Ryde
N.S.W., Australia 2113

INTRODUCTION

The odoriferous skin glands of the wild rabbit, *Oryctolagus cuniculus*, which function for conspecific communication purposes have been the subject of intensive ethological and chemical studies. Volatile constituents of the anal and chin glands have been found to convey information concerning territoriality, whilst the odor of the secretion of inguinal glands conveys information about the identity and sex of an individual animal (Mykytowycz, 1965, 1966a,b; Hesterman and Mykytowycz, 1982). This paper reviews our knowledge of the chemistry of secretions from these glands.

ANAL GLANDS

In the field, rabbits deposit fecal pellets coated with the secretion from their anal glands on special sites which are commonly known as 'dung hills'. The odors emanating from the 'dung hills' not only repel conspecific intruders into a territory, but more importantly reinforce the confidence of the residents (Mykytowycz, 1966a). The chemical composition of the anal gland secretion has been examined particularly intensively (Goodrich *et al.*, 1978, 1981a). In early work macerated anal glands were used for studies of the chemical composition of smell derived from this organ (Goodrich *et al.*, 1978). More recently however, the examination of volatiles given off by fecal pellets have been studied (Goodrich *et al.*, 1981a), since these are a more convenient source of odor of the secretion from anal glands for chemical and ethological studies.

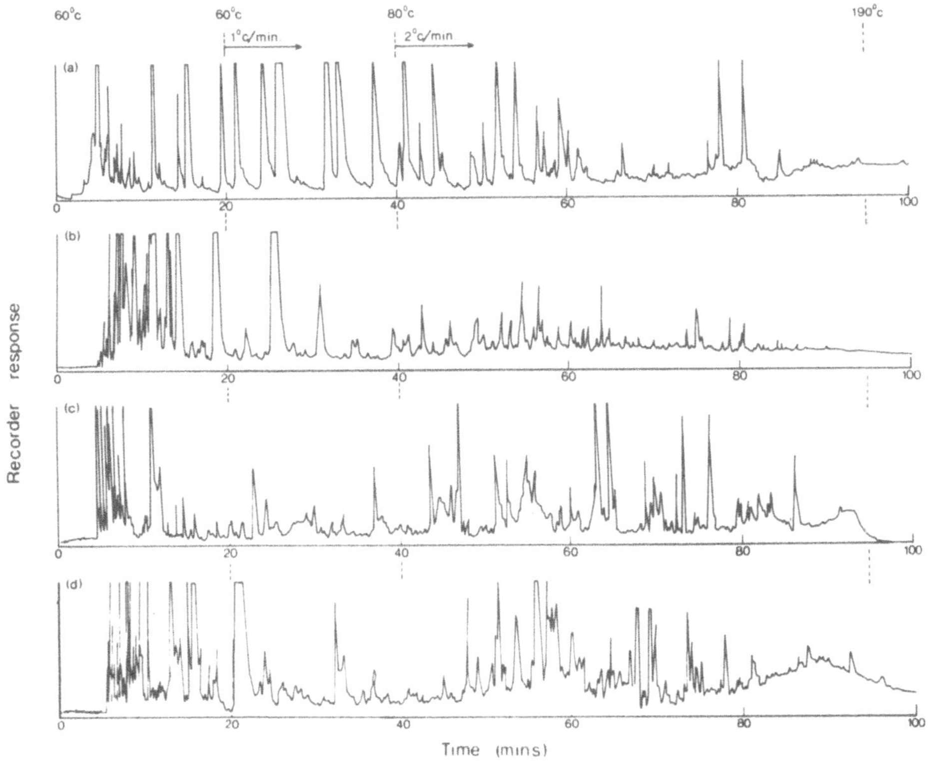

Fig. 1. Gas chromatograms on FFAP SCOT column of headspace volatiles above odoriferous secretions from adult male rabbits, Oryctolagus cuniculus. (a) Anal gland homogenate. (b) Fecal pellets. (c) Chin gland secretion. (d) Inguinal gland secretion.

The volatiles given off by macerated anal glands and fecal pellets (headspace volatiles) were collected on Chromosorb 105 and then introduced through a modified gas chromatographic inlet onto a glass SCOT column coated with a polar phase (Carbowax 20M or FFAP). The gas chromatograms of these headspace collections (Figs. 1a and 1b) indicate the complexity of the odor associated with the anal gland secretion.

To facilitate the search for behaviorally significant components in the odors, adult male rabbits were positioned to sniff the effluents from the gas chromatographic column (Fig. 2) and responses of their heart rate to the eluted components were recorded (Hesterman et al., 1976; Goodrich et al., 1978, 1981a). Thus rabbits were used to detect those constituents of their own odors which may be behaviorally important.

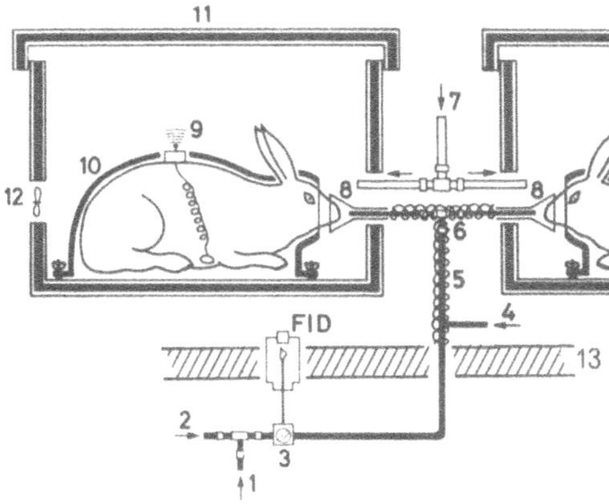

Fig. 2. Diagram showing two rabbits, Oryctolagus cuniculus located to sniff the gas chromatographic column effluent with simultaneous recording of heart rate. 1 = Capillary column outlet; 2 = make-up N_2; 3 = column effluent splitter; 4 = make up air for dilution of effluent; 5 = heated delivery tube for presentation of volatile components to rabbits; 6 = T-piece union for splitting effluent between rabbits; 7 = precooled air to each rabbit; 8 = glass funnels; 9 = miniature radio transmitters; 10 = fiberglass casts for restraining rabbits; 11 = soundproofed boxes for housing rabbits; 12 = exhaust fan; 13 = gas chromatographic oven lid.

Guided by changes in the rabbits' heart rates, zones in the chromatograms were selected for further studies. Fractions of the odor corresponding to these zones were trapped, rechromatographed on a capillary column of lower polarity (Silicone SF96), and rabbits again employed to monitor the effluent in order to pinpoint components which elicit most consistently changes in cardiac response.

Gas chromatography-mass spectrometry of the constituents of the odor of anal glands and fecal pellets revealed the presence of several classes of organic compounds (see Table 1). Whilst some of these compounds were found to induce changes in heart rate in male rabbits, their behavioral role has not yet been tested.

To test the behavioral significance of volatiles present in the odors associated with the anal gland secretion, an experimental technique was developed involving a specially designed chamber (Goodrich et al., 1981b; Hesterman et al., 1981).

The rationale for this new method was based on information obtained earlier that when two rabbits are placed in a new

Table 1. Compounds identified in the odor of anal glands and fecal pellets of the rabbit, *Oryctolagus cuniculus*

Anal glands

Aldehydes

3-Methylbut-2-enal	*n*-Decanal
6-Methylhept-5-enal	*cis*-Undec-4-enal
n-Octanal	*n*-Undecanal
Non-2-enal	*cis*-Dodec-5-enal
6-Methyloctanal	*trans*-Dodec-5-enal
n-Nonanal	9-Methylundecanal
Dec-2-enal	10-Methylundecanal
Citronellal	n-Dodecanal
7-Methylnonanal	Tridec-5-enal
8-Methylnonanal	10-Methyldodecanal

Ketones

6-Methylhept-5-en-2-one

Alcohols

2-Methylbut-2-en-1-ol	*n*-Heptanol
3-Methylbut-2-en-1-ol	Oct-1-en-3-ol
2-Methylbutan-1-ol	*n*-Octanol
3-Methylbutan-1-ol	Linalool
n-Hexanol	

Fecal Pellets

Hydrocarbons

Two monoterpenes	$C_{15}H_{30}$
n-Decane	*n*-Pentadecane
n-Dodecane	$C_{16}H_{32}$
n-Tridecane	$C_{16}H_{34}$
n-Tetradecane	*n*-Hexadecane
β-Gurjenene	

Table 1 (continued)

Alcohols	
2-Methylpropan-I-ol	*n*-Hexanol
n-butanol	*n*-Heptanol
3-Methylbut-2-en-1-ol	2-Phenylethanol
3-Methylbutan-1-ol	*n*-Octanol
n-Pentanol	*n*-Nonanol
Ketones	
Pentan-2,3-dione	Octan-3-one
Heptan-2-one	2,2,6-Trimethylcyclohexanone
6-Methylhept-5-en-2-one	Nonan-2-one
6-Methylheptan-2-one	
Carboxylic acids	
Ethanoic acid	*n*-Butanoic acid
Propanoic acid	
Esters	
Ethyl propanoate	Ethyl hexanoate
Methyl butanoate	*n*-Butyl pentanoate
Ethyl butanoate	*n*-Pentyl butanoate
Ethyl pentanoate	*n*-Butyl hexanoate
3-Methylbutyl ethanoate	*n*-Hexyl butanoate
n-Propyl butanoate	Ethyl 2-phenylpropanoate
Aryl compounds	
Phenol	*p*-Cresol
Styrene	Naphthalene
o-Cresol	2,6-Di-*tert*-butyl-*p*-cresol
Heterocyclic compounds	
Pyridine	2-*n*-Pentylfuran
Trimethylpyrazine	2-Ethyl-3,5,6-trimethyl-pyrazine
Tetramethylpyrazine	
Others	
Dimethyl disulfide	Trichloroethylene

environment together with an odor used for territorial marking obtained from only one of them, the owner of this odor dominates the other one (Mykytowycz et al., 1976).

In a series of tests a pair of male rabbits was introduced into the test chamber (Fig. 3) which contained one of the following odors derived from one of them: the total odor collected from above fecal pellets; fractions prepared by either chemical scrubbing, or preparative gas chromatography of the volatiles.

Chemically-scrubbed fractions were prepared by washing the volatiles with distilled water, 1M hydrochloric acid, or 1M sodium hydroxide. The communicative power of the volatiles remaining after washing with distilled water was only slightly weaker than that of the untreated odor. However scrubbing with either acid or alkali completely eliminated the behaviorally-active constituents. This suggests that both acidic and basic components are essential for the formation of the territorial message contained in the odor from anal glands. The effects of the different scrubbing procedures on the composition of the volatiles are illustrated in the gas chromatograms presented in Fig. 4.

A more precise technique for fractionation of the odor constituents of fecal pellets was achieved by preparative gas chromatography (Fig. 5). After a preliminary clean-up by scrubbing with distilled water, the volatiles were trapped on Chromosorb 105, transferred to a Silicone SF96 trap, fractionated on a GC capillary column, retrapped on SF96 from the column effluent and introduced through a heated valve into the test chamber (Goodrich et al., 1981b). The bioassay showed that fraction 1 (see Fig. 5) was ineffective in the development of the donors' territorial confidence, whilst fraction 2 was less effective than the total effluent or the unfractionated odor. Fraction 3 however, retained the characteristic properties of the total odor (unpublished data). The zones covered by the respective fractions were selected on the basis of heart rate response data (Goodrich et al., 1981a). Fractions 2 and 3 contained the volatiles of the total effluent which elicited the most numerous heart rate responses. These results indicate the usefulness of the heart rate technique as a method of screening complex odors for components of possible behavioral importance.

These studies demonstrate that gas chromatographic techniques can be used for the fractionation of complex mammalian odors with precision, and the trapped fractions rechromatographed and retrapped without loss of activity. The fractions can be stored for later use, remixed, or transported from one laboratory to another. Combined with a descriminative bioassay such procedures facilitate the identification of specific combinations of volatile substances involved in the formation of olfactory signals.

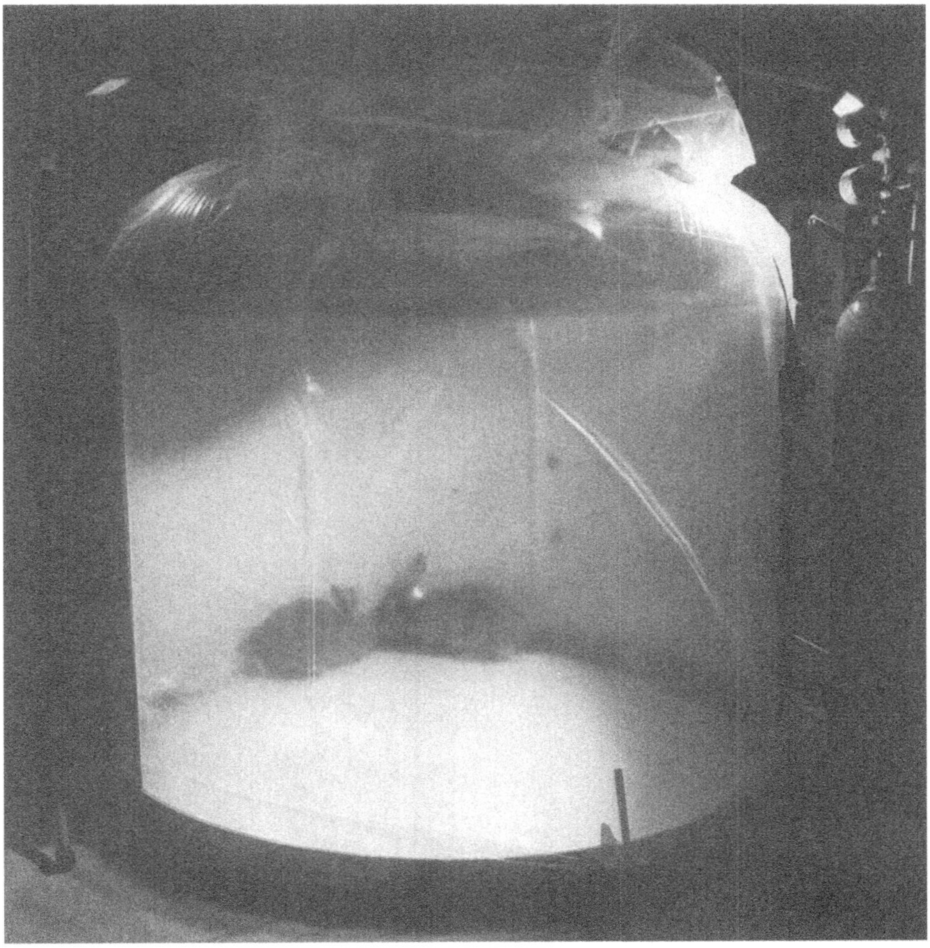

Fig. 3. The totally enclosed chamber used for testing the behavioral reaction of rabbits, Oryctolagus cuniculus to the volatiles from fecal pellets and related fractions, and chin gland secretion.

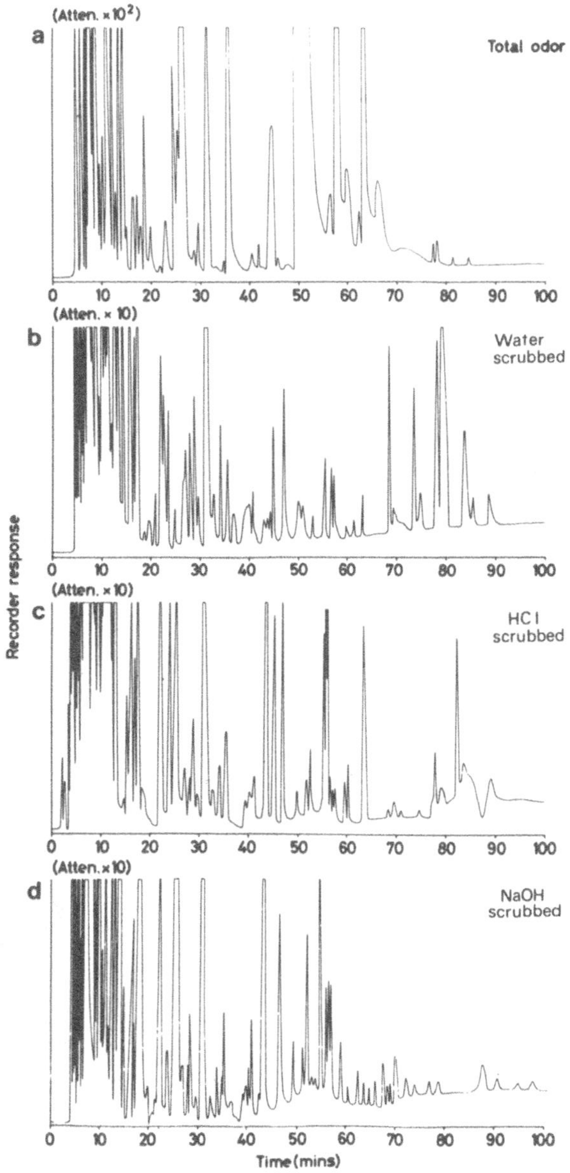

Fig. 4. Gas chromatograms on FFAP SCOT column of the headspace volatiles from the fecal pellets of a male rabbit, Oryctolagus cuniculus, showing the effects of different scrubbing reagents: (a) total volatiles; (b) distilled water; (c) 1M hydrochloric acid; (d) 1M sodium hydroxide. Note that the signal attenuation in (a) is one tenth that for the other chromatograms.

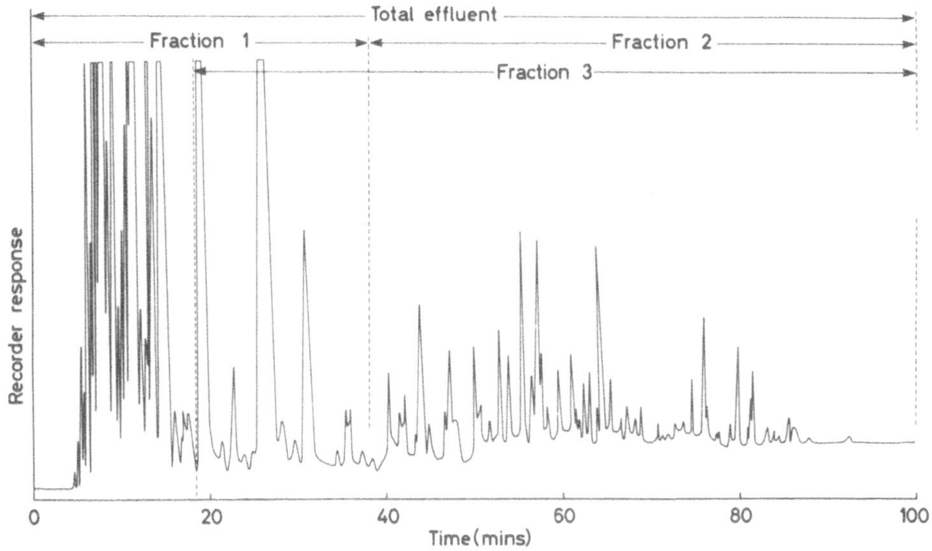

Fig. 5. Gas chromatogram on FFAP SCOT column of the non-water-soluble volatiles from the fecal pellets of a male rabbit, Oryctolagus cuniculus, desorbed from a Silicone SF96 trap. The zones included in fractions 1, 2 and 3 are indicated.

CHIN GLANDS

The submandibular or chin gland of the rabbit, as the name implies is located under the chin. In the course of chinning, secretion is deposited deliberately on various objects within the rabbits' environment. This gland is much larger in males than in females. It has been experimentally established that testosterone stimulates and oestrogen inhibits its secretory activity and the intensity of chinning in both wild and domestic rabbits (Mykytowycz, 1965, 1968; Wales and Ebling, 1971).

Preliminary chemical examination of the chin gland secretion identified proteins as the major constituents (Goodrich and Mykytowycz, 1972). Both qualitative and quantitative differences were observed for the protein components and variations related to sex as well as individual identity of the animals were demonstrated (Fig. 6).

The chin gland differs from the anal and inguinal glands in not having an odor detectable by the human nose. The 'rabbity' odors of secretions from these last two glands are associated with free lipids. However these are present in negligible quantities in the chin secretion. Instead high molecular weight, non-volatile constituents predominate. The observation however that pure chin gland secretion obtained directly from the gland ducts is detected olfactorily by rabbits (Mykytowycz, 1975) prompted a more detailed chemical study (Goodrich, in prep.).

Headspace volatiles from above chin gland secretion were collected as previously described on Chromosorb 105 sequentially after one, two, three and ten days. The volatiles collected in the four traps produced gas chromatographic profiles which were qualitatively similar and of approximately equal intensities. It is likely that there would be an advantage for the owners of a given territory that such olfactory signals should persist in the environment. A typical gas chromatogram of chin gland secretion is shown in Fig. 1c.

The volatile organic compounds present in the headspace of the chin gland secretion were also examined using gas chromatography-mass spectrometry. The identity of the major constituents is shown in Table 2. Aromatic compounds were the main components. This is in contrast to that which has been found for the anal gland and fecal pellet volatiles (Goodrich et al., 1978, 1981a), where aliphatic compounds predominated. These results point to the likelihood of a difference in the message carried by the odor of the anal and chin glands, although both glands function for territorial purposes.

To test the behavioral significance of the volatiles emanating from chin gland secretions the same test chamber technique as used in bioassays of the odor of anal glands was used (see Fig. 3). It was found however that the confidence of donor rabbits was not influenced when the air space of the chamber was filled uniformly with total odor of chin gland secretion. The confidence-giving effect was observed however, when the odor from the secretion was introduced continously into the test chamber through a number of outlets protruding through the floor. This observation illustrates again the importance of selecting an appropriate method for presentation of odor in testing the reaction of an animal to an olfactory cue.

Fractionation according to molecular size of the macromolecular components present in the chin secretion was carried out through a series of ultrafiltration membranes (XM 100A, CM 50, PM 10, and UM 05; Amicon Corporation), and by chromatography on G-25 Sephadex (Pharmacia Fine Chemicals).

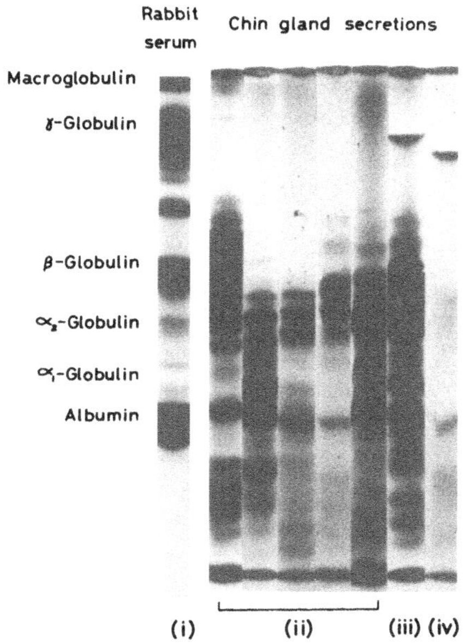

Fig. 6. Disc electrophoresis on polyacrylamide gels of protein components of chin gland secretions from individual rabbits, *Oryctolagus cuniculus*. (i) Pooled rabbit serum; (ii) chin gland secretions from individual males; (iii) pooled chin gland secretion from males; (iv) pooled chin gland secretion from females.

Gas chromatographic analysis of the headspace volatiles from all the fractions gave similar profiles of the components present. This suggests that volatiles in the chin gland secretion are bound to the high molecular weight constituents such as proteins, which may act to slowly release a territorial message. The involvement of macromolecules in vertebrate olfactory communication has previously been reported by many workers (Goodrich and Mykytowycz, 1972; Berüter *et al.*, 1973; Novotny *et al.*, 1980; Singer *et al.*, 1980; Vandenbergh, 1980).

Table 2. Compounds identified in the odor of chin gland secretion and inguinal pouch sebum of the rabbit, *Oryctolagus cuniculus*

Chin gland secretion	
Aliphatic compounds	
n-Butanoic acid	*n*-Nonanal
n-Undecane	*n*-Decanal
n-Tridecane	
Aromatic compounds	
Ethyl benzene	Acetophenone
Naphthalene	Dimethylacetophenone*
Benzaldehyde	Ethylacetophenone*
Benzdialdehyde*	Phenol
Dimethylbenzaldehyde*	2,6-Di-*tert*-butyl-*p*-cresol
Ethylbenzaldehyde*	Dichlorobenzene*
Vinylbenzaldehyde*	
Inguinal pouch sebum	
Ethanoic acid	3-Methylbutanoic acid
Propanoic acid	*n*-Pentanoic acid
2-Methylpropanoic acid	4-Methylpentanoic acid
n-Butanoic acid	*n*-Hexanoic acid

* Identity of isomer unknown.

The presence of proteins also provides suitable conditions for the survival of microorganisms. Therefore it is essential that in studies of the chemistry of some gland secretions the role of microorganisms in shaping of the olfactory message should also be considered. Current experiments in which germ-free whole chin gland secretions were inoculated with bacterial strains isolated from the chin region of rabbits show changes in the composition of the volatiles (Goodrich and Merritt, unpublished).

INGUINAL GLANDS

Information about ownership of a given space is only one of the messages that the rabbit has to convey to its conspecifics. There are many others, and also other sources of odor which function specifically for the purpose of communication. For instance, the inguinal gland secretion is used for individual identification (Mykytowycz, 1966b). The glands consist of two parts - a superficial white sebaceous part, and an adjacent more deeply seated brownish sudoriferous portion - both secreting into hairless skin pouches located in the groins, where a brownish sebum with the characteristic 'rabbity' odor is formed.

To demonstrate that the inguinal gland odor conveys information on the identity of individual rabbits, members of stabilised groups were smeared with inguinal gland secretion from unfamiliar individuals. When these smeared rabbits were returned to their usual groups they were attacked by their pen-mates, sometimes very severely. In contrast smearing with anal or chin gland secretion, urine, macerated inguinal glands or a commercially-produced perfume did not cause serious disruption to the social relationships within the groups (Hesterman and Mykytowycz, 1982). Such reaction of rabbits to the inguinal secretion suggests that this could be exploited for the bioassay of the total odor as well as fractions of the secretion.

Fractionation of the highly complex odor of sebum from the inguinal pouches (Fig. 1d) resulted in losses of behaviorally-active constituents. To test the behavioral function of a separate fraction of the total inguinal odor it is essential that this material be in such a form that it can be applied physically to the rabbits' body. The high volatility of some of the components require that fractions be incorporated into a suitable fixative. However such substances, which can release test odors in a similar way as occurs naturally, are not yet available.

The failure of macerated inguinal glands to release aggression suggests again that bacterial action on the sebum in the inguinal pouches may be essential for the formation of meaningful odor signals (Albone and Perry, 1976).

The contents of inguinal pouches of the wild rabbit were examined for the presence of microorganisms and volatile fatty acids. Organisms belonging to four genera have been isolated (Merritt _et al._, in press). Table 2 lists the volatile fatty acids which have been isolated. Quantitative analysis of the acids in the sebum showed considerable variation between individual animals. Ethanoic acid and 3-methylbutanoic acid were the most abundant. In a liquid medium containing one of these acids,. the growth of microorganisms was only slightly affected. It became inhibited however, when both of these acids were present. The synergistic inhibitory effect of the acids suggests the existence of an additional mechanism for shaping odor profiles of the inguinal pouches. Inclusion of other acids as well as other pouch constituents may increase the titre of the inhibitory effect and the resultant composition of the final odor.

CONCLUSIONS

The results of chemical studies of the secretions from anal, chin and inguinal glands of the rabbit, Oryctolagus cuniculus, provide us with a deeper insight into the composition of mammalian odors generally. Odors used by mammals for communication consist of complex mixtures of organic compounds, but only fractions of these may be required for the production of specific behavioral responses.

Until now chemists have mainly attempted to identify behaviorally-active constituents of total odors. However the role of components which appear not to be directly involved in signalling must also be explained. For instance, the interaction of various non-volatile constituents such as proteins in chin gland secretions, and hydrocarbons and triglycerides in sebum from inguinal pouches of rabbits require special consideration. The role of microorganisms in shaping of odor profiles must also be explored more fully.

To determine the behavioral importance of the chemical components in an odor mixture, it is essential to use reliable bioassays to test them. In designing a valid bioassay it is necessary to understand thoroughly the behavior of a given species and particularly how odor signals are produced, transmitted and perceived. The responses of mammals to odor signals depend not only on their chemical compositions, but also on the way in which they are presented during experimentation.

The study of the role of odor in communication of vertebrates is an interdisciplinary task. To ensure progress, collaboration is essential not only between chemists and ethologists, but also with specialists in other disciplines.

ACKNOWLEDGEMENTS

I wish to thank Mr E.R. Hesterman, Mr G. Merritt and Dr R. Mykytowycz for their collaborative contributions and helpful comments on this manuscript.

REFERENCES

Albone, E.S., and Perry, G.C., 1976. Anal sac secretion of the red fox, Vulpes vulpes; volatile fatty acids and diamines. Implications for a fermentation hypothesis of chemical recognition, J. Chem. Ecol. 2:101.

Berüter, J., Beauchamp, G.K., and Muetterties, E.L., 1973, Complexity of chemical communication in mammals: urinary components mediating sex discrimination by male guinea pigs, Biochem. Biophys. Res. Comm. 53:264.

Goodrich, B.S., and Mykytowycz, R., 1972, Individual and sex differences in the chemical composition of pheromone-like substances from the skin glands of the rabbit, Oryctolagus cuniculus, J. Mammal. 53:540.

Goodrich, B.S., Hesterman, E.R., Murray, K.E., Mykytowycz, R., Stanley, G., and Sugowdz, G., 1978, Identification of behaviorally significant volatile compounds in the anal gland of the rabbit, Oryctolagus cuniculus, J. Chem. Ecol. 4:581.

Goodrich, B.S., Hesterman, E.R., Shaw, K.S., and Mykytowycz, R., 1981a, Identification of some volatile compounds in the odor of fecal pellets of the rabbit, Oryctolagus cuniculus, J. Chem. Ecol. 7:813.

Goodrich, B.S., Hesterman, E.R., and Mykytowycz, R., 1981b,. Effect of volatiles collected above fecal pellets on behavior of the rabbit, Oryctolagus cuniculus, tested in an experimental chamber. II. Gas chromatographic fractionation of trapped volatiles, J. Chem. Ecol. 7:947.

Hesterman, E.R., Goodrich, B.S., and Mykytowycz, R., 1976, Behavioral and cardiac responses of the rabbit, Oryctolagus cuniculus, to chemical fractions from anal gland, J. Chem. Ecol. 2:25.

Hesterman, E.R., Goodrich, B.S., and Mykytowycz, R., 1981, Effect of volatiles collected abov fecal pellets on behavior of the rabbit, Oryctolagus cuniculus, tested in an experimental chamber. I Total volatiles and some chemically prepared fractions, J. Chem. Ecol. 7:795.

Hesterman, E.R., and Mykytowycz, R., 1982, Misidentification by wild rabbits, Oryctolagus cuniculus, of group members carrying the odor of foreign inguinal gland secretion. I. Experiments with all-male groups, J. Chem. Ecol. 8:419.

Merritt, G.C., Goodrich, B.S., Hesterman, E.R., and Mykytowycz, R., In press, The microflora and volatile fatty acids present in the inguinal pouches of the wild rabbit, Oryctolagus cuniculus, in Australia, J. Chem. Ecol.

Mykytowycz, R., 1965, Further observations on the territorial function and histology of the submandibular cutaneous (chin) glands in the rabbit, Oryctolagus cuniculus (L.), Anim. Behav. 13:400.

Mykytowycz, R., 1966a, Observations on odoriferous and other glands in the Australian wild rabbit, Oryctolagus cuniculus (L.) and the hare, Lepus europaeus P. I. The anal gland, CSIRO Wildl. Res. 11:11.

Mykytowycz, R. 1966b, Observation on odoriferous and other glands in the Australian wild rabbit, Oryctolagus cuniculus (L.) and the hare, Lepus europaeus P. II. The inguinal glands, CSIRO Wildl. Res. 11:49.

Mykytowycz, R., 1968, Territorial marking by rabbits, Sci. Amer. 218:116.

Mykytowycz, R., 1975, Activation of territorial behaviour in the rabbit, *Oryctolagus cuniculus*, by stimulation with its own chin gland secretion, in: "Olfaction and Taste, V", D.A. Denton and J.P. Coghlan, eds., Academic Press, New York.

Mykytowycz, R., Hesterman, E.R., Gambale, S., and Dudzinski, M.L., 1976, A comparison of the effectiveness of the odors of rabbits, *Oryctolagus cuniculus*, in enhancing territorial confidence, J. Chem. Ecol. 2:13.

Novotny, M., Jorgenson, J.W., Carmack, M., Wilson, S.R., Boyse, E.A., Yamazaki, K., Wilson, M., Beamer, W., and Whitten, W.K., 1980, Chemical studies of the primer mouse pheromones, in: "Chemical Signals : Vertebrates and Aquatic Invertebrates", D. Müller-Schwarze and R.M. Silverstein, eds., Plenum Press, New York.

Singer, A.G., Macrides, F., and Agosta, W.C., 1980, Chemical studies of hamster reproductive pheromones, in: "Chemical Signals : Vertebrates and Aquatic Invertebrates", D. Müller-Schwarze and R.M. Silverstein, eds., Plenum Press, New York.

Vandenbergh, J.G., 1980, The influence of pheromones on puberty in rodents, in: "Chemical Signals : Vertebrates and Aquatic Invertebrates", D. Müller-Schwarze and R.M. Silverstein, eds., Plenum press, New York.

Wales, N.A.M., and Ebling, F.J., 1971, The control of the apocrine glands of the rabbit by steroid hormones, J. Endocr. 51:763.

THERMAL AND OSMOLARITY PROPERTIES OF PHEROMONAL COMMUNICATION IN THE GERBIL, MERIONES UNGUICULATUS

Del D. Thiessen
Department of Psychology
University of Texas
Austin, Texas 78712

Arthur E. Harriman
Department of Psychology
Oklahoma State University
Stillwater, Oklahoma 74078

INTRODUCTION

Integumental or glandular release of chemosignals is constrained by metabolic capacity and core temperature (Thiessen, 1977). Cellular metabolism and heat production conjointly determine the nature and quantity of chemicals released as signals, and also set limits on sensory sensitivity and motor efficiency (Campbell, 1969; Heinrich, 1979; Paulus & Reisch, 1980). Indeed, a strong case could be made that homeostatic processes and chemocommunication evolved together (Stoddart, 1980). In any event, the relationship has important implications for advance in the understanding of animal behavior.

Our present intent is to demonstrate that olfactory communication among Mongolian gerbils, Meriones unguiculatus, is associated with both thermoregulation and osmoregulation. Emphases are on two systems that have been implicated in pheromonal communication, namely the Harderian gland and the ventral abdominal scent gland (Thiessen & Yahr, 1977). Harderian gland secretions, spread throughout the pelage during an autogroom, have diverse functions. Chiefly, the exudate insulates the gerbil against cold and wet environments, increases the absorption of radiant energy, and decreases the ability to withstand water deprivation under conditions of low relative humidity. The ventral scent gland, among its other effects (Thiessen & Yahr, 1977), appears to be a heat-exchange mechanism for the body. There is evidence that the animal thermoregulates by bringing the ventral scent gland into contact with thermal objects in its environment. The female, in addition, apparently uses the gland during lactation to transfer body heat to her pups. Thus, the pheromonal activities of the two glands appear to be constituents of a more general homeostatic system.

THE HARDERIAN GLAND

The Harderian gland is located in the orbit of the eye of most terrestrial vertebrates, with the exception of the highest primates (Kennedy, 1970). In the Mongolian gerbil, as well as other rodent species, the gland periodically discharges secretions into the Harder-lacrimal canal and out the external nares of the nose. The secretions also reach the mouth, presumably by way of the nasal palatine duct. The anatomical site of the gland, the secretory tracts and the area of the body initially covered by secretions are presented in Figure 1.

The secretion is a complex mixture of lipids and pigments which can be seen only because the protoporphyrin pigment in the mixture

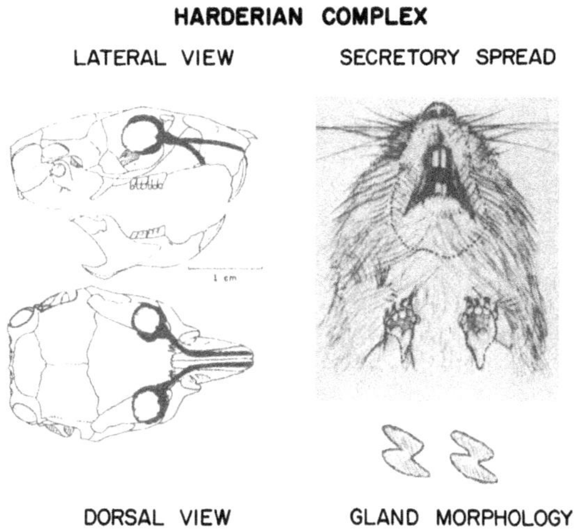

Fig. 1. Drawings indicating relative position of Harderian glands, Harder-lacrimal canals, and the primary points of spread of Harderian material following an autogroom. The z-shaped Harderian glands are also indicated. The drawings are a composite of diagrams by Gulotta (1971) and Goodwin (personal communication).

fluoresces bright red under ultraviolet light stimulation. The Harderian secretion is released to the external nares during a thermoregulatory autogroom and is spread widely over the face and hair. Major steps in the autogroom are pictured in Figure 2, and the relationship between changes in body temperature and grooming is seen in Figure 3. Elevations in body temperature whether caused by social interactions or in other ways, evoke an autogroom and the release of Harderian secretions to the external nares. The animal mixes these secretions with saliva and spreads the secretions around the face and over the pelage. The saliva has three functions: (1) evaporative cooling, (2) interaction with Harderian material to destroy its fluorescent and pheromonal qualities, and (3) service as a social attractant (Block et al., 1980).

While the Harderian spread has a number of functions, only some of these have been clearly determined (Campbell, 1969; Heinrich, 1979, Paulus, 1980). Among the latter is the finding that the secretory material provokes mutual investigation of nose and mouth areas among interacting adults. Immediately following a groom, animals approach each other and investigate the nasal and facial areas. The probability of mutual investigation is highest just after the groom and decreases within seconds thereafter. The decrease in investigation correlates with the disappearance of facial fluorescence. This relationship is illustrated in Figure 4. An autogroom in one

Figure 2. Photographs indicating the major movements of an autogroom.

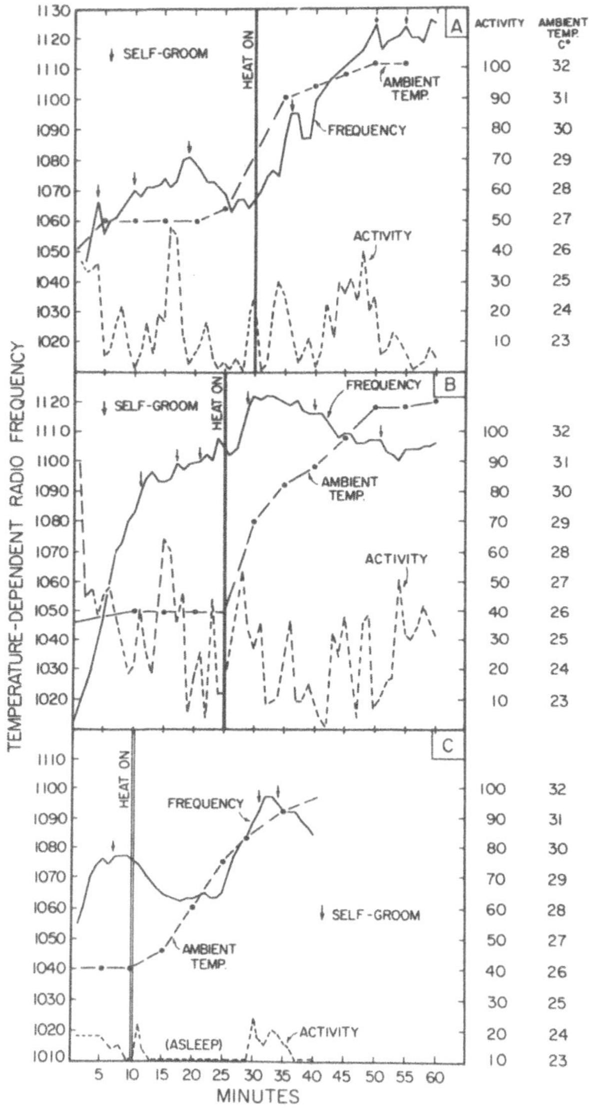

Fig. 3. Relationships among ambient and body temperature, and autogrooms (↓) and general activity. Body temperature is represented by AM radio signals from implanted biotelemetry Mini-mitters, and not absolute body temperature (Thiessen, Graham, Perkins and Marcks 1977).

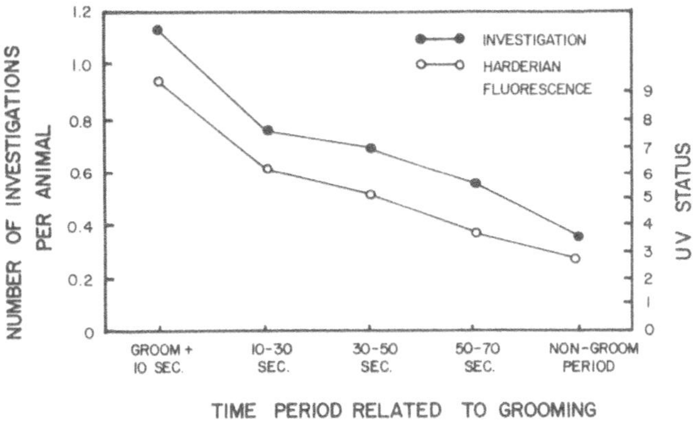

Fig. 4 Relationship of time and autogrooming to the number of facial investigations by conspecifics and the spread of Harderian material on the face, as indexed by the intensity of fluorescence (UV Status) (Thiessen, 1977).

animal often triggers an autogroom in a conspecific, however, the effect is reduced by Harderianectomy (Goodwin & Regnier, unpublished).

We have also found that, prior to weaning, pups are attracted to the parental nest because of Harderian odors (Thiessen, 1982, in press). Further, Harderianectomy, which reduces the interaction among adults as well as between offspring and parents, probably signifies by exclusion that the pheromone triggers aggregation. Additionally, Payne (1977, 1978) has reported that the Harderian components of female hamsters reduce male aggression and act as a sexual signal.

The foregoing discussion points to possible interactions among Harderian secretion, pheromone influence and thermoregulation. Certainly the release and spread of the material is linked to increases in body temperature and to saliva spread. The relationship is more extensive, however. We have recently found that Harderian lipids on the pelage protect the animal against cold and wet stress (Thiessen & Kittrell, 1980). Thus, Harderianectomy reduces an animal's ability to maintain its body temperature during exposure to a wet and cold (3-5°C) environment (see Figure 5). This effect can be duplicated with animals which have had their hair lipids removed by shampooing. Apparently, the Harderian lipids placed throughout the pelage by grooming provide an insulative barrier against potentially hypothermic environments.

Moreover, Harderian spread of lipids and pigments darkens the pelage and increases the amount of radiant energy absorbed (Thiessen et al., 1981). Figure 6 shows the relative degree of hair reflectance across the visual spectrum for intact and Harderianectomized animals. Further, Figure 7 shows that the percent increase in body temperature due to radiant heat exposure is less for the lighter Harderianectomized animals. We have found that either allowing

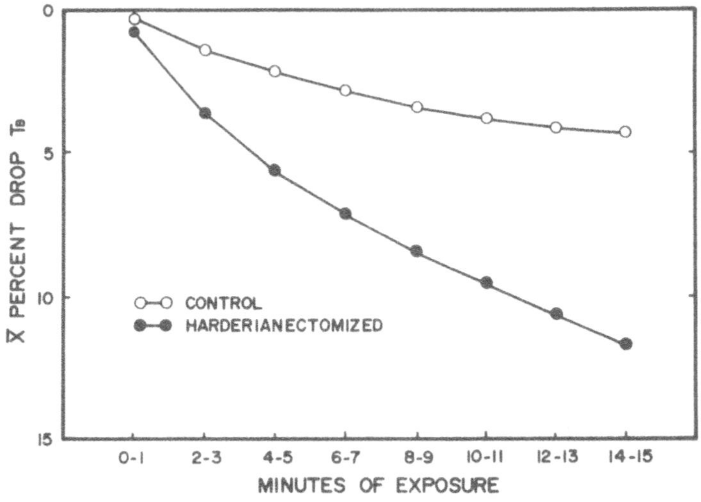

Fig. 5. Body temperature decline in Harderianectomized and control gerbils following a one sec ice water dip and 15 min of exposure to three 5°C. Shampooing out the fur lipids has a similar effect (Thiessen & Kittrell, 1980).

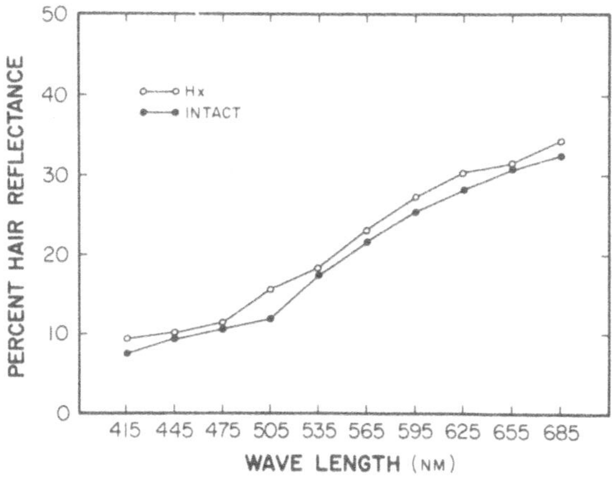

Fig. 6. Reflectance of hair for Harderianectomized (Hx) and intact animals within the visual spectrum (Thiessen et al., 1981).

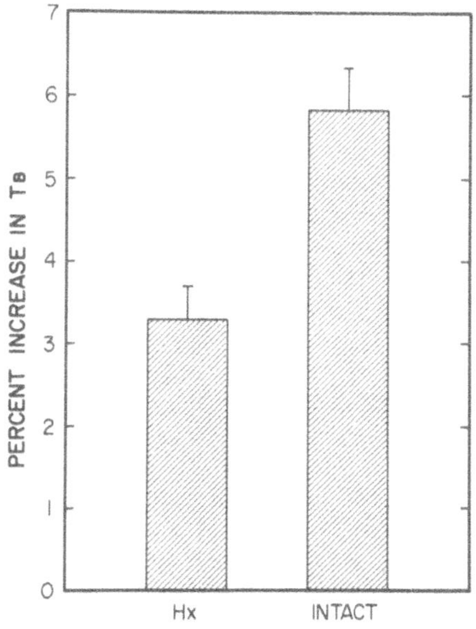

Fig. 7. Percent increase in body temperature for Harderianectomized (Hx) and intact animals (Thiessen et al., 1981).

animals extended opportunities to sandbathe or shampooing animals to remove pigments and lipids have comparable effects on hair reflectance and on temperature response to radiant heat. It seems probable, therefore, that one thermal function of Harderian material is to darken the pelage and increase the absorption of radiant energy. The increased heat absorptive capacity of the pelage is of obvious value to Meriones unguiculatus, a species that is diurnal during colder months of the year (Randall & Thiessen, 1980). Sandbathing lightens the pelage and decreases the absorption of radiant energy. Consequently, the Harderian spread and sandbathing act oppositely in the control of body temperature.

In addition to the Harderian effects on communication and thermoregulation, there are osmolarity consequences. Intact male animals not allowed to sandbathe (i.e., living on floors made of wire mesh screen) show greater losses of body weight during water deprivation than do either Harderianectomized animals or intact animals allowed to sandbathe (living on a substrate of ground corn cobs). At 23.0°C, the mean difference in weight loss after four days of water deprivation is over three grams (about 25.0%). The reason for the greater loss of body weight among animals which have their Harderian glands intact but which are not permitted to sandbathe may be due to the condition of the hair. About seven days after loss of opportunity to "sandbathe," gerbils develop a darker, matted, greasy, and wet-appearing coat with many small areas of skin showing around clumps of hair (Thiessen & Kittrell, 1980). Animals living on ground corn cobs, sand or other bedding material, like those that are Harderianectomized, continue to show a dry, lighter and fluffy coat which covers the entire skin. Increased loss of body water through insensible skin evaporation among the former animals may account for this difference in body weights. Hence, "sandbathing" may not only be used to lighten the pelage and thus to increase radiant energy reflectance, but it may also help condition the hair to provide a more insulative covering of the skin. Thus, like clothing for humans, the pelage serves the interests of homeostasis by altering the transfer rate of heat from skin to environment.

We have also reasoned that, if water evaporation from the skin among intact animals was more rapid than for Harderianectomized animals, the loss of body weight during water deprivation would be greater under conditions of lower relative humidity. When both intact and Harderianectomized animals were water-deprived for four days while housed on wire mesh floors, and thus prevented from sandbathing, the difference in weight loss was minimal at a relative humidity of 70%, but increased at 50% and increased even more at 30%. These data are shown in Table 1. Thus Harderian material on the hair increases evaporative water loss, but only in conditions of low relative humidity.

This finding obtains, as it now appears, independently of the

TABLE 1

Mean Body Weight Losses During Four Days of Water Deprivation by Harderianectomized and Intact Gerbils Tested Under Different Levels of Relative Humidity

Animal Condition	Level of Relative Humidity 70% $\bar{X}$ S.E.	50% $\bar{X}$ S.E.	30% $\bar{X}$ S.E.
Intact	14.5±0.58	14.8±0.43	15.5±0.58
Harderianectomized	13.6±0.92	12.8±0.61	13.0±0.61
±Difference (g)	0.9	2.0	2.5

ambient temperature. Intact and Harderianectomized gerbils are exposed to environmental temperatures of 5°C and of 30°C, which differ in each case between high (80.0%) and low (30.0%) relative humidity, during counterbalanced four-day water deprivation periods. Body weight losses were significantly greater among the intact animals than among the operated gerbils at 30% relative humidity, but not at 80% relative humidity. The temperature level prevailing during the tests could not be associated with the difference in rate of body weight loss by the two groups during water deprivation. Through use of four-day water deprivation periods body weight losses were kept below 15% which is the point where most animals develop dehydration hyperthermia (Richards, 1973). We intend to determine whether build-up of Harderian spread on the gerbil's pelage advances onset of this hyperthermia.

Conclusions drawn from the studies discussed above are that the Harderian gland is used in pheromonal communication and thermoregulation and is important in osmoregulation by feral gerbils. The release of Harderian material, when triggered by an elevation in body temperature, acts as an aggregating signal. At the same time, its spread throughout the pelage can insulate the animal against hypothermic environments and result in the absorption of higher levels of radiant energy. Under some conditions of ambient humidity, the Harderian spread may also facilitate evaporation of water from the skin. Under different environmental circumstances, a reduced amount of Harderian material in the pelage may be beneficial, as on bright hot days when the relative humidity is low. On these occasions, it would be adaptive to lighten the pelage to reflect more radiant energy and so to protect against water loss. The different adaptations could be achieved by a reduced spread of Harderian lipids and pigments and/or an increased intensity of sandbathing. Whatever the case, one can expect complex interactions among signaling, spread of Harderian secretions, sandbathing, body temperature and body fluids. We are continuing to study these interactions.

THE VENTRAL SCENT GLAND

Ventral scent gland marking in Meriones unguiculatus is associated with sexual dimorphism, aggression, social class, and variations in environmental stimuli (Agren & Meyerson, 1977; Cheal, 1979; Daly, 1977; Pettijohn & Barkes, 1978; Thiessen & Yahr, 1977). The gerbil shares these characteristics in common with many other mammalian species (Beauchamp, 1974; Drickamer et al., 1973; Eisenberg & Kleiman, 1972; Epple, 1981; Goodrich & Mykytowycz, 1972; Johnson, 1973; Müller-Schwarze, 1971; Ralls, 1971; Schultze-Westrum, 1969; Steiner, 1974; Thiessen & Rice, 1976). The complex physiological control of the gerbil scent gland and of scent marking depends upon gonadal steroids acting on the gland tissue and on restricted areas of the central nervous system (Swanson, 1980; Thiessen & Yahr, 1977; Turner, 1975; Yahr, 1976).

To return to an earlier point, we believe that the abdominal scent gland of the gerbil may have a thermoregulatory as well as signaling function. The gerbil is very sensitive to internal and external thermal variations (McManus & Mele, 1969; Mele, 1972; Robinson, 1959; Steffen & Roberts, 1977). Moreover, the gland pad is one of the few bare areas of its body where heat exchange with the environment can occur. Conceivably, when an animal marks an object it may transfer body heat to the environment or, conversely, acquire heat from the environment. Of course, the direction of heat flow would depend on the thermal gradient between the animal and its environment. Body heat transferred to the substrate would also increase the volatility of the deposited sebum.

Two recent experiments support this notion. The first shows that there is a specific heat transfer through the ventral pad of the adult male. The second suggests that the ventral gland of the lactating female acts as a "brood patch" which elevates the body temperature of nurselings and facilitates maturation.

In the first experiment, adult males with or without their ventral glands (surgically removed under nembutal anesthesia) were tested for marking in a Plexiglas arena in which a single marking peg was positioned at the center of the floor. A thermometer, brought through the transparent floor and fixed at the surface of the peg, measured the temperature change resulting from each mark. Body temperature was recorded from interperitoneal telemetry units (Minimitter) at intervals during the test. After one week, animals were again placed in the arena. On this occasion, a "floating glass peg" was attached to a weighing scale below the apparatus. The pressure of each mark was recorded to the nearest milligram. Animals were video-taped from beneath the apparatus, and illustrations were prepared from the tapes.

Drawings of the ventral surface of an animal moving across the floor and marking the peg are shown in Figure 8. Illustrated are the extension of the body and the consequent application of pressure to the peg which parts the shingled layers of hair covering the gland. In result, the entire gland surface is brought into contact with the peg.

Figure 9 gives the mean temperature exchanges at the surface of the peg for the two groups. The mean peg temperature increased 0.64°C during a mark for animals with ventral glands and 0.33°C for animals without ventral glands. The difference cannot be attributed to a difference in the pressure applied to the peg, as the two groups showed no systematic variation on this measure (see Figure 10.) Ventral marking frequencies and core body temperatures did not differ significantly for the groups, although glandless animals tended to show a steeper rise during the first minutes of the test (see Figure 11).

The data suggest that there is a significant thermal effect related to ventral marking. Subsequent experiments will attempt to confirm a thermoregulatory role.

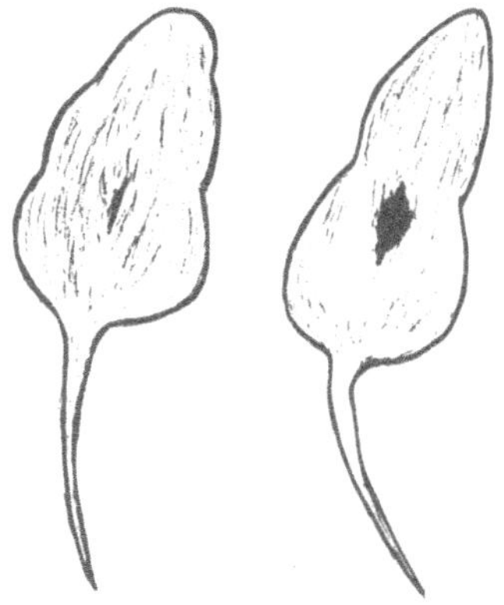

Fig. 8. Schematic view of ventral scent gland in non-marking movement (left) and during a mark (right). Body temperature and pressure applied to substrate expose a 1 X 2 cm gland surface. Drawings taken from video pictures from beneath Plexiglas floor.

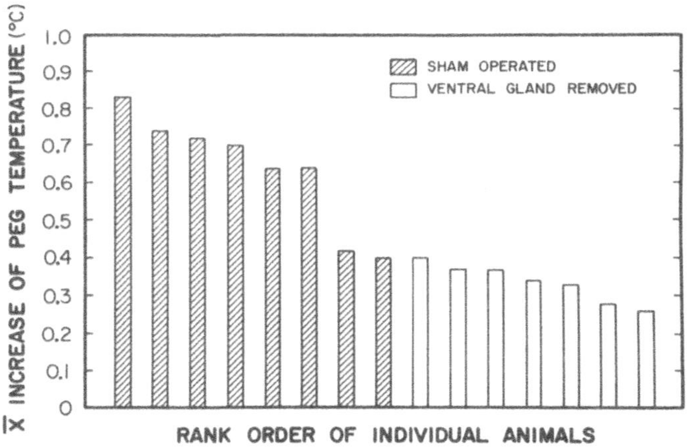

Fig. 9. Rank order of individuals according to average peg temperature increase during a 30-min test period.

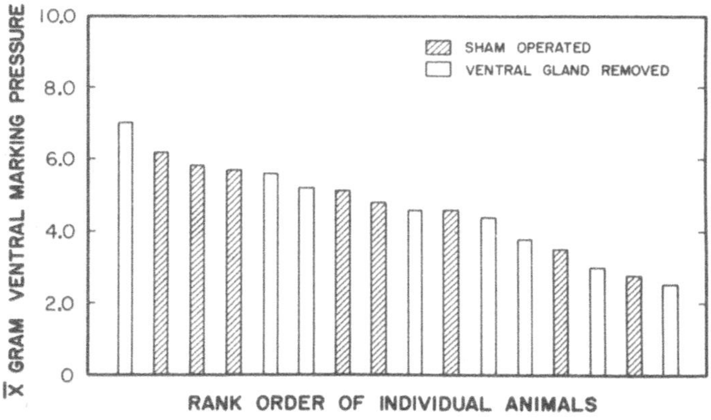

Fig. 10. Rank order of individuals according to average pressure applied to marking peg during a 30-min test period.

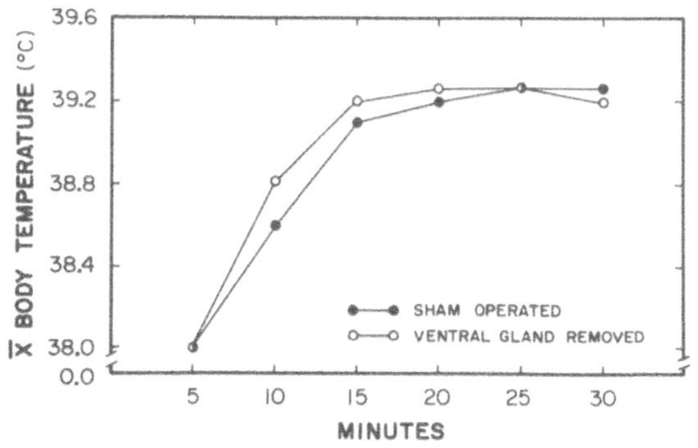

Fig. 11. Average body temperature changes at 5-min intervals during a 30-min marking test.

The second illustration of a thermal function for the ventral scent gland concerns lactation in females. Two startling changes occur in the postpartum female - ventral gland size is increased and marking frequency is elevated (Wallace et al., 1973). Gland size enlargement for a lactating female is illustrated in Figure 12. During this period, both gland size and marking frequency approach size and marking levels found in adult males.

The ventral gland enlargement has apparent implications for pup development (Kittrell et al., 1982). Offspring reared with mothers which had their ventral glands removed prior to pregnancy developed at a slower rate than did offspring with normal mothers. As indicated in Table 2, pups from glandless animals showed a delayed righting response, reduced ultrasound frequency, and lower body weights. The ventrum temperature of glandless mothers was also lower.

These results indicate a reduced transfer of maternal heat for mothers without scent glands. During the first few days of life, when the pups are essentially ectothermic, the reduced heat transfer could critically restrict growth rate. We recognize that the loss of sebum odors or a change in maternal behavior could contribute to the results. Nevertheless, pilot data indicate that mothers with and mothers without ventral glands spend equivalent amounts of time with their litters.

TABLE 2

Developmental Differences Between Pups

Reared by Mothers with or without Ventral Glands

Variable Studied ($\bar{X}$ Value)	Pups with Glandless Mothers	Pups with Intact Mothers	F
No. of ultrasounds (2 min) ($\bar{X}$ from day 2-20)	66.37	99.02	14.60***
Body Weight (day 10)	7.73	9.06	3.43*
Righting Latency (Sec) (Day 10)	36.62	11.67	6.61**
Maternal ventral temperature ($0^{o}C$; $\bar{X}$ from days 2-20)	35.86	36.61	11.93**

* p<.10>.05
** p<.05
*** p<.01

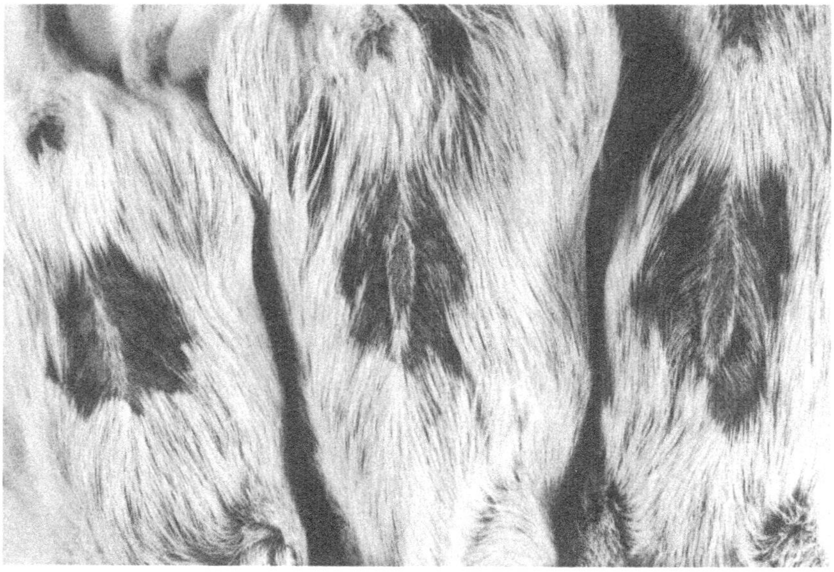

Fig. 12. Ventral gland enlargement of lactating female (middle) relative to non-lactating female (right) and male (left) (Kittrell et al., 1982).

CONCLUSIONS

The investigations referred to in this chapter are intended to show that chemocommunication functions within a more general regulatory system. Specifically, we have attempted to draw associations between pheromonal communication in the Mongolian gerbil and homeostatic changes in thermoregulation and water conservation. Two systems of pheromone communication involving Harderian gland secretions and ventral scent gland marking show such relations.

Harderian secretion is sequenced by variations in body temperature and is associated with the use of hair lipids for insulation in hypothermic environments and with color changes of the pelage that modify the amount of radiant energy absorbed. Harderian lipids on the hair also affect how much body water is lost under desiccating environments. Excess body lipids coupled with low relative humidity exaggerate the loss of body water.

The ventral scent gland may be a "heat window" to the environment. Through scent marking the animal may be able to adjust its body temperature, either by transmitting body heat to the environment or by absorbing environmental heat. We have demonstrated that the scent pad conducts substantial heat during each ventral mark. The lactating female has an enlarged gland pad and an elevated gland pad temperature. Our experiments indicate that heat transfer through the gland may facilitate the rate of pup development. The effect is similar to that of a ventral brood patch in birds.

The demonstrations are not definitive, but they do open the way for investigations that may implicate "scent glands" in a variety of functions with conservative value for the individual and/or for the species. Communication may exemplify but one such function. Indeed, natural selection may act primarily to effect physiological homeostasis and may only secondarily amalgamate signaling capabilities to these processes. An illustration of this last possibility is seen with respect to some organized skin glands, especially those found in primitive species, which conceivably have no signaling function at all, even though the glands are actively rubbed against substrates. Their function could be entirely physiological We suspect that marking behavior influences body heat and fluid content in more advanced species as well. The study of marking from this perspective stands to reveal how homeostatic reactions interact with information transfer.

Footnote

Experiments referred to in this chapter were funded by NIMH Grant MH 14076, awarded to D. D. Thiessen. We wish to thank Tim Barth, Dr. Melanie Kittrell, Dr. Barbara Gregg, Merri Pendergrass and Kendell Thiessen for their many contributions.

REFERENCES

Agren, G., & Meyerson, D. J., 1977, Influence of gonadal hormones and social housing conditions on agonistic, copulatory, and related sociosexual behavior in the Mongolian gerbil (Meriones unguiculatus). Behavl. Proc., 2: 265-282.
Beauchamp, G. K., 1974, The perineal scent gland and social dominance in the male guinea pig. Physiol. Behav., 13: 669-673.
Block, M. L., Volpe, L. C., & Hayes, M. J., 1980, Saliva as a chemical cue in the ontogeny of social behavior. Submitted to Science, 1980.
Campbell, H. W., 1969, The effects of temperature on the auditory sensitivity of lizards. Physiol. Zool., 42 (2): 183-210.
Cheal, M. L., 1979, Stimulis-elicited investigation in Apomorphine-treated gerbils. Behavl. & Neural Biol., 27: 157-174.
Daly, M., 1977, Some experimental tests of the functional significance of scent-marking by gerbils (Meriones unguiculatus). J Comp. Physiol. Psychol., 91 (5): 1082-1094.
Drickamer, L. C., Vandenbergh, J. G., & Colby, D. R., 1973, Predictors of dominance in the male golden hamster (Mesocricetus auratus). Anim. Behav., 21: 557-563.
Eisenberg, J. F., & Kleiman, D. G., 1972, Olfactory communication in mammals. In: Annual Review of Ecology and Systematics, R. F. Johnston, P. W. Frank, & C. D. Michener, eds., Annual Rev., Cal.
Epple, G., 1981, Effects of prepubertal castration on the development of the scent glands, scent marking and aggression in the saddle back tamarin (Saguinus fuscicollis, Callitrichidae, Primates). Horm. & Behav. 15 (1) : 54-67.
Goodrich, B. S., & Mykytowycz, R., 1972, Individual and sex differences in the chemical composition of pheromone-like substances from the skin glands of the rabbit, Oryctolagus cuniculus. J. Mammal., 53 (3): 540-548.
Gulotta, E. F., 1971, Mammalian species (Meriones unguiculatus). The Am. Soc. Mammal., 3: 1-5.
Heinrich, B., 1979, Bumblebee Economics, Harvard University Press, Cambridge.
Johnson, R. P., 1973, Scent marking in mammals. Anim. Behav., 21 (3): 521-536.
Kennedy, G. Y., 1970, Harderoporphyrin: A new porphyrin from the Harderian gland of the rat. Comp. Biochem. Physiol., 36: 21-36.
Kittrell, E. M. W., Gregg, B. R., & Thiessen, D. D., 1981, Brood patch function for the ventral scent gland of the female Mongolian gerbil, Meriones unguiculatus. Devl. Psychobiol., in press (1981).
McManus, J. J. & Mele, J. A., 1969, Temperature regulation in the Mongolian gerbil, Meriones unguiculatus. Bull., 14 (1-2): 21-22.

Mele, J. A., 1972, Temperature regulation and bioenergetics of the Mongolian gerbil, Meriones unguiculatus. Am. Midl. Nat., 87: 272-282.

Müller-Schwarze, D., 1971, Pheromones in black tailed deer (Odocoileus hemionus columbianus). Anim. Behav., 19: 141-152.

Paulus, K., & Reisch, A. M., 1980, The influence of temperature on the threshold values of primary tastes. Chem. Sen., 5 (1): 11-21.

Payne, A. P., 1977, Pheromonal effects of Harderian gland homogenates on aggressive behavior in the hamster. J. Endocr., 73: 191-192.

Payne, A. P., 1978, Attractant properties of the Harderian gland and its products on male golden hamsters of differing sexual experience. Proc. Soc. Endocr., 8-30/9-2.

Pettijohn, T. F., & Barkes, B. M., 1978, Surface choice and behavior in adult Mongolian gerbils. Psychol. Rec., 28: 299-303.

Ralls, K., 1971, Mammalian scent marking. Science, 171: 443-449.

Randall, W., Elbin, J., & Swenson, R., 1974, Biochemical changes involved in a lesion-induced behavior in the cat. J. Comp. Physiol. Psychol., 86: 747-750

Richards, S. A., 1973, Temperature Regulation, Springer-Verlag, New York.

Robinson, P. F., 1959, Metabolism of the gerbil, Meriones unguiculatus. Science, 130: 502-503.

Schultze-Westrum, T. G., 1969, Social communication by chemical signals. In: Olfaction and Taste, C. Pfaffmann, ed., Rockefeller Univ. Press, New York.

Steffen, J. M., & Roberts, J. C., 1977, Temperature acclimation in the Mongolian gerbil (Meriones unguiculatus): Biochemical and organ weight changes. Comp. Biochem. Physiol., 58B: 237-242.

Steiner, A. L., 1974, Body rubbing, marking and other scent-related behavior in some ground squirrels (Sciuridae), a descriptive study. J. Canadian Zool., 52: 889-906.

Stoddart, D. M., 1980, Aspects of the evolutionary biology of mammalian olfaction. In: Olfaction in Mammals, D. M. Stoddart, ed., Academic Press, London.

Swanson, H. H., 1980, Social and hormonal influences on scent marking in the Mongolian gerbil. Physiol. Behav., 24: 839-842.

Thiessen, D. D., 1977, Thermoenergetics and the evolution of pheromone communication. In: Progress in Psychobiology and Physiological Psychology, J. M. Sprague, A. N. Epstein, eds., Academic Press, New York.

Thiessen, D. D., 1982, The thermoenergetics of communication and social interactions among Mongolian gerbils, Meriones unguiculatus. In Symbiosis, H. Moltz & L. Rosenblum eds., Plenum, NY

Thiessen, D. D., Graham, M., Perkins, J., & Marcks, S., 1977, Temperature regulation and social grooming in the Mongolian gerbil (*Meriones unguiculatus*). *Behav. Biol.*, 19: 279-288.

Thiessen, D. D., & Kittrell, M. W., 1980, The Harderian gland and thermoregulation in the gerbil (*Meriones unguiculatus*). *Physiol. Behav.*, 24: 417-424.

Thiessen, D. D., Pendergrass, M., & Harriman, A. E., 1981, The thermoenergetics of coat color maintenance by the Mongolian gerbil, *Meriones unguiculatus*. *Ther. Biol.*, in press.

Thiessen, D., & Rice, M., 1976, Mammalian scent gland marking and social behavior. *Psychol. Bull.*, 83 (4): 505-539.

Thiessen, D. D., & Yahr, P., 1977, *The Gerbil in Behavioral* Investigations. University of Texas Press, Austin.

Turner, J. W., Jr., 1975, Influence of neonatal androgen on the display of territorial marking behavior in the gerbil. *Physiol. Behav.*, 15 (3): 265-271.

Wallace, P., Owen, K., & Thiessen, D. D., 1973, The control and function of maternal scent marking in the Mongolian gerbil. *Physiol. Behav.*, 10: 463-466.

Yahr, P., 1976, The role of aromatization in androgen stimulation of gerbil scent marking. *Horm. & Behav.*, 7: 259-265.

THE EVOLUTION OF ALARM PHEROMONES

Paul J. Weldon

Department of Zoology
University of Tennessee
Knoxville, TN

It is generally agreed that most signals evolved from aspects of the signal emitter not originally used for signalling. Discussants of ritualization, for example, often state that visual displays have been built up from or enhanced by locomotory, preening or other movements previously devoid of iconic value. Some authors portray receivers as playing a passive role in signal installment. The assertion by Dawkins and Krebs (1978) that "the actor is selected to manipulate the behaviour of the reactor" epitomizes such a portrayal because it is implied that receivers already have perceptual-behavioral tendencies which the emitters exploit.

An alternative way in which a signal may come into being is through evolutionary adjustments on the part of the receiver, wherein perceptual and behavioral responses are made to pre-existing emissions from an emitter. This process is uniquely illustrated in parsimonious scenarios for the establishment of alarm chemical signals.

"Alarm pheromone" and "alarm substance" are terms generally used to denote chemicals from one organism which elicit anti-predator responses in conspecifics. The citation of defense against predators furnishes a touchstone against which to evaluate the function of behaviors elicited by these chemicals. Further, the events associated with predator attack mark the release of _the_ chemicals to which anti-predator responses by receivers would be appropriate. Attention is given here to two widespread mechanisms of alarm chemical release: accompanying defensive discharges and by damage of tissue.

Ghent (cited in Blum and Brand, 1972) established the double-edged defensive role of citronellal as a predator repellent and alarm pheromone for the ant _Acanthomyops claviger_. This aldehyde is aversive to arthropod offenders, and it elicits aggressive alarm in which members of a colony rally to the defense of conspecifics. Ghent hypothesized that the alarm properties of citronellal were acquired secondarily to those of predator repellency.

Blum and Brand (1972) adduce comparative evidence furthering the plausibility of Ghent's hypothesis, suggesting a mechanism of broad scope. Their examination of predator repellent compounds of non-social insects reveals that many of the chemicals involved are identical to those used by social insects for alarm. They conclude that "the presence of compounds identical to hymenopterous alarm pheromones in arthropod taxa (e.g., opilionids) which are considerably more ancient than these social insects, clearly demonstrates that the ability to synthesize these compounds was present in arthropod stock before these insects arose... Once solitary arthropods possessed the ability to synthesize defensive compounds which were both fairly volatile and stimulatory olfactants, it only remained for the more recently derived social insects to secondarily utilize these olfactants for communicative functions." Volatility may fulfill one of the requirements for chemicals subserving pheromonal roles. Also important in this case is that these chemicals are discharged upon predator attack, thereby providing cues if its imminence to which conspecifics have auspiciously become attuned. Perhaps some chemicals, not themselves repellent but discharged along with those that are, also act as alarm chemicals.

A widely occurring mechanism through which alarm chemicals are released involves the abrasion or tearing of the skin or exoskeleton of the emitter. This is known for anemones (Howe and Sheikh, 1975), marine and freshwater gastropods (Atema and Stenzler, 1977; Snyder, 1967), echinoderms (Snyder and Snyder, 1970), treehoppers (Nault et al., 1974), fishes, amphibians (Pfeiffer, 1974; Smith, 1977), birds (Jones and Black, 1979) and rodents (Mackay-Sim and Laing, 1981).

It is more than coincidental to the establishment of alarm chemical signalling that the damage of tissue, an act tightly and universally linked to acts of predation, is also the mechanism through which alarm chemicals are released. What renders the chemicals involved here ideal alarm signals is the occasion upon which these, or chemicals from which they have evolved, become available - predators at work. To paraphrase Snyder (1967), perhaps we are not dealing with a question of evolving an alarm substance but of evolving a response to metabolites of requisite properties that had or still have an entirely different function.

Smith (1977) proposes a rationale for the evolution through kin selection of abrasion released chemicals (evolved expressly perhaps) for alarm in fishes. He cites the costs inherent in the growth and maintenance of the specialized club cells containing the alarm substances <u>and</u> the unlikelihood of the sender ever benefitting from the release of such chemicals as evidence for altruism. Both he and Atema and Stenzler (1977) contrast fish and snail alarm systems, citing the fish system as potentially altruistic because fishes have skin cells specialized to contain alarm chemicals. The snails presumably do not, thus their system is not, as Atema and Stenzler put in quotes, "intentional." In support of a similar point, Smith refers to failed attempts to elicit fright reactions in fishes with extracts from skin regions without club cells. He states that fishes have progressed beyond the stage of responding to unspecialized metabolites, "if they ever went through it."

Kin selection acting upon alarm chemical systems may be manifest through special storage organs. But kin selection could also act to increase the production of chemicals without modes of sequestration, or it could foster other specializations to enhance alarm chemical release. Further, whilst kin selection may account for the storage of alarm chemicals in fishes, this hypothesis contributes little to an understanding of the inception of signalling. The biosynthesis of chemicals and the ability to respond to them - both highly complex processes - most likely appear in a stepwise fashion, no matter what level of selection acts later. Failure to elicit alarm reactions from skin areas unspecialized to house alarm chemicals may be due to inadequate chemical concentrations or to the evolutionary divergence of alarm chemicals from those from which they were derived. Interesting results may be forthcoming when alarm chemicals are identified and compared to metabolites from regions of the body not specialized to contain alarm chemicals, or compared to those from homologous tissues of related taxa without alarm chemical systems.

The idea that metabolites achieve roles as chemical releasing stimuli through chemosensory and behavioral adjustments enacted by receivers is not a new one. The pioneering chemical ecologist C. E. Lucas (1944), building upon inferences from physiologists, stated that "much of the history of evolution has concerned the development by living things of responses to metabolites, sometimes their own and sometimes produced by others." Kairomones, for example, a type of interspecific chemical releaser, may also become established through perceptual and behavioral ploys on the part of allospecific receivers (Weldon, 1980). Alarm pheromones, and perhaps other types of intraspecific releasers, originate through a process similar to that of kairomones. Any serious consideration of this analogy between the origins of intra- and interspecific chemical releasers should serve to caution against the leisurely acceptance of claims that (i) receptors are endowed for the reception of pheromones by enzymes of pheromone synthesis (Blum, 1981; pg. 497) or (ii) pheromone receptors appear as a result of the externalization of internal receptors for hormones and the like (Kittredge and Takahashi, 1972). If allospecific receivers can home in on emitters without head starts (i.e., pheromone synthetic enzymes or interoceptors), there is no reason to believe that conspecific receivers rely upon special preadaptations to do so.

REFERENCES

Atema, J., and Stenzler, D., 1977, Alarm substance of the marine mud snail, Nassarius obsoletus: Biological characterization and possible evolution, J. Chem. Ecol. 3:173-187.

Blum, M. S., 1981, "Chemical Defenses of Arthropods," Academic Press, New York.

Blum, M. S., and Brand, J. M., 1972, Social insect pheromones: Their chemistry and function, Am. Zool. 12:553-576.

Dawkins, R., and Krebs, J. R., 1978, Animal signals: Information or manipulation?, in: "Behavioural Ecology: An Evolutionary Approach," J. R. Krebs and N. B. Davies, eds., Sinauer Ass., Inc., Sunderland, Mass., 282-309.

Howe, N. R., and Sheikh, Y. M., 1975, Anthopleurine: A sea anemone alarm pheromone, Science 189:386-388.

Jones, R. B., and Black, A. J., 1979, Behavioral responses of the domestic chick to blood, Behav. and Neural Biol. 27:319-329.

Kittredge, T. S., and Takahashi, F. T., 1972, The evolution of sex pheromone communication in the Arthropoda, J. theor. Biol. 35: 467-471.

Lucas, C. E., 1944, Excretions, ecology and evolution. Nature 153: 378-379.

Mackay-Sim, A., and Laing, D. G., 1981, Rats' responses to blood and body odors of stressed and non-stressed conspecifics, Physiol. Behav. 27:503-510.

Nault, L. R., Wood, T. K., and Goff, A. M., 1974, Treehopper (Membracidae) alarm pheromones, Nature 249:387-388.

Pfeiffer, W., 1974, Pheromones in fish and amphibia, in: "Pheromones," M. C. Birch, ed., Frontiers in Biology, vol. 32, North-Holland Pub. Co., Amsterdam, 269-296.

Smith, R. J. F., 1977, Chemical communication as adaptation: Alarm substance of fish, in: "Chemical Signals in Vertebrates," D. Müller-Schwarze and M. M. Mozell, Plenum, New York, 303-320.

Snyder, N. F. R., 1967, An alarm reaction of aquatic gastropods to intraspecific extract, Cornell Univ. Agric. Exp. Station Memoir 403:1-122.

Snyder, N. F. R., and Snyder, H. A., 1970, Alarm response in Diadema antillarum, Science 168:276-278.

Weldon, P. J., 1980, In defense of "kairomone" as a class of chemical releasing stimuli, J. Chem. Ecol. 6:719-725.

ACKNOWLEDGEMENTS

M. S. Blum, G. M. Burghardt and R. J. F. Smith graciously offered comments on an earlier draft of this paper. This was written while the author was supported by an NSF grant (BNS-78-14196) to G. M. Burghardt and a Hilton Smith Fellowship from the University of Tennessee.

INVESTIGATIONS INTO THE ORIGIN(S) OF THE FRESHWATER ATTRACTANT(S) OF THE AMERICAN EEL

Peter W. Sorensen

Graduate School of Oceanography
University of Rhode Island
Narragansett, RI 02882

INTRODUCTION

Anguillid eels are thought to spawn in mid-ocean gyres and after spending a year or more as drifting larvae metamorphose into elvers (juveniles). Elvers possess a strong tendency to swim into fresh water. As this tendency is not shown by olfactory-ablated animals (Hain, 1975), or in charcoal-filtered, aged, or well water (Creutzberg, 1961; Miles, 1968) a chemical attractant is thought responsible. Because elvers do not consistently prefer the water from which they were collected the attractant is thought to be innately recognized (Miles, 1968). Teichmann (1957) showed eels to detect odorants at concentrations of 1:2.9 x 10^{-18} molar, giving them the best sense of smell among the fishes.

Although olfaction is important to the migratory behavior of many fishes, virtually nothing is known about the nature of the odorants involved. Walker and Hasler (1949) demonstrated that trained minnows could distinguish between rinses of different aquatic weeds and suggested plants to be important to the home stream "bouquet" recognized by returning adult salmon. Atema et al. (1973) suggested that low molecular weight acids and bases emanating from decaying organic matter might be a component of the home stream odor recognized by spawning alewives (*Alosa pseudoharengus*). Bodznick (1978) showed calcium to be a nonessential component of the odor of lake water recognized by migrating sockeye salmon fry.

Hain (1975) and the author have assayed the relative attractivity of washings of various flora, fauna, sediments, and organic matter to migrating elvers. Hain (1975) found several aquatic weeds to be mildly attractive and suggested an important role for leaf detritus. The author has confirmed these findings and has also found stream bed gravel and stones, dead alewife eggs, and river bank mud to be attractive. Both workers found crushed aquatic weeds to be repulsive. This paper discusses experiments performed to test the possibility that the attractivity of these items is attributable to a film of chemical compounds and/or microorganisms found on them.

METHODS

Water attractivity was assayed with a Y-maze constructed of glass and tygon tubing wrapped with black tape. Test waters were introduced through two darkened entrapment bottles located at the end of the Y. Groups of 25 elvers were collected just above the freshwater interface of a local stream and introduced into the base arm. Ten minutes later their positions were noted. The apparatus was drained, the positions of the arms reversed, and the procedure repeated with the same group of elvers. A replicate experiment was usually performed. Elvers found in the entrapment bottles were considered to have made a choice.

Control experiments have shown the apparatus to be unbiased and elvers to choose independently of one another. Results were analyzed according to two criteria: 1) whether a majority of elvers made a choice, and 2) whether there was a preference for a water type at $p=.05$. The latter was determined by the sign test ($p=q=.5$). Individual tests were compared using the chi square statistic. Previous work had shown deionized charcoal-filtered water to exert no influence, and it was used for the washings and as the blank against which the washings were tested. All washed materials were obtained from the stream from which the elvers were collected.

Gravel, 1-2 cm. in diameter, was collected from the stream bed, rinsed free of detritus with deionized water, and blotted dry. Three hundred milliliters (by displacement) of this gravel was then placed into 10 liters of deionized water for 15 minutes and the water assayed. This was repeated with gravel from the same location which had been soaked in chlorox for 4 hours and thoroughly rinsed.

In Rhode Island the alewife spawning run coincides with the elver run. Fifty grams of dead spawned alewife eggs were collected, rinsed free of detritus, blotted dry, and placed into 10 liters of deionized water for 2.5 hours before being assayed. This procedure was repeated with eggs freshly removed from sacrificed fish.

One hundred fifty grams of decaying leaf detritus (mostly from oak and beech trees) was collected from the stream bed, rinsed with deionized water, blotted dry, and placed into 10 liters of deionized water for 2 hours. This rinse was poured through cheesecloth and assayed against deionized water which also had been poured through cheesecloth. This procedure was repeated for the same weight of dead dry leaves collected from the nearby forest floor and living leaves picked from trees. To test whether the attractant(s) were entirely attributable to microbial decomposition a 150 gram sample of leaf detritus was boiled, autoclaved (230°F 20 lb./in^2), rinsed for eight hours, autoclaved, and rinsed again. This sample was then placed into 10 liters of deionized for 2.5 hours, strained through cheesecloth and assayed.

It was hypothesized that if microbes were responsible for the attractant, dry (unattractive) leaves should become attractive when placed into a nutrient medium innoculated with stream water. Twenty four grams of rinsed dry leaves were placed into 10 liters of artifical pond water (.5mM NaCl, .05mM KCl, .4mM $CaCl_2$, and .2mM $NaHCO_3$). Nitrogen (NH_4Cl) and phosphate ($FePO_4$) were added to create a molecular ratio of 50C:10N:1P and the resulting mixture was autoclaved. Three batches were made. When cool, two bottles were inoculated with 1.5 gr of fresh leaf detritus and 100 ml of fresh stream water. A control sample was inoculated with the same ingredients which had been autoclaved. All samples were aerated at room temperature through sterile tubing equipped with millipore filters. One experimental bottle was assayed the first day. After one week, half of each of the remaining samples was assayed against deionized water. After two weeks the remaining samples were assayed against each other.

RESULTS

The values represent the number of elvers which made that particular choice.

TABLE 1. STREAM GRAVEL

	Exper.	Blank	No Choice
Gravel	46*+	11	31 } *
Chlorox-treated Gravel	23	36	22 }

TABLE 2. ALEWIFE EGGS

	Exper.	Blank	No Choice
Fresh Alewife eggs	6*	9	75 } *
Dead Alewife eggs	36*	7	50 }

TABLE 3. LEAF ASSAYS

	Exper.	Blank	No choice
Dry Forest Leaves	5	7	32
Live Forest Leaves	23	14	57
Fresh Leaf Detritus	24*+	2	16
Autoclaved Rinsed Detritus	31*+	13	42

TABLE 4. CULTURE EXPERIMENT

	Exper.	Blank	No Choice
1 day: Exp. Leaf Mixture	2	22*	64
1 week: Exp. leaf Mixture	18*	4	64
Control leaf Mixture	17*	4	57
2 weeks:	Exper.	Control	No Choice
	51*+	25	50

KEY

* significant at p<.01 + Majority made a choice

DISCUSSION

Gravel washings were attractive because of an exterior film which was removed by treatment with a strong base. Decaying alewife eggs were attractive while fresh eggs were not, suggesting that microbial degradation might be the cause. These and other experiments have consistently found decaying leaf detritus to be strongly attractive. Because sterilized rinsed leaf detritus was attractive, and sterile dry leaves became so, the attractant is thought to occur in leaves. Sterile leaves became more attractive when cultured with fresh stream water than with autoclaved stream water, suggesting that microorganisms either accelerate the release of the attractant or synthesize it themselves. Evidence that only the outside coating of living aquatic weeds is attractive suggests that epiphytic microbes and/or a film of organic matter is responsible.

REFERENCES

Bodznick, D. 1978. Calcium ion: an odorant for natural water discrimination and the migratory behavior of sockeye salmon. J. Comp. Physiol. 127 157-166.

Creutzberg, F. 1961. On the orientation of migrating elvers (*Anguilla vulgaris* Turt.) in a tidal area. Netherlands J. Sea Res. 1(3) 257-338.

Hain, J.H.W. 1975. Migratory orientation in the American eel. Unpublished Ph.D. thesis. University of Rhode Island. Kingston, Rhode Island. 126p.

Miles, S.G. 1968. Rheotaxis of elvers of the American eel (*Anguilla rostrata*) in the laboratory to water from different streams in Nova Scotia. J. Fish Res. Bd. Can. 25(3) 1591-1602.

Walker, T.J. and A.D. Hasler. 1949. Detection and discrimination of odors of aquatic plants by the bluntnose minnow (*Hyborhynchus notatus*). Physiol. Zoo. 22 45-63.

This work was supported by N.M.F.S. Grant no. NA-81-FB-A-00301.

A PREGNANCY BLOCK RESULTING FROM MULTIPLE-MALE COPULATION OR EXPOSURE AT THE TIME OF MATING IN DEER MICE (PEROMYSCUS MANICULATUS)

Donald A. Dewsbury

Department of Psychology
University of Florida
Gainesville, FL 32611 U.S.A.

Deer mice, Peromyscus maniculatus, are known to engage in multiple-male copulations in the field. Litters conceived in the field are known to be of multiple paternity (Birdsall and Nash, 1973). In a subsidiary analysis of data from several experiments, Dewsbury and Baumgardner (1981) found evidence that females mating with more than one male may have a decreased probability of pregnancy compared to females receiving similar amounts of stimulation from one male. Three experiments were designed to verify and explore this phenomenon.

In the first experiment, 36 blonde, female deer mice mated in three tests in each of which they received three complete ejaculatory series. In one test they mated undisturbed with a blonde male. In a second test they received two ejaculations from a blonde male and one from a wild-type male. In an effort to control for disturbance and delay, females in the third condition were removed from a blonde male after two ejaculations and returned to the same male after 20 min to receive a final ejaculation. Females mating with one male, with or without delay, were significantly more likely to be pregnant than those mating with two males. There were no significant differences across groups with respect to the number of intromissions received, gestation period, or litter size.

In Experiment 2, 29 blonde females completed three tests in which they first copulated for three ejaculatory series with a blonde male. They then spent 2 hr in a divided cage, the other side of which a.) was empty, b.) contained a strange blonde male, or c.) contained a strange wild-type male. Significantly more females were pregnant after being placed opposite the empty cage than when across from either a blonde or wild-type male for 2 hr.

The latter two conditions did not differ.

In Experiment 3, 27 wild-type females copulated for three ejaculatory series with a wild-type male. They then were placed for a period of 2 hr alone in the cage of either the male with which they had just mated or a different wild-type male. Significantly more females became pregnant after exposure to the cage and bedding of the familiar than the strange male.

Together, these data indicate the existence of a blockage of pregnancy in female deer mice if they mate with, or are exposed to, more than one male around the time of copulation. A chemical cue is implicated. Although the effect is not large, it is stable across various treatments and variations. No effects on litter size were detected; the effect was all-or-none.

The phenomenon bears at least a superficial resemblance to the Bruce effect in that in both types of study there is a failure of pregnancy after exposure to a strange male. This phenomenon differs in that a) exposure occurs around the time of copulation, b) exposure to the first male is relatively brief, and c) exposure to the second male is brief. These parameters have been manipulated in studies of the Bruce effect, suggesting that the two effects may have the same basis. On the other hand, prairie voles, *Microtus ochrogaster*, appear to show a Bruce effect (Stehn and Richmond, 1975) and not a multi-male pregnancy block (Dewsbury and Baumgardner, 1981). Further work is needed to determine whether or not the same physiological processes are responsible for both types of pregnancy block.

There appears to be no support for the notion that heterospermic insemination increases fertility (see Beatty, 1970).

It is generally agreed that if the Bruce effect occurs in the field, it is likely that it reflects the takeover of a deme by a strange male. The necessity of such an interpretation may simply be a function of the design of experiments on the Bruce effect. As copulations are not observed, strange males must be introduced some time after copulation. The present phenomenon may reflect behavior known to occur in the field (i.e., one female mating with more than one male) (Birdsall and Nash, 1973). It is possible that females elect to forego reproduction for a brief period when several males mate--perhaps because this indicates a time of social instability. Analagous phenomena have been reported in Norway rats and chimpanzees (Calhoun, 1962; Tutin, 1979).

REFERENCES

Beatty, R. A., 1970, The genetics of the mammalian gamete. *Biol.*

Revs., 45:73.
Birdsall, D. A., and Nash, D., 1973, Occurrence of successful multiple insemination of females in natural populations of deer mice (Peromyscus maniculatus), Evolution, 27:106.
Calhoun, J. B., 1962, "The Ecology and Sociology of the Norway Rat," U.S. Dept. of Health, Education, and Welfare, Bethesda, MD.
Dewsbury, D. A., and Baumgardner, D. J., 1981, Studies of sperm competition in two species of muroid rodents, Behav. Ecol. Sociobiol., 9:121.
Stehn, R. A., and Richmond, M., 1975, Male-induced pregnancy termination on the prairie vole, Microtus ochrogaster, Science, 187:1211.
Tutin, C. E. G., 1979, Mating patterns and reproductive strategies in a community of wild chimpanzees (Pan troglodytes schweinfurthii), Behav. Ecol. Sociobiol., 6:29.

OLFACTORY COMMUNICATION IN KANGAROO RATS (*D. MERRIAMI*)

Jan A. Randall

Department of Biology
Central Missouri State University
Warrensburg, Missouri 64093

Kangaroo rats (family Heteromyidae) form a major component of mammalian fauna in the arid deserts of western North America. They have long been considered solitary rodents with a low threshold for the exhibition of agonistic behaviors (Eisenberg, 1963). Although much is known about their ecology, little is known about their social systems and communication. Four years of my research on olfactory communication in the Merriam's kangaroo rat, *Dipodomys merriami*, indicate that olfactory signals are a major means of communication in this species. Signals originate from at least two sources: sandbathing deposits and urine marks. These odors communicate species and sex, coordinate reproduction, and help to maintain social order.

SANDBATHING

Kangaroo rats sandbathe in a series of side rolls and ventral rubs followed by grooming and scratching behaviors. Eisenberg (1963) described the sandbathing activity of kangaroo rats and proposed that the behavior had two functions: (1) care of the body surface, and (2) olfactory communication. *D. merriami* sandbathe to remove excess oils from the fur and clean their pelage in a series of rubbing and grooming behaviors (Randall, 1981a). *D. merriami* also appear to leave olfactory signals at sandbathing loci. In the laboratory, *D. merriami* sandbathed at high frequencies and preferred the sandbathing loci of conspecifics to those of another *Dipodomys* species. Males preferred the sandbathing loci of both sexes, and females preferred only females (Randall, 1981b). Laine and Griswold (1976) found in bannertail kangaroo rats (*D. spectabilis*) that the sandbathing loci of females attracted males, and both sexes sandbathed at high frequencies at male loci. Kangaroo rats, therefore, appear to communicate species and sex by olfactory signals left at sandbathing sites.

The source of the olfactory signal at sandbathing areas is probably from sebaceous glands associated with hair follicles, and from a specialized dorsal gland. Male D. merriami exhibit larger dorsal glands than females, and the period of greatest gland size and secretion coincides with reproductive intervals. Function of the gland depends on androgens, as the glands of newly castrated males rapidly decreased in size. Gland sizes of the castrated animals were significantly smaller than glands of intact males in only one week after castration (Lepri, 1981). Furthermore, testosterone implants significantly increased the size of the dorsal gland in castrated males in five weeks after implantation (Randall, unpublished data).

Because kangaroo rats are nocturnal and solitary (males and females occupy separate burrows), olfactory signals seem a likely means of coordinating reproduction. Accordingly, castrated and intact male D. merriami were tested for their responses to sandbathing loci of four scent donors: intact males, castrated males, estrous females, and anestrous females. All female donors had been ovariectomized and had received replacement estrogen by either an oil injection or silastic implants. Testing procedures were similar to those reported in Randall (1981b). Results showed that castrated and intact males responded similarly to sandbathing loci of all four classes of scent donor. All sandbathing loci were equally attractive (Lepri, 1981). (This experiment was replicated and yielded the same results (Lepri and Randall, in prep.).) Reproductive status, therefore, is not communicated to males at sandbathing loci. Female responses to sandbathing loci of animals of different hormonal condition remain to be tested, and the function of the dorsal gland is unresolved.

In the field, D. merriami scent mark throughout the night by periodically rubbing their bodies in numerous pits scattered throughout their home ranges. Sandbathing with grooming components occurs early in the evening and after release from a trap (Randall, unpub. obs.). This species does not appear to defend a territory. Instead, animals occupy home ranges with male areas overlapping those of females. Overlap increases during times of reproduction (O'Farrell, 1980), and animals encounter each other frequently in these areas of high density (Randall, unpubl. obs.). I have mapped sandbathing loci in the field by using color-coded flags. Data are preliminary, but males seem to sandbathe at higher frequencies than females in these areas of overlap. Kangaroo rats also inspect the sandbathing sites of other animals and periodically sandbathe on top of a mark or in proximity to it. Both males and females sandbathe in the vicinity of their burrows. Besides probable communication of species and sex identity, individual identity might be communicated at sandbathing sites. Field observations seem to support results from the laboratory that sandbathing is an important means of olfactory communication in kangaroo rats.

URINE MARKS

Urine plays a major role in olfactory signaling and in synchronizing reproduction in many rodent species (Aron, 1979). *D. merriami* are desert-adapted rodents that greatly limit urine production when no water source is available (Schmidt-Nielsen, 1964). In years when no rain fell in the desert, and no winter annuals germinated, *D. merriami* failed to reproduce (Beatley, 1969). When they consumed green vegetation, however, reproduction occurred, thus, reproduction is linked to the consumption of green vegetation in this species. As kangaroo rats begin to consume green vegetation, urine production probably increases. Because animals could quickly communicate in their urine physiological changes associated with reproduction, a series of experiments was begun to ascertain whether urine communicates changes in reproductive status in *D. merriami*.

Estrous and anestrous females and intact and castrated males were tested (Randall, in prep.) for their preferences to pairings of various combinations of water and the urine from intact males, estrous females, castrated males, and anestrous (ovariectomized) females. Both intact males and estrous females exhibited a preference for urine from intact males. Estrous females showed a highly significant preference for intact male urine when it was paired with urine from castrated males. Intact males showed no attraction to the urine from estrous females, and castrated animals exhibited no preference for any of the stimulus choices.

Johnston (1979) suggested that females in solitary, well-dispersed species with relatively short periods of receptivity might tend to advertise their upcoming reproductive status early in order to ensure the attraction of a mate. The information that signals the female's reproductive state could be in proestrous, instead of estrous, urine. Accordingly intact and castrated males were tested for their responses to urine that was collected from females at different stages of the reproductive cycle: proestrus, estrus, and anestrus. Urine was collected from naturally cycling females, and the stage of reproduction was verified by vaginal smears. Even though intact males tended to spend more time at a sample of proestrous urine, no significant effects were found. Estrous urine once again appeared totally unattractive to intact males (Randall, unpubl. data). Female urine, therefore, does not appear to signal reproductive status to male *D. merriami*. The urine from intact males seems to contain information for coordinating reproduction in this species.

FEMALE CHOICE-MALE COMPETITION

Field observations, coupled with laboratory findings, suggest that reproduction in D. merriami conforms to a female choice-male competition model. In the field, D. merriami males enter the home range of an estrous female and attempt to mate with her (Randall, unpubl. obs.). Males have been observed fighting with each other during this time, presumably in competition for the female. In all observed matings (N=4), the male followed the female and attempted to mount when she stopped and remained still. In almost every case, the female instigated termination of the mount by moving away. Estrous females may seek out intact males in response to encountering their urine in areas of home range overlap. Because males compete for estrous females, the urine of intact males may communicate the presence of a potential competitor to an investigating male in reproductive condition. Results from a pilot study indicate that male D. merriami are attracted to the vaginal odor of estrous females. A male may be unaware of the exact reproductive status of a female until he actually smells her vaginal area.

Kangaroo rats afford a unique opportunity to obtain an integrated picture of how chemical signals function within the ecological context of natural communities of small mammals. Sandbathing deposits and urine marks seem to convey a variety of messages that probably function to minimize aggression and to maintain social order in D. merriami, especially during reproduction.

REFERENCES

Aron, C., 1979, Mechanisms of control of the reproductive function by olfactory stimuli in female mammals, Physiol. Rev., 59:229.
Beatley, J. C., 1969, Dependence of desert rodents on winter annuals and precipitation, Ecology,50:721.
Eisenberg, J. F., 1963, The behavior of heteromyid rodents, Univ. California Publ. Zool., 69:1.
Johnston, R. E., 1979, Olfactory preferences, scent marking, and "proceptivity" in female hamsters, Horm. Behav., 13:21.
Laine, J., and Griswold, J. G., 1976, Sandbathing in kangaroo rats (Dipodomys spectabilis), J. Mammal., 57:408.
Lepri, J. J., 1981, Hormonal regulation of sandbathing in male kangaroo rats (Dipodomys merriami)., M.S. thesis, Warrensburg.
O'Farrell, M. J., 1980, Spatial relationships of rodents in a sagebrush community, J. Mammal., 61:589.
Randall, J. A., 1981a, Comparison of sandbathing and grooming in two species of kangaroo rat, Anim. Behav., 29:1213.
Randall, J. A., 1981b, Olfactory communication at sandbathing loci by sympatric species of kangaroo rat, J. Mammal., 62:12.
Schmidt-Nielsen, K., 1964, "Desert Animals," Oxford Univ. Press.

ODOR PREFERENCES OF YOUNG RATS: PRODUCTION OF AN ATTRACTIVE ODOR BY MALES

Richard E. Brown

Department of Psychology
Dalhousie University
Halifax, Nova Scotia, Canada B3H 4J1

INTRODUCTION

Infant rats prefer the odors from lactating females to those from males, non-lactating females or non-odorized stimuli (Leon, 1974; Brown, 1978). There are strain differences and dietary factors in these odor preferences. Long-Evans hooded rat pups show a greater attraction to the odors of non-lactating females than do albino pups (Galef and Muskus, 1979), and pups prefer the odors of lactating females eating the same diet as their own dam to the odors of females on another diet (Leon, 1975; Galef, 1981).

Nulliparous females placed with newborn pups show an increase in prolactin levels and maternal behavior and produce a maternal odor (MO) but males do not produce a "maternal odor" even though they show "maternal behavior" (Leidahl and Moltz, 1975). Since males injected with bile from lactating females produce an odor which is as attractive as the maternal odor (Moltz and Leidahl, 1977); it was hypothesized that males which ingest feces from lactating females might also produce an odor which is attractive to infant rats.

EXPERIMENT 1

Adult male Long-Evans hooded rats were fed 40 g of food per day. This consisted of crushed Purina lab chow (No. 5001) mixed with water and feces from lactating female rats. Two males each received 0, 4, 8, or 12 grams of feces mixed into their food (0%, 10%, 20%, 30% of total). Feces were taken from females between 12 and 19 days post partum and males received the feces diet for 7 days. On day 8, 30 rat pups (16 female and 14 male),

18 to 20 days of age from 4 different litters reared with their mother alone were given three preference tests between pairs of male odors. These odors consisted of feces from males fed 0 (MO), 4 (M4), 8 (M8) and 12 grams (M12) of feces from lactating females.

Tests were conducted in a plastic arena measuring 26.5 x 15 cm and 12.5 cm high. At each end of the arena was a circular dish (7.5 cm in diameter x 1.5 cm high) filled with fresh pine shavings (as used in the home cages). In each dish were placed 4 to 6 grams of male feces. Each odor preference test was 2 minutes in duration and the time spent on each half of the arena was recorded using two stopwatches (forced choice paradigm).

The results of these tests are shown in Figure 1. Preferences of male and female pups did not differ and their results are pooled on each test. Odors of feces from M4 males were not preferred to those of MO males (correlated samples t-test; $t = 0.639$); nor were odors of M8 males preferred to those of MO males ($t = 0.631$); but the odors from M12 males were preferred to those of MO males ($t = 3.032$, $df = 29$, $p < .01$). Thus, male rats which consume a diet consisting of 30% feces from lactating females for 1 week produce an odor which is more attractive to rat pups than the odor of male rats consuming only rat food.

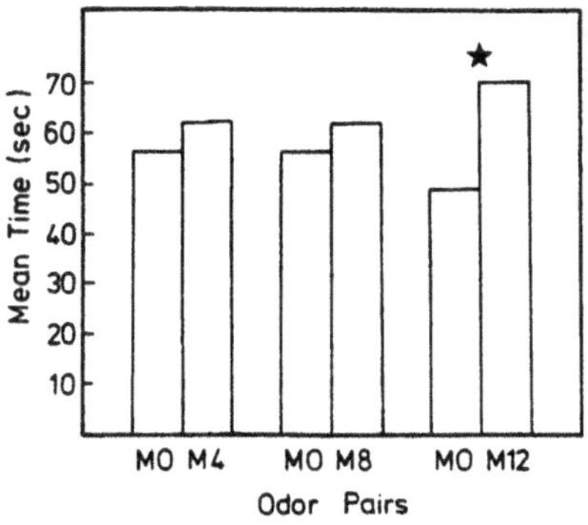

Fig. 1. Mean time spent on the half of the arena having NO, M4 M8 or M12 odors. Significant differences are indicated by a star.

EXPERIMENT II

In order to compare the attractiveness of male odors with the odors of lactating females, rat pups were tested for preferences among all pairs of 4 odors: feces from their own mother (OM); feces from adult males (MO); feces from adult males fed 12 grams of female feces per day (30% of diet) for 1 week (M12) and no feces odor (NO) which consisted only of wood shavings.

Eighteen male and 12 female pups from 3 litters reared with their mother alone were tested between 18 and 20 days of age. The apparatus and procedure were the same as those used in Experiment I except that each rat pup was used in six odor preference tests.

The results of these tests are shown in Figure 2. OM odors were preferred to NO odor ($t = 5.528$, $df = 29$, $p < .001$); M12 odors were preferred to NO odor ($t = 7.106$, $df = 29$, $p < .001$); and OM odors were preferred to MO odors ($t = 2.666$, $df = 29$, $p < .05$). MO odors were not preferred to NO odor ($t = 0.392$); OM were not preferred to M12 ($t = 0.335$) and M12 odors were not preferred to MO ($t = 1.519$). This last result was surprising in light of the results of Experiment I and was examined more carefully. Males preferred M12 odors to MO ($t = 2.910$, $df = 17$, $p < .01$) but females did not ($t = 0.753$, $df = 11$). Over all, of 30 subjects 20 preferred M12 odors, 9 preferred MO and 1 had no preference. A sign test

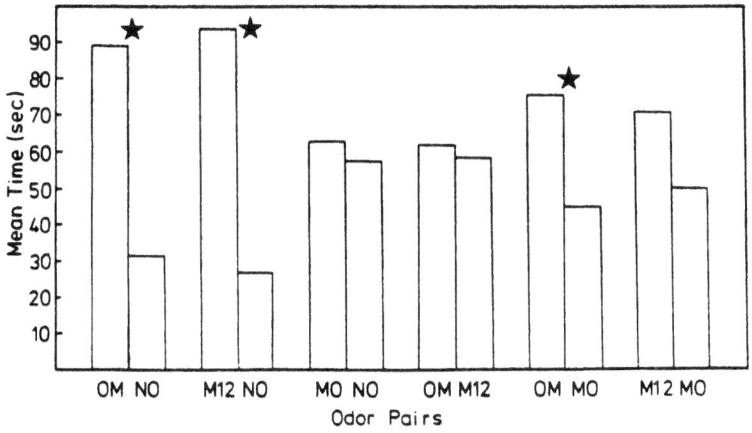

Fig. 2. Mean time spent on the half of the arena having NO, OM, M12 or MO odors. Significant differences are indicated by a star.

indicates preference for the M12 odor ($Z = 1.857$, $N = 29$, $p = 0.032$). Based on these results, I conclude that the M12 odors are as attractive as OM odors and are preferred to both MO odor and NO odor. Thus the preference scale is: OM = M12 > MO = NO.

DISCUSSION

The results of these experiments indicate that males which ingest the feces of lactating females at the time that these females are producing maternal odors (14 - 20 days post partum) produce an odor which is as attractive to infant rats as that of a lactating female. This result is similar to that obtained by injecting bile from lactating females into males. Whether the male manufactures a specific attractant odor or merely passes on the substances from the female feces is unknown. Consuming feces of lactating females may alter the pH or the microflora of the male's gut (Lee and Moltz, 1980; Moltz and Lee, 1981) or act merely as a laxative, causing a higher rate of emission of male caecotrophe (Leon, 1978).

REFERENCES

Brown, R.E. 1978. Rearing environment and the generalization of odor preferences in infant rats. Paper presented at the third ECRO Congress, Pavia, Italy.

Galef, B.G., Jr. 1981. Preference for natural odors in rat pups: Implications of a failure to replicate. Physiol. Behav., 26: 783-786.

Galef, B.G., Jr., and Muskus, P.A. 1979. Olfactory mediation of mother-young contact in Long-Evans rats. J. Comp. Physiol. Psychol., 93:708-716.

Lee, T.M., and Moltz, H. 1980. How rat young govern the release of a maternal pheromone. Physiol. Behav., 29:983-989.

Leidahl, L.C., and Moltz, H. Emission of the maternal pheromone in the nulliparous female and failure of emission in the adult male. Physiol. Behav., 14:421-424.

Leon, M. 1974. Maternal pheromone. Physiol. Behav., 13:441-453.

Leon, M. 1975. Dietary control of maternal pheromone in the lactating rat. Physiol. Behav., 14:311-319.

Leon, M. 1978. Emission of maternal pheromone. Science, 201: 938-939.

Moltz, H., and Lee, T.M. 1981. The maternal pheromone of the rat: Identity and functional significance. Physiol. Behav., 25: 301-306.

Moltz, H., and Leidahl, L.C. 1977. Bile, prolactin, and the maternal pheromone. Science, 196:81-83.

RATE AND LOCATION OF SCENT MARKING BY PIKAS DURING THE BREEDING SEASON

Carron Meaney

Department of EPO Biology
University of Colorado
Boulder, Colorado

INTRODUCTION

This study presents evidence that scent marking may play a role in breeding by pikas (*Ochotona princeps*). Pikas are small lagomorphs inhabiting talus slopes in montane regions of western North America. Individuals maintain territories which overlap to a variable extent. The breeding season extends from April through June and varies with specific location (Smith, 1980). Pikas have apocrine sebaceous cheek glands (Harvey and Rosenberg, 1966) and can discriminate individual cheek gland secretions (Meaney, 1982).

METHODS

Field observations were made on Niwot Ridge (35 km west of Boulder, Colorado) from June through August in 1979 and 1980 and from mid-April to mid-June in 1981 on a one-hectare site with an average resident population of 14 animals. Pin flags were placed at 5-meter intervals to form a grid for noting animal locations. Locations of animals (location points) were recorded every five minutes, and locations of cheek marks (cheek-mark points) were recorded when they occurred from June through August. Convex polygons enclosing 80% of the location points (closest to the geometric center defined by all of the individual's location points) were generated for each animal (Michener, 1979). Cheek-mark points were superimposed on the polygons. The rate of cheek marking during 15-minute observation periods was recorded from mid-April through August. These data were separated into eight two-week periods (April 16-31, May 1-15, June 1-15, 16-30, July 1-15, 16-31, August 1-15, 16-31, for the eight periods, respectively).

Table 1. Size of 80% polygon (m^2), total overlap area (m^2), and percent overlap for June-August (±SEM). Kruskal-Wallis analyses of variance were used in monthly comparisons.

	June	July	August
No. of pikas (locations)	10 (392)	23 (1128)	23 (1095)
Mean Area of 80% polygons	681 ± 106.9	625 ± 162.5	317 ± 37.0*
Mean Overlap Area	436 ± 89.6	240 ± 45.0	99 ± 20.2***
Mean Percent Overlap	68%	50%	29% **

* p<.05; ** p<.01; *** p<.005

SEASONAL VARIATIONS

The latter part of the breeding season (June), was characterized by a larger mean area encompassed within the 80% polygons, a greater overlap with other individuals' polygons, and a greater percent overlap when compared to July and August. These three measures decreased significantly through the summer (Table 1). There also was significant variation in rates of adult cheek marking per 15 minutes with a peak from early May (period 2) through mid-June (period 3) ($\bar{x}$±SEM: Period 1 = 1.1±.20, period 2 = 2.2±.51, period 3 = 2.3±.33, period 4 = 1.0±.22, period 5 = 1.7±.31, period 6 = 1.3±.20, period 7 = 0.9±.12, period 8 = 1.7±.27, and $F = 3.29$, $df = 7$, $p<.005$) (Figure 1). Adult males marked at a significantly higher rate ($\bar{x}$ = 1.8±.14) than females ($\bar{x}$ = 1.1±.11)($F = 17.24$, $df = 1$, $p<.005$). Although these differences are small they are statistically significant.

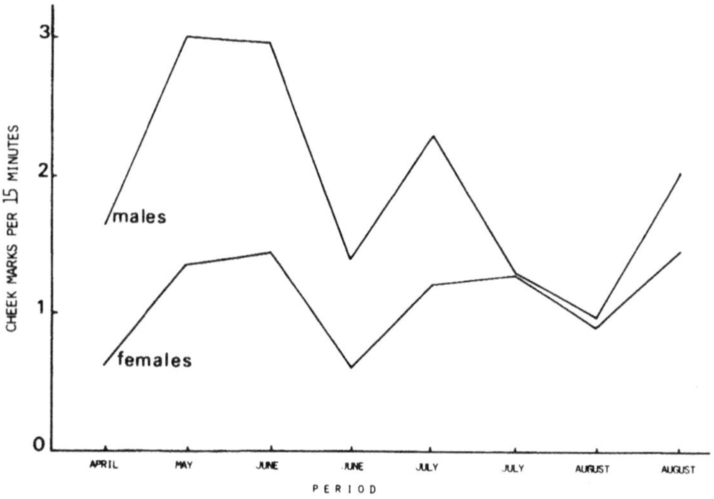

Figure 1. Rate of adult cheek marking from mid-April through August. No. of observation periods = 559. No. of animals = 18.

Table 2. Number of adult check marks located in male and female 80% polygons, compared to an expected distribution generated from the observed sex ratio. n is the number of animals. G-test.

Sex	Month	n	Marks in Male Polygons	(Expected)	Marks in Female Polygons	(Expected)	Significance
Male	June	4	0	(26)	52	(26)	***
	July	5	2	(20)	44	(26)	***
	August	4	8	(20)	45	(33)	***
Female	June	4	23	(12)	1	(12)	***
	July	7	47	(27)	16	(36)	***
	August	4	18	(11)	13	(20)	*

* p<.05
*** p<.005

MARKS IN NEIGHBORING POLYGONS

Do pikas concentrate their marks in neighbors' polygons? The proportion of all location points that fell in the area of overlap was used to generate expected frequencies of cheek mark points in the overlap area. A comparison of observed to expected frequencies revealed that a greater number of marks were placed in the area of overlap than expected (30/37 cases or 81%). Of those cases (30) in which observed marking frequencies were greater than expected, 67% (20) were statistically significant.

The occurrence of cheek marks in polygons of same- and opposite-sexed individuals was compared to an expected frequency based on the sex ratio which was obtained from a count of all animals whose 80% polygons overlapped. Any mark that fell into both a male and female polygon was not counted. Adult males and females as a group (Lewis and Burke, 1949) placed significantly more marks in polygons of opposite-sexed individuals than expected (Table 2). This effect is most pronounced in June. The concentration of scent marks in polygons of opposite-sexed individuals may be due to the spatial distribution of pikas (Smith, 1980; Svendsen, 1979). Males (n=6) had a significantly greater area of overlap with females ($\bar{x}$=303.2m^2±127.51) than with other males ($\bar{x}$=20.5m^2±10.53)(Wilcoxon matched-pairs signed-rank test, Z=-1.99, p<.05). Females (n=8) overlapped more with males ($\bar{x}$=221.2m^2±93.11) than with other females ($\bar{x}$=92.2m^2±52.95) but this difference was not significant (Z=-0.98, p<.05).

Pikas do not place a greater number of marks in polygons of opposite-sexed individuals than would be expected from the proportion of location points. The number of marks deposited in male or female polygons by each sex was compared to an expected number generated from the proportion of location points in polygons of each sex. The values for the months were pooled due to low expected frequencies. No significant difference was found between the pooled values ($G=.39$, $n=13$, $p>.05$ and $G=2.02$, $n=15$, $p>.05$ for males and females, respectively). As a group they marked in polygons of same- and opposite-sexed individuals in accordance with their own occurrence there.

CONCLUSION

These data show that the rate of marking on Niwot Ridge is elevated for both sexes during the period approximating the middle of the breeding season. Although a greater number of marks are deposited in polygons of opposite- than in those of same-sexed individuals, the picture is confounded by the fact that pikas also overlap more with polygons of opposite-sexed individuals. The location of scent marks during the breeding season is probably due to space utilization patterns rather than to selective placement of scent marks in polygons of opposite-sexed individuals. Nonetheless, there is a significant increase in the area of overlap during the breeding season, and the net effect is to bring pikas into greater contact with the odors of potential mates.

Pikas can discriminate individual odors (Meaney, 1982) and they are highly aggressive. Reduction of aggression by olfactory familiarity, as Daly (1977) has proposed for gerbils, is a possible mechanism by which scent marking could facilitate breeding in pikas. Information on variation in rates of marking in different parts of the polygon could be valuable in further testing the hypothesis of a breeding-facilitation function of marking in pikas.

REFERENCES

Daly, M. 1977. Some experimental tests of the functional significance of scent marking by gerbils (Meriones unguiculatus). J. Comp. Physiol. Psych., 91:1082-1094.

Harvey, E.B. and L.E. Rosenberg. 1960. An apocrine gland complex of the pika. J. Mamm., 41:213-219.

Lewis, D. and D.J. Burke. 1949. The use and misuse of the Chi-square test. Psych. Bull., 46:433-489.

Meaney, C. 1982. Discrimination of individual odors in pikas (Ochotona princeps). In preparation.

Michener, G. 1979. Spatial relationships and social organization of adult Richardson's ground squirrels. Can. J. Zool., 57:125-139.

Smith, A.T. 1980. Territoriality and social behavior of Ochotona princeps. Proc. of the World Lagomorph Conference (K. Myers, ed.).

Svendsen, G.E. 1979. Territoriality and behavior in a population of pikas (Ochotona princeps). J. Mamm., 60:324-330.

EFFECTS OF URINE ON THE RESPONSE TO CARROT-BAIT IN THE EUROPEAN WILD RABBIT

Diana Bell, Stephen Moore, David Cowan*

School of Biological Sciences, University of East Anglia
Norwich, Norfolk, NR4 7TJ, England
*M.A.F.F., Worplesdon Laboratory, Nr. Guildford, Surrey

INTRODUCTION

Despite success with insects, the applied use of chemical signals in the management of mammalian pest species remains to be fully explored. Shumake (1977) suggests that such chemical signals might be used (1) to disrupt an animal's reproductive processes (through odour 'primer' effects, see Aron,1979); (2) to attract it or repel it from an area (lure/repellent); or (3) to encourage or discourage it from feeding (bait enhancer/food repellent). Müller-Schwarze (1971) for example found that black-tailed deer could be discouraged from feeding by applying conspecific metatarsal secretions to the undersurface of their feeding bowls.

Poisoned carrot-bait is used as a technique for controlling populations of European wild rabbit *Oryctolagus cuniculus* (L). in Australia and New Zealand. In the light of those sex-specific attractant and repellent qualities accorded to rabbit urine by Bell (1980; in press (a) and (b)) the present study set out to determine the effects of the presence of conspecific urine on the effectiveness of carrot-baiting in this species. The first experiment monitored the long-term response to urine-treated and untreated carrots, as percentage consumed (% take), and the second monitored the behavioural response to these baits by direct observation.

Experiment 1

Method

This study was conducted in an area of the University campus known to support a colony of free-living wild rabbits. Three piles of sliced carrots were provided at set sites 10 m apart, at 1700 h on 20 consecutive nights. Each pile consisted of 400g of carrot on Day 1 and 800g on subsequent days. Two of the piles were sprayed with 50 ml. of urine pooled from either adult male or female conspecifics and third left untreated (control). The urine was collected daily from 4 laboratory-caged, sexually mature, Dutch-strain rabbits of each sex.

Any remaining carrots were weighed at 0900 h each morning to provide daily measures of % take for each carrot treatment. The location of each carrot pile was shifted 5 m on Day 11. The siting of the untreated control was swapped with that of the female-urine treatment of Day 19 and again with the male-urine treatment on Day 20. Rainfall and daily min./max. temperatures were recorded throughout the 20 day period.

Results

The % take of the male urine-treated carrots over 20 nights was significantly lower than that of either the female urine-treated carrots or the untreated controls ($p<0.002$ Mann-Whitney U-test) with means (± SD) of 45.11 ± 24.60; 78.69 ± 16.67 and 85.22 ± 13.56 (control) respectively (Fig. 1). The difference between the % takes for female urine-treated and control piles was not statistically significant.

These differences were apparent during the 10 day periods both before and after re-siting of carrot piles on Day 11, suggesting that were not site-specific effects. The % take of untreated control, male- and female-urine treatments each showed a significant positive correlation with minimum night temperature (r_s values = .40 $p<0.05$; .71 $p<0.01$ and .54 $p<0.01$ respectively - Spearman Rank test). It is interesting to note that the rainfall on Day 6 was accompanied by an increase in % take of the male urine-treated carrots (90%) together with a decrease of the amount of untreated controls consumed (75%).

Experiment 2

Method

The second investigation used a colony of 15-19 individually tagged adult wild rabbits living in a 270 m^2 grassland enclosure at Gt. Witchingham Wildlife Park, Norfolk. Nine 200g piles of sliced carrots were sequentially set out as 3 replicates of 3 treatments on each of 5 separate days, i.e. two piles were sprayed with 10 ml. of either male or female urine and the third pile (control) left untreated. Piles were set 3 m. apart avoiding sites of focal

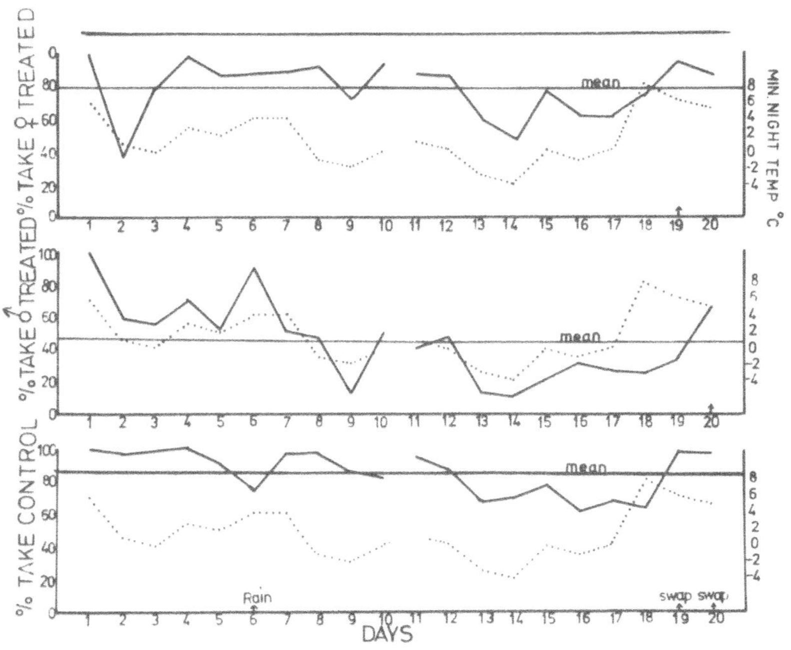

Fig. 1. Per cent take urine-treated and control carrots.

..... minimum night temperature
DAY 11 - each pile shifted 5m.
DAY 19 SWAP - ♀ treated with control.
DAY 20 SWAP - ♂ treated with control.

activity e.g. latrine areas. Behavioural responses to treated/untreated piles were recorded until all the carrots had been consumed, or the rabbits ceased feeding.

Results

Male and female rabbits made a total of 57 and 74 visits to carrot piles, i.e. an average of 7.1 and 10.6 visits/individual for males and females respectively. The overall number of visits made to carrot piles by males and females varied significantly between treatments (χ^2 = 12.08, df = 2, p<0.01 - Table 1), with control piles receiving the most visits (N = 61) and male urine-treated piles the least number of visits (N = 29).

The probability that an initial approach to a carrot pile would be followed by feeding also varied significantly between treatments. Again a lower percentage of visits to male- and female-urine treated piles led to feeding (75.9% and 82.9% respectively) than visits to the untreated controls (98.4%). This difference was also apparent when visits by males and females were analysed separately (Table 1).

The emerging pattern of apparent preference for control and female urine-treated carrots over those sprayed with male urine, was further supported by marked differences between treatments in the duration of initial investigation of the bait, prior to feeding or rejection. When the results for male and female visitors are considered together, the mean duration of initial investigation of the male and female treatments (9.66 and 3.32 secs) differed significantly from each other ($p<0.001$ - t-test) and from the initial investigation of controls (1.69 secs) - $p<0.001$ and $p<0.01$ respectively.

The duration of initial investigation of male-treatments appeared to be longer in visits by males than by females ($\bar{x}$ = 12.57 and 6.93 secs). Interestingly, the duration of initial investigation of female-treatments and controls differed significantly for female visitors ($p<0.05$) but not for males.

DISCUSSION

These results suggest that the attractiveness of carrot-bait may be influenced by the presence of conspecific urine. Experiment 1 found that the consumption of carrots treated with male urine, was consistently lower than that of carrots treated with female urine or untreated controls. This discriminatory response was maintained throughout a 20 bait-night period with the exception of one night of heavy rainfall when it seems probable that the male urine-treatment was washed away.

The percentage take of each carrot treatment was positively correlated with minimum night temperature. It has been suggested that this relationship reflects the preference for unfrozen over frozen carrots found in other carrot-bait studies (Cowan pers. comm.).

Further support for an apparent preference for untreated or female urine-treated carrots over those sprayed with male urine was provided by corresponding differences in the number of bait visits, the duration of initial investigation prior to feeding/rejecting the bait and also the percentage of visits followed by feeding (Experiment 2). Bell (1977; in press, b) attributed an analogous preference for oestrous female over adult male urine found in two-choice preference tests, to the presence of androgen - dependent

Table 1. Behavioural response to treated/untreated carrot-bait

	Treatment			
	♀ urine	♂ urine	control	p†
Total number of visits				
by ♂ and ♀	41	29	61	**
by ♂	16	14	27	N.S.
by ♀	25	15	34	*
Number (and %) of visits followed by feeding				
by ♂ and ♀	34 (82.9)	22 (75.9)	60 (98.4)	***
by ♂	13 (81.3)	11 (78.6)	27 (100)	*
by ♀	21 (84.0)	11 (73.4)	33 (97.06)	**
Duration (secs) of initial investigation (mean ± S.D.)		** (♀ urine – control)		
	*** (♀ urine – ♂ urine)		*** (♂ urine – control)	
by ♂ and ♀	3.32 ± 3.32	9.66 ± 8.48	1.69 ± 1.94	
		N.S. (♀ urine – control)		
	** (♀ urine – ♂ urine)		*** (♂ urine – control)	
by ♂	3.00 ± 3.06	12.58 ± 10.64	1.48 ± 2.05	
		* (♀ urine – control)		
	* (♀ urine – ♂ urine)		*** (♂ urine – control)	
by ♀	3.52 ± 3.53	6.93 ± 4.71	1.50 ± 1.86	

*** $p < 0.001$, ** $p < 0.01$, * $p < 0.05$ } t-test

† probability derived from chi-squared test (2 d.f.)

male-aversive qualities in the urine of males and male-attractant qualities in that of oestrous does. Male aversion to the urine of conspecific males may also vary in relation to the social status of the urine donor, with the urine of high-ranking males apparently more aversive than that of low-ranking individuals (Bell, in press a).

The results of the present study suggest that these aversive urinary qualities may also reduce the consumption of previously attractive food items over relatively long-term baiting periods. The possible application of such food repellent effects to problems of crop damage control, clearly warrants further investigation.

REFERENCES

Aron, C., 1979, Mechanisms of control of the reproductive function by olfactory stimuli in female mammals, Physiol, Rev. 59: 229-284.

Bell, D.J. 1977, Aspects of the social behaviour of wild and domesticated rabbits, *Oryctolagus cuniculus* (L). Unpubl. Ph.D. Thesis: University of Wales.

Bell, D.J., 1980, Social olfaction in lagomorphs, Symp. Zool. Soc. Lond., 45: 141-163.

Bell, D.J., in press (a), Chemical communication in the European rabbit: urine and social status, in: Proceedings of the World Lagomorph Conference, Guelph, 1979, K. Myers and C.A. McInnes. eds.

Bell, D.J., in press (b), Lagomorph scents, in: Social Odours in Mammals, D. MacDonald and R. Brown, eds., Oxford University Press, Oxford.

Müller-Schwarze, D., 1971, Pheromones in black-tailed deer (*Odocoileus hemiones columbianus*), Anim. Behav., 19: 141-152.

Shumake, S.A., 1977, The search for applications of chemical signals in wildlife management, in: Chemical signals in vertebrates, D. Müller-Schwarze and M.M. Mozell, eds., Plenum Press, New York.

AN INVESTIGATION INTO THE 'BRUCE EFFECT' IN DOMESTICATED RABBITS

Diana Bell and Cindy Reece

School of Biological Sciences
University of East Anglia
Norwich NR4 7TJ, England

INTRODUCTION

Exposing a newly mated female to a 'strange' male (i.e. a male other than the original stud) has been found to block pregnancy in several rodent species. First reported in a spontaneously ovulating species, the mouse *Mus musculus* (Bruce, 1959), this so-called 'Bruce effect' has since been demonstrated in induced ovulators such as the field vole *Microtus agrestis*, meadow vole *Microtus pennsylvanicus*, and praire vole *Microtus ochrogaster* (Clulow & Clarke, 1968; Milligan, 1976; Clulow & Langford, 1971; Stehn & Richmond, 1975).

In female *Mus* pregnancy-blockage can only be induced prior to implantation of the fertilized egg (Bruce, 1961) while female prairie voles remain vulnerable until the fairly late stages (around 15th day) of their 21-23 day gestation (Stehn & Richmond, 1975). The effect may occur even after relatively brief pairings with the original 'stud'; Milligan (1979) for example, found a 55% blockage rate among females exposed to a 'strange' male after only 1 hour's contact with the original stud some 2 days earlier.

The existence of similar primer effects in the European rabbit has not been explored despite the well-documentated use of pheromonal signals by this species (e.g. Mykytowycz, 1974; Bell, 1980). On discovering that commercial rabbit farmers often 'check-mate' females with a different buck several days after the initial mating, we set out to determine whether such practice influenced littering rates amongst the breeding stock. Previously mated females were exposed to either the original 'stud' or a 'strange' male, during both the pre- and post-implantation phases of pregnancy.

METHOD

The experiments were conducted at a commercial rabbit farm, Norfolk Rabbits Ltd., Attleborough, Norfolk, during the spring of 1981. The rabbits were a New Zealand White strain bred at the farm. They were housed in unheated farm sheds in individual wire-mesh cages (0.85 x 0.45 x 0.35 m high) under a 14 hr/10 hr light/dark cycle and provided with an *ad libitum* supply of rabbit chow and water. 36 females of known fertility, were paired with 'stud' males according to the farm's standard husbandry procedure, i.e. each female was taken to the male's cage, the first copulation and ejaculation observed and the pair then left together for one hour before returning the female to her home-cage. 19 of these females were paired with either the original 'stud' (controls) or a 'strange' male (experimentals) 5 days later and another 17 similarly paired with the original 'stud' or a 'strange' male 10-15 days after their initial pairing. The females behaviour was observed during the first 10 minutes of these second, hour long pairings.

If the females had conceived during their first pairings, they should have been in pre- and post-implantation phases of pregnancy respectively, at the time of second pairings 5 and 10-15 days later (Brambell, 1948).

All males were selected on the basis of proven fertility and distant caging from any female partner. The paternity of resulting litters was determined by back-dating from birth date.

RESULTS

Prenatal mortalities (3 'experimentals' and 3 'controls') reduced the number of females to 15 in both the 5 and 10-15 day groups. Of the 25 litters produced, 24 resulted from the females first pairings and only one (in a 10-15 day control) from second matings. Within groups re-mated 5 or 10-15 days after first pairings there were no significant differences between 'experimental' and 'control' females in terms of either: a) the percentage of does littering, or b) the percentage of does 'accepting' the buck during the first 10 mins of their second pairing (Table 1). The apparent tendency for females to accept either male more readily 10-15 days after first matings (53%) than after 5 days (26.3%) was not statistically significant.

DISCUSSION

Pregnant female rabbits paired with a 'strange' male either 5 or 10-15 days post-coitum showed levels of male 'acceptance' and litter production similar to those of inseminated females re-exposed to the original 'stud' male, under commercial rabbit farm conditions.

TABLE 1.

Effects of 're-mating' with the original (controls) or a 'strange' (experimentals) male 5 or 10-15 days after initial mating.

Treatment	Number (and %) of does littering	Number (and %) of does showing ready acceptance [a] of male on re-mating
5 day controls	5/7 (71.4%)	2/9 (22.2%)
5 day experimentals	8/8 (100%)	3/10 (30%)
10-15 day controls	5/7 (71.4%)	4/8 (50%)
10-15 day experimentals	6/7 (85.7%)	5/9 (55.6%)

[a] i.e. copulation and ejaculation within first 10 mins of second pairing.

The present study was designed to determine the consequences of 'check-mating' breeding stock under commercial rabbit farm conditions. The results cannot therefore be taken as evidence for the absence of male-induced pregnancy-blocking effects in this species. It is possible, for example, that the 'odour milieu' permeating these mixed rabbit sheds somehow masked the females' ability to distinguish between 'strange' and 'familiar' males. Also, the males and females here were of the same inbred strain, a factor found to reduce the probability of pregnancy-blocking in mice (Parkes & Bruce, 1961). Furthermore, in a deliberate attempt to follow the farm's regular husbandry procedure as closely as possible, several female variables found to influence susceptibility to pregnancy-blockage in other species were not standardized in the present study, e.g. age (Chipman & Fox, 1966), number of previous litters (Terman, 1969) and concurrent lactation (Bruce & Parkes, 1961). The effects of these factors are currently being explored however, in laboratory-based investigations of the pregnant rabbit's response to male conspecifics.

In summary, the results of this applied investigation suggest that within the context of a commercial rabbit farm, pregnancy completion rates are unaffected by the husbandry practise of 'check-mating' females with a second 'strange' male during the pre- or post-implantation phases of pregnancy.

REFERENCES

Bell, D.J., 1980, Social olfaction in lagomorphs, Symp. Zool. Soc. Lond., 45: 141-163.

Brambell, F.W.R., 1948, Prenatal mortality in mammals, Biol. Rev., 23: 370-407.

Bruce, H.M., 1959, An exteroceptive block to pregnancy in the mouse, Nature (Lond.), 184: 105.

Bruce, H.M., 1961, Time relationships in the pregnancy block induced in mice by strange males, J. Reprod. Fert., 2: 138-142.

Bruce, H.M., 1963, Olfactory block to pregnancy among grouped mice, J. Reprod. Fert., 6: 451-460.

Bruce, H.M. and Parkes, A.S., 1961, The effect of concurrent lactation on olfactory block to pregnancy in the mouse, J. Endocrinology, 22: 6-7.

Chipman, R.K. and Fox, K.A., 1966, Factors in pregnancy blocking, age and reproductive background of females: numbers of strange males, J. Reprod. Fert., 12: 399-403.

Clulow, F.V. and Clarke, J.R., 1968, Pregnancy block in *Microtus agrestis*, an induced ovulator, Nature (Lond), 219: 511.

Clulow, F.V. and Langford, P.E., 1971, Pregnancy block in the meadow vole *Microtus pennsylvanicus*, J. Reprod. Fert., 24: 275-277.

Milligan, S.R., 1976, Pregnancy blocking in the vole, *Microtus agrestis*. I Effect of the social environment, J. Reprod. Fert., 46: 91-95.

Milligan, S.R., 1979, Pregnancy blockage and the memory of the stud male in the vole, *Microtus agrestis*, J. Reprod. Fert., 57: 223-225.

Mykytowycz, R., 1974, Odours in the spacing behaviour of mammals, in: "Pheromones", M.C. Birch, ed., North-Holland, Amsterdam.

Parkes, A.S. and Bruce, H.M., 1961, Olfactory stimuli in mammalian reproduction, Science, 134: 1049-1054.

Stehn, R.A. and Richmond, M.G., 1975, Male induced pregnancy termination in the prairie vole, *Microtus ochrogaster*, Science, N.Y., 187: 1211-1213.

Terman, C.R., 1969, Pregnancy failure in prairie deermice related to parity and social environment, Anim. Behav., 17: 104-108.

INDIVIDUAL DISCRIMINATION ON THE BASIS OF URINE IN DOGS AND WOLVES

Donna S. Brown and Robert E. Johnston

Department of Psychology
Cornell University
Ithaca, NY 14853

The importance of olfactory communication among dogs and wolves is renowned, but human beings still rarely fathom how pervasive such communication is. A glimpse into this world is suggested by the observation that members of a typical wolf pack scent mark or inspect a scent mark once every two minutes of the day (Peters and Mech, 1975). The functions of scent marks among dogs and wolves are legion, and may well involve subtleties that humans have not yet imagined. It is however known that dogs can discriminate males from females on the basis of urine odors and that males are preferentially attracted to the odor of urine from estrous as opposed to non-estrous females (Dunbar, 1978). Both naturalistic observations of dogs and wolves and preferences of female dogs for individual males suggest the ability of individuals of both species to recognize other individuals (e.g., Mech, 1970; LeBoeuf, 1967). Such observations also suggest the importance of scent cues, and it has been shown that male dogs can distinguish between their own odor and that of another male (Dunbar and Carmichael, 1981). We are not aware, however, of any proof that individuals can discriminate two other individuals on the basis of odor cues.

We investigated the ability of both dogs and wolves to distinguish between urine from two conspecifics by the use of an habituation paradigm. For both species the urine of one individual was presented to a subject once per day until the subject habituated, that is until a low, asymptotic level of investigation was reached. The odor of a different individual was then presented. A change in the level of investigation indicates that the animal noticed the difference.

Twenty-four sexually mature and 11 prepubertal female beagles were used in the first experiment. They were housed individually in cages measuring 1 x .5 x .5 meters and were maintained on dry dog food and water. All subjects were anestrous during the course of the study. Each female was first presented with an unscented paper towel once a day for seven days, then with a paper towel with urine from one of sixteen stimulus females for seven days and finally with a towel with urine from a second female for seven days. The stimuli were presented for two minutes and then removed; the time spent sniffing the stimuli were timed with a stopwatch.

Female dogs in both groups clearly habituated to each stimulus over the course of several days (Fig. 1); comparisons of responses on the first and last days of each condition were all significantly different (Wilcoxan Signed-Ranks test, $p<.05$ or less for all comparisons). Females also showed significant increases in investigation when a urine sample was first presented (comparison of day 7 and 8) and when the urine sample was changed (comparison of days 14 and 15; all comparisons $p<.05$ or less). Thus the females clearly discriminated the odors of urine from two other females. There was a tendency for the prepubertal females to spend more time investigating the urine odors than the sexually mature females but this difference was not statistically significant (Mann-Whitney U test).

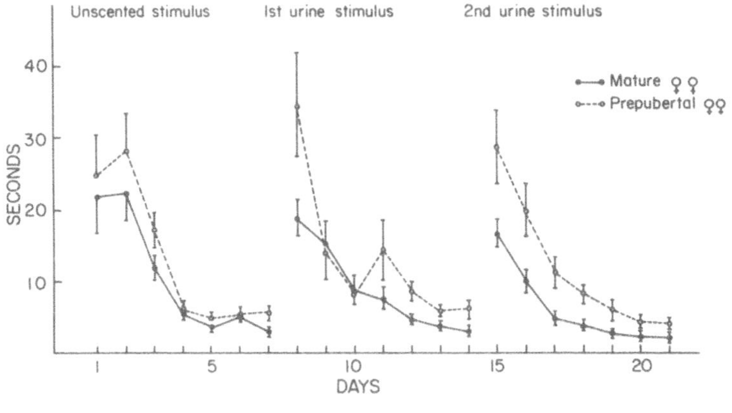

Figure 1. Mean (±SE) number of seconds females spent investigating control or urine stimuli on each test day. N = 24 mature females; N = 11 prepubertal females.

Nine captive wolves (8 Canis lupus lycaon, 1 Canis lupus occidentalis) were used in the second study. They were kept in two outdoor enclosures of approximately .5 acres each that were separated by 50 meters. Pack A contained two male and two female littermates, 2 years of age, descended from the alpha pair of pack B; they had been separated from their parents since 5 wks of age. Pack B contained two male and two female littermates, five years of age, and one six year old male of the other subspecies. The animals were fed fresh meat on a regular basis. Urine samples were obtained by observing males urinate and collecting a dish of fresh urine and snow. Each pack was presented first with a dish of snow alone for ten days, then a dish of snow with fresh urine from a male of the other pack for ten days, and finally snow and fresh urine of a second male of the other pack for 10 days. The total time that all the wolves in a pack spent investigating the stimuli was measured and recorded.

Data from pack B are shown in Fig. 2. Although the data are not amenable to statistical analysis there is little doubt that the wolves distinguish the urine of different individuals, since their investigation time increased dramatically when a new urine stimulus was presented. The data from pack A were almost exactly the same;

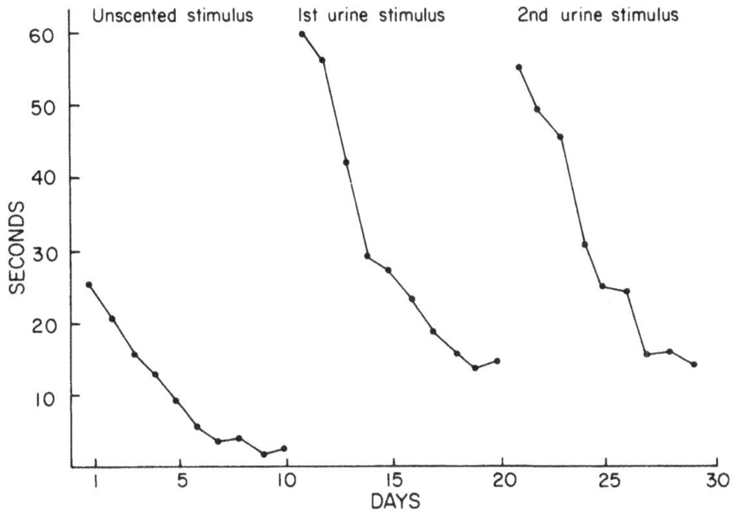

Figure 2. Total time spent investigating control and urine stimuli by a pack of wolves (Pack B, 3 males, 2 females).

most importantly on the last day of presentation of the first urine stimulus they sniffed it 21 seconds and on the next day sniffed the second urine stimulus for 46 seconds.

The results of these experiments demonstrate that both wolves and dogs spontaneously discriminate between urine from two individuals of the same sex and species. These data do not prove that either species actually uses this information in their daily social encounters, but it is hard to imagine that they wouldn't since they are so attentive to the differences. As noted above individuals of both species seem to treat other conspecifics as recognizable individuals, and the present data also hint at an extraordinary memory for the individuality of urine odors, since the test situation always involved comparing a stimulus with the memory of the odor presented the day before. This ability suggests a highly evolved sensitivity to subtle differences in odor quality, especially since the individual odors discriminated by both species came from close relatives and/or inbred strains and from individuals that were maintained on the same diet. Whether dogs or wolves are more sensitive to subtle differences in social odors than similar differences in non-social odors has yet to be determined, but in any case it is unlikely that such a highly developed ability would go unused in the daily life of either dog or wolf.

REFERENCES

Dunbar, I. F., 1978, Olfactory preferences in dogs: the response of male and female beagles to conspecific urine, Biol. Behav., 3:273.

Dunbar, I. and Carmichael, M., 1981, The response of male dogs to urine from other males, Behav. Neural. Biol., 31:465.

LeBoeuf, B. J., 1967, Interindividual associations in dogs, Behaviour, 29:268.

Mech, L. D., 1970, "The Wolf," Natural History Press, New York.

Peters, R. P., and Mech, L. D., 1975, Scent marking in wolves, Amer. Sci., 63:628.

THROAT-RUBBING IN RED HOWLER MONKEYS (*ALOUATTA SENICULUS*)

Ranka Sekulic and John F. Eisenberg

Department of Zoology
University of Maryland
College Park, MD

National Zoological Park
Smithsonian Institution
Washington, D.C.

INTRODUCTION

Olfactory communication in New World primates may be more developed than in most Old World monkeys and apes (Epple and Lorenz, 1967; Eisenberg, 1977). All the New World genera have glands in the sternal and/or gular region. The purpose of this paper is to describe throat-rubbing in wild red howler monkeys (*Alouatta seniculus*).

METHODS

Throat-rubbing was recorded during a one-year study at Hato Masaguaral, Guarico State, Venezuela (Sekulic, 1981 in press). For recent information about the howler population see Rudran (1979). The study troops were found in a seasonally flooded open woodland, in groups of 4-17 individuals with a mean of 8.9 (n=20) and a mode of nine (Rudran, 1979). Each troop usually included two to four adult females and one to two adult males. Most maturing males and some females emigrated from their natal troops.

Four contiguous troops and solitary individuals were observed during 10-hour periods (0700-0500 h), as well as during frequent early morning and late afternoon contacts; one other troop was contacted on irregular basis. Systematic sampling of throat-rubbing is based on 644.25 hours of contact between 31 January and 9 August 1980. The expected rate of marking for each sex is based on the number of hours of contact (e.g., if two females were observed for three hours, this was equal to six female contact hours).

RESULTS

Throat-rubbing was generally performed on the underside of a large branch while suspended by the limbs and tail. The neck area was moved forwards and backwards for 1-2 or more minutes. The throat is at times rubbed over a tree trunk, or over a branch above where the animal is stationed.

A total of 127 cases of throat-rubbing were observed by 27 individuals. Based on 23 10-hour samples, the marking was most frequently performed during the active period, before 1100 h (x^2= 15.65, d.f.=1, p<0.001). Most of the animals observed marking (24) were members of five troops. Three (two subadult females and one subadult male) were solitary. The behavior was generally performed by adults and subadults (Table 1). Marking by juveniles usually followed adults. The youngest animal observed displaying this behavior was an 11-month-old juvenile male. A total of 31 cases (24.4%) were followed by the same behavior in nearby individuals. Of these, 10 (32.3%) were performed over the location where the previous animal had rubbed.

Circumstances associated with throat-rubbing are shown in Table 2. In the majority of cases (93.7%), the marking was clearly associated with a hostile reaction toward conspecifics, and was accompanied by howling and/or piloerection. Three cases followed disturbance by other species, including humans. Assuming equal probability of approach by the two sexes, there was no difference among males and females in response to male approach (x^2=0.12, d.f.= 1, p>0.05), but females rubbed significantly more than males following female approach (x^2=11.00, d.f.=1, p<0.001).

Table 1. Age and Sex Classification of Individuals Performing Throat-Rubbing.

Sex	Adult	Subadult	Juvenile
0	29	30	3
0	31	31	3
Total	60	61	6

Table 2. Classification of Throat-Rubbing by Context

Context		
♂ Approach	16 (12.6%)	11 (8.7%)
♀ Approach	18 (14.1%)	33 (26.0%)
♂ and ♀ Approach	10 (7.9%)	7 (5.5%)
Troop Encounter	16 (12.6%)	8 (6.3%)
Disturbance	2 (1.6%)	1 (0.8%)
No Evident Stimuli	1 (0.8%)	4 (3.1%)
Total	63 (49.6%)	64 (50.4%)

Overall, there was a tendency for the males to rub at a higher rate (0.039/h) than females (0.028/h; x^2=3.26, d.f.=1, $0.1>p>0.05$). The highest rate of throat-rubbing (0.150/h), however, was found among solitary females. More than half of the observed markings (58.3%) were associated with the presence of one solitary female (fS), who was observed most of the time in the range of one troop and was often followed by subadult troop males (Sekulic, in press). Rubbing rates by the fS, troop males and troop females were significantly higher on days when fS was proceptive (0.354/h; 0.202/h; 0.281/h, respectively) than on days when she was not proceptive (0.122/h; 0.027/h; 0.048/h, respectively; $p<0.05$, Binomial Test).

CONCLUSION

Throat-rubbing of *Alouatta seniculus* conforms in many respects to the sternal marking described in several captive Callitrichids and in wild titi monkeys (*Callicebus moloch*; Mason, 1966). Unlike scent marking rates among monogamous Callitrichids (Epple, 1970; Box, 1975; Mack and Kleiman, 1978), in *A. seniculus* there is a tendency for the males to rub at a higher rate than the females. As in captive *A. palliata* (Eisenberg, 1976), *A. seniculus* marking was associated with a high level of arousal and was not performed at any particular location within the home range. Throat-rubbing rates of the solitary *A. seniculus* female, although low compared with the rates of marking among captive Callithrichids, were significantly higher than rates among reproductive troop females. These results suggest that, unlike some Callitrichids (e.g., golden lion tamarins, *Leontopithecus rosalia*; Kleiman and Mack, 1980), subordinate *A. seniculus* are not behaviorally inhibited from scent marking.

ACKNOWLEDGEMENTS

We are most grateful to Cecilia and Tomas Blohm for allowing us to conduct research at Hato Masaguaral. The study was supported by the Smithsonian's Environmental Science grant to J. F. Eisenberg.

REFERENCES

Box, H.O., 1975, Quantitative behavioral studies of C. jacchus, Primates, 16:155.

Eisenberg, J. F., 1976, Communication mechanisms and social integration in the black spider monkey, Ateles fusciceps robustus, and related species, Smithsonian Contrib. Zool.,213.

_____, 1977, Comparative ecology and reproduction of New World monkeys, in: "Biology and Conservation of the Callitrichidae," D. G. Kleiman, ed., Smithsonian Institution Press, Washington.

Epple, G., 1970, Quantitative studies on scent marking in the marmoset (Callithrix facchus), Folia primatol,13:48.

_____, and Lorenz, R., 1967, Vorkommen, Morphologie und Funktion der Sternaldrüse bei den Platyrrhini, Folia primatol., 7:98.

Kleiman, D. G., and Mack, D. S., 1980, Effects of age, sex and reproductive status on scent marking frequencies in the golden lion tamarin, Leontopithecus rosalia, Folia primatol., 33:1.

Mack, D. S., and Kleiman, D. G., 1978, Distribution of scent marks in different contexts in captive lion tamarins Leontopithecus rosalia, in: "Biology and Behavior of Marmosets," H. Rothe, H. J. Wolters, and J. P. Hearn, eds., Eigenverlag, Gottingen, W. Germany.

Mason, W. A., 1966, Social organization of the South American monkey, Callicebus moloch; a preliminary report, Tulane Studies Zool., 13:23.

Rudran, R., 1979, The demography and social mobility of red howler (Alouatta seniculus) population in Venezuela, in: "Vertebrate Ecology in the Northern Neotropics," J. F. Eisenberg, ed., Smithsonian Inst., Washington.

Sekulic, R., 1981,"The Significance of Howling in the Red Howler Monkeys," Ph.D. dissertation, Univ. Maryland.

_____, in press, Behavior and ranging patterns of a solitary female red howler (Alouatta seniculus), Folia primatol.

AUTHOR INDEX

Abbot, J. 207
Adams, D.B. 265
Adey, W.R. 105
Agosta, W.C. 180, 186, 187,285
Agren, G. 300
Albone, E.S. 287
Albrecht, E.D. 141
Aleksiuk, M. 57
Alexander, B.K. 196, 245
Alexander, G. 262
Alexander, R.D. 265
Allison, A.C. 27, 29
Altieri, R. 239
Altman, J. 168, 171
Alvarez, J. 162, 165
Aonuma, S. 141
Arnold, A.P. 124, 128, 129,131
Aron, C. 170, 173, 266, 323, 333
Aschoff, J. 161
Asofsky, R. 141
Atema, J. 310, 311
Attili, G. 266
Atz, J.W. 28
Auffenberg, W. 32
Baker, R.H. 197
Baker, R.R. 6, 7
Bakker, R.T. 36
Banks, E.M. 135, 187
Barlow, G.W. 8
Bargmann, W. 105
Barkes, B.M. 300
Barrette, C. 262
Bartke, A. 139, 144, 187
Bashinski, H. 7
Baumann, F. 9, 16, 17
Baumgardner, D.J. 317, 318
Baxter, G. 34
Beach, F.A. 246, 264
Beamer, W. 285
Beatley, J.C. 323
Beauchamp, G.K. 29, 67, 73, 75, 76, 79, 80, 90, 91, 142, 145, 187, 265, 266, 285, 300
Behrends, P. 237
Bell, D.J. 333, 339
Bellamy, D. 183
Bellows, R.A. 155
Bellringer, J.F. 73, 252
Benimoff, N. 79
Bennett, A.F. 25
Benshoof, L. 265
Beresford, W.A. 111 unpublished
Bernhard 186
Berry, K.H. 33
Bertmar, G. 27
Bertram, B.C.R. 146
Berüter, J. 75, 187, 285
Best, E.N. 159
Biaty, E. 141
Birdsall, D.A. 317, 318
Black, A.J. 310
Blackwell, R.L. 155
Blanchard, F.C. 57
Blanchard, F.N. 57
Blask, D.E. 113
Blass, E.M. 34, 35
Bleier, R. 111
Block, M.L. 32, 293
Blum, M.S. 309 - 311
Blum, S.L. 120, 144
Bodner, M. 196
Bodznick, D. 313
Bogert, C.M. 18
Boland, B.D. 246, 247
Bonsall, R.W. 264
Boonstra, R. 146, 229, 237
Boot, L.M. 137, 171
Booth, D.A. 265
Bowen, D.V. 187
Bowers, J.M. 196, 245
Box, H.O. 349
Boyden, A. 28

Boyse, E.A. 207, 285
Boyse, J. 207
Brambell, F.W.R. 340
Brand, J.G. 67, 75, 79, 80, 90
Brand, J.M. 309, 310
Breen, M.F. 189
Brill, A.A. 264
Brink, A.S. 25, 32, 33, 35
Broadwell, R.D. 67, 110
Brock, O.G. 17
Brom 186
Broman, I. 27
Bronson, F.H. 135, 137, 139 - 142, 153, 155, 161, 183, 184, 197, 198, 203 - 206, 208, 259, 262, 267
Brooks, P.H. 155
Brooks, R.J. 135, 146
Brooksbank, B.W.L. 265
Broom, R. 30, 100
Brown, J.H. 19
Brown-Grant, K. 165
Brown, R.E. 181, 182, 245, 265, 325
Bruce, H.M. 140, 145, 339, 341
Brüggemann, I. 237, 238
Buck, E. 160, 171
Buck, J. 160, 171
Buck, K.T. 265
Buffler, G. 165
Bullock, T.H. 161
Bunnell, B. 246, 247
Burfening, P.J. 155
Burghardt, G.M. 1, 6, 7, 26-29, 37, 45, 47, 63, 64, 90
Burke, D.J. 331
Burrell, L. 120
Burris, M.J. 155
Butcher, R.L. 162
Butler, H. 265
Byatt 186
Cain, W.S. 266
Calhoun, J.B. 318
Campbell, A. 159
Campbell, H.W. 291, 293
Capone, M.R. 229, 237
Cardé, R.T. 195
Carmack, M. 241, 242, 285
Carmichael, M. 343
Caroom, D. 141
Carpenter, G. 260
Carr, W.J. 181, 245
Carregal, E.J.A. 108
Carroll, D. 25, 265
Carter, C.S. 187
Cavanaugh, P.H. 263
Cegavske, C.E. 120
Champlin, A.K. 137, 144
Chapman, V.M. 137, 142
Cheal, M.L. 259, 300
Chee, P. 73, 171, 252
Chenoweth, P.J. 91
Chipman, R.K. 141, 144, 341
Chiszar, D. 1-3, 6-9, 13, 15, 17, 18
Chitty, D. 223
Chorover, S.L. 187
Christian, J.J. 145
Chudinov, P.K. 25, 32, 35
Clabough, J.W. 249
Clancy, A.N. 126
Clark, D.L. 7, 17, 45
Clark, I.J. 96
Clarke, J.R. 135, 145, 339
Clee, M.D. 137
Cluclow, F.V. 339
Clulow, F.K. 135, 145
Cluver, M.A. 30, 31
Coblentz, B.E. 94, 96, 99
Cohen, J. 57
Colbert, E.H. 37
Colby, D.R. 136, 140, 141, 155, 300
Cole, D.J. 155
Collins, W.E. 162
Comfort, A. 263, 264
Commins, D. 120, 130
Conway, C.H. 267
Cope, E.D. 18
Coppock, C.E. 91
Coquelin, A. 139
Corker, C.S. 165
Costanzo, D.J. 245
Covacevich 1, 17, 18
Cowan, W.M. 110
Cowles, R.B. 6, 28, 63, 64
Cowley, J.J. 265
Creutzberg, F. 313
Crews, D. 32, 57, 120

Criss, B.R. 75
Crompton, A.W. 25, 30
Csernus, V.J. 126
Curry, J.J. 168
Czaplicki, J.A. 34
Dafny, J. 114
Dagg, A.I. 91
Dalterio, S. 139
Daly, M. 229, 237, 300, 332
D'Angelo, W. 139, 187
Darby, E.M. 187
Dawkins, R. 309
Darwin, C. 35
Davidson, J.M. 164
Davidson, W. 95
Davies, P. 265
Davies, V.J. 183
Davis, B.J. 67, 108
DeWar, A.D. 137
Debski, E.A. 67, 73, 252
Deceverski, V. 114
Deck, J. 267
DeCoursey, P. 160
Denenberg, V.H. 144
Desjardins, C. 155, 204
Devor, M. 89, 187
Dewsbury, D.A. 246, 247, 317, 318
Dezell, H.E. 135
Dhariwal, A.P.S. 164
Dice, L.R. 197
Dissinger, M.L. 181
Dills, S.S. 141, 142, 153
Distel, H. 68
Dixon, A.F. 267
Dixon, A.K. 179, 180, 184, 185
Dizinno, G. 180, 183-186
Dobrogosz, W.J. 141, 142, 153
Dominguez, R. 164
Dominic, C.J. 140
Dorner, G. 121
Doty, R.L. 181, 182, 196, 259, 263-266
Doty, R.W. 110
Dowling, H.G. 18
Dressler, M. 265
Drickamer, L.C. 136, 138, 141, 170, 185, 203, 229, 237, 300
Dudzinski, M.L. 280
Dullemeijer, P. 1, 9, 17
Dunbar, I. 343
Duvall, D. 3, 8, 13, 25-30, 32-36
Ebling, F.J. 283
Edinger, H.M. 111
Eckmann, K.W. 252
Eichenbaum, H. 252
Eisenberg, J.F. 29, 33, 198, 300, 321, 347, 349
Eleftheriou, B.E. 135
Ellis, H. 264
Ellis, L.B. 144
Emery, D.E. 89
Engen, T. 260
Epple, G. 67, 196, 262, 300, 347, 349
Espmark, Y. 245
Estes, R.D. 27, 29, 89, 91, 92, 99, 100
Evans, C.M. 142, 184, 185
Evans, H.M. 161, 165
Evans, L.T. 34
Everett, J.W. 164, 165, 168
Fan, Z. 237
Faux, S.F. 229, 237
Feder, H.H. 29, 67, 142
Fernandez, F. 139, 187
Field, F.H. 187
Field, P.M. 121, 126
Fields, R.B. 67, 79, 90
Finlayson, J.S. 141, 142, 153
Firbas, W. 29
Fishman, W.H. 141
Fitch, H.S. 34
Fitzgerald, K.M. 165
Fleming, A. 252
Fleming, S. 73
Foote, R.H. 96
Forbes, J.M. 156
Fox, K.A. 145, 341
Frakel, A.J. 139
Francoeur, D. 267
Freeman, M.E. 162
Frumin, N. 28, 45, 67
Fugo, N.W. 162
Galef, B.G. 325
Galton, P.M. 37
Galvin, R. 35, 36
Gambale, S. 280
Gamble, J. 245

Garstka, W.R. 32, 57
Gaudelman, R. 144
Gaughwin, M.D. 89, 99
Gaunt, S.L. 141
Gawienowski, A.M. 181, 182
Geist, V. 99
Gerall, A.A. 89, 168
Gerisch, G. 160
Gerling, S.A. 144
Gern, W. 34
Gerzog-Thomas, D.A. 32
Giantonio, G.W. 89
Gillingham, J.C. 6, 7, 17, 45
Gilmore, R.W. 264
Globus, A. 127
Gloor, P. 105
Godowicz, B. 142, 144
Godfrey, J. 196
Goff, A.M. 310
Golan, L. 9, 13, 15
Gold, A.R. 265
Goldberg, N. 250
Goldfoot, D.A. 264-266
Golob, N.F. 196
Goodall, J. 265
Goodrich, B.S. 275-277, 280, 283-287, 300
Goodwin, M. 119, 292, 296
Gorbman, A. 120
Gordon, J.H. 121, 126, 131
Gorski, R.A. 120, 121, 126, 131
Gove, D. 67
Gow, C.E. 33
Goy, R.W. 89
Graham, C.A. 267
Graham, M. 294
Grant, P.R. 207
Grau, G. 267
Graves, B.M. 35, 36
Greene, H.W. 28, 37
Greene, W.A. 96
Greenstein, J.A. 142
Greg, B.R. 303, 304
Gregory, P.T. 57
Grine, F.E. 33
Griswold, J.G. 321
Guerillot, C. 114
Guild, S.R. 107
Guillette, L.H.,Jr. 26, 34, 36
Gullotta, E.F. 292
Gurevich, B. Kh. 160
Gustafsson, J.A. 265
Gut, J. 237, 238
Haag, C. 184
Hain, J.H.W. 313
Hall, K.R.L. 265
Hall, W.C. 67, 110
Halpern, M. 27-29, 45, 46, 49, 56-58, 62, 63, 66, 67,76,79
Halpin, Z.T. 245
Hamilton, J.B. 250
Handelmann, G. 171
Hansson, L. 237
Harding, C. 129
Harding, J.W. 108
Harriman, A.E. 296, 297
Harrington, J.P. 267
Harris, V.S. 89
Hart, B.L. 75, 89, 91-95, 98, 99, 120
Harvey, E.B. 329
Hasler, A.D. 313
Hasler, J.F. 135
Haugen, C.M. 89
Hausfater, G. 168, 265
Hayashi, B. 186
Hayes, M.J. 32, 293
Heckman, S. 236, 237
Heiligenberg, W. 161
Heimer, L. 67, 89
Heinrich, B. 291, 293
Heller, S.B. 29, 45, 63, 67
Herrick, C.J. 67, 110
Hess, E.H. 260
Hesterman, E.R. 275-277, 280, 284, 285, 287
Heth, G. 196
Hill, J. 141
Hodgdon, H.E. 236
Hofer, M.A. 34, 35
Hold, B. 261, 263, 266
Hong, K.M. 262, 263
Holt, J.A. 145
Hoogland, J.L. 207
Hoover, J.E. 170
Hoope, P.G. 141, 142, 144
Hoppensteadt, F. 161
Hopson, J. 36, 37, 263
Hotton, N.H. 31
Howard, W.E. 207

Howe, N.R. 310
Hrdy, 146
Huber, G.C. 107
Hughes, P.E. 155, 156
Humphrey, S.R. 242
Humphreys, E.M. 137
Hurley, H.J. 264
Issacson, R.L. 105
Izard, M.K. 154-156
Jackson, J.C. 120, 130
Jacobs, V.L. 91
Jacobson, C D. 126, 127
Jacobson, L. 26
Jalife, J. 161
James, W.H. 265
Janak, J. 265
Jenkins, F.A. 25, 30
Jerhoff, B. 184
Jerison, H.J. 26
Jesmajian, S. 180
Johns, M.A. 28, 29, 67, 142, 173
Johnson, A.L. 18, 265
Johnson, P.A. 180
Johnson, R.P. 300
Johnston, R.E. 75, 180, 186, 187, 245, 250, 262, 323
Jones, R.B. 310
Jones, R.E. 34
Jones, T.O.A.C. 94, 95
Jorgenson, J.W. 241, 242, 285
Kahmann, H. 45, 63
Kakihana, R. 144
Kamel, F. 139
Kaneko, N. 67, 73, 252
Kaplan, J. 260 - 262
Kappers, J.A. 113
Katz, Y. 73
Kayne, H. 260
Keiser, E.D. 56, 67
Keller, J.B. 161
Kelly, M.J. 114
Kemp, T.S. 31
Kampson, D.A. 223
Kennedy, G.Y. 292
Kannedy, J.F. 142, 184
Kessler, S. 144
Keverne, E.B. 67, 73, 252, 264
Kilpatrick 183
Kimura, T. 186
King, G.M. 30, 31
King, M.B. 25, 34
Kirkpatrick, J.F. 184
Kirk-Smith, M. 265
Kirkwood, R.N. 155, 156
Kitching, J.W. 33
Kittredge, T.S. 311
Kitrell, E.M.W. 296, 298, 303, 304, 306
Klauber, L.M. 7
Kleiman, D.G. 29, 33, 300, 349
Klemm, W.R. 91
Kligman 264
Klopfer, P.H. 245
Knapp, H. 67
Knappe, H. 91
Knight, L. 2
Kochkian, C.D. 141
Komisaruk, B.R. 29, 67, 142
Korf, H.W. 114
Kost, T.A. 141, 142, 153
Kowell, A.P. 67
Kraeplin, G. 168
Krames, L. 245
Kratzing, J. 89
Krebs, C.J. 229, 237
Krebs, J.R. 309
Krehbiel, R.H. 171
Krstic, R. 114
Kruger, L. 105
Krzanowska, H. 144
Kubie, J.L. 4, 28, 29, 45-49, 56-58, 62, 63, 66, 67, 75, 76, 79, 80, 90
Kübler, F. 160
Kuehn, R.E. 246
Kuno, Y. 264
Kwan, M. 187
Labov, J.B. 146
Ladewig, J. 75, 89, 91-93,98,99
Lage, C.D. 114
Laine, J. 321
Laing, D.G. 310
Lamond, D.R. 171
Landauer, M.R. 181, 187, 250
Langford, P.E. 135, 339
Larsell, O. 107
Larsson, K. 89, 120
Lauderdale, J.W. 156
LaVelle, A. 127
Law, J.M. 63

Lawless, H.T. 266
LeBoeuf, B.J. 343
Lee, T.M. 328
Leffray, P. 114
Le Gros Clark, W.E. 111
Leidahl, L.C. 183, 325
Leimeister, J.L. 155
LeMagnen, J. 181
Leon, M. 34, 170, 171, 183, 262, 325, 328
Leonard, B.E. 107, 111, 113
Lepri, J.J. 322
Leshner, A.I. 189
Levine, J.M. 263
Lewis, D. 331
Lincoln, G.A. 95
Lindzey, G. 120, 222
Lipsitt, L.P. 260
Liu, S. 250
Loeb, L.S. 181
Lombardi, J.R. 140, 141, 155
Long, C.A. 35
Long, J.A. 161, 165
Lorenz, R. 347
Lott, D.F. 99
Lucas, C.E. 311
Luderschmidt 250
Luine, V. 129
Lund, N.L. 89
Lydell, K. 181, 182
MacFarlane, A. 259, 260
Mack, D.S. 349
Mackay-Sim, A. 310
Mackintosh, J.H. 142, 179, 180, 184, 185
MacLean, P.D. 26, 37, 105
Macrides, F. 67, 79, 108, 139, 180, 186, 187, 285
Madison, D.M. 32, 34
Mainwarning, W.I.P. 127
Malchow, D. 160
Mallory, F.F. 135, 145, 146
Marchlewska-Koj, A. 141, 142
Marcks, S. 294
Margolis, F.L. 108
Marinari, K.T. 170
Marques, D.M. 67, 73, 79, 92, 253
Marsden, H.H. 142, 161
Martin, I.G. 76
Martys, M. 89
Maruniak, J.A. 139, 183, 184, 203
Marx, H. 1, 18, 19
Mascitti, T.A. 110
Mason, W.A. 349
Massey, A. 145, 153
Mathieson, B.J. 207
Matthes, E. 75
Matthews, M.R. 111
Mayer, A.D. 29, 67, 142
Mazder, E. 229, 237
McBurney, D.H. 263
McCarron, D. 34
McClintock, M.K. 153, 159, 160, 165, 168-171, 173, 267
McClung, R. 114
McCrady, D.E. 168
McCusky, S. 168
McDowell, S.B. 45
McEwen, B.S. 129
McGrew, W.C. 267
McIntosh, T.K. 136, 138, 185
McManus, J.J. 300
Means, A.R. 127
Meaney, C. 329, 332
Mech, L.D. 343
Mele, J.A. 300
Meredith, M. 4, 6, 27, 29, 67, 73, 91, 92, 253
Merritt, C.G. 286, 287
Meyer, D.R. 105
Meyer, P.M. 105
Meyer, R.P. 110
Meyerson, D.J. 300
Michael, R.P. 264
Michener, G. 329
Mike, V. 207
Miles, S.G. 313
Miline, R. 114
Miller, T. 1, 9, 13, 15, 17, 18
Milligan, S.R. 339
Mingyue, Wang 215
Minton, S.A. 19
Mitchell, D. 33
Mock, E.J. 139
Moe, G.K. 161
Mogil, L. 96
Møller, M. 114
Molnar, R.E. 37

Moltz, H. 170, 183, 262, 325, 328
Montagna, W. 250
Montgomery, W.I. 230
Moody, E.L. 156
Moore, C.P. 161
Moore, J.D. 261
Moore, R.E. 196
Morgantaler, A. 120
Morin, L. 165
Moulton, D.G. 108, 168, 265, 266
Mozell, M.M. 259
Mueller-Schwarze, D. 95, 96, 98, 99, 196, 236, 237, 239-241, 259, 264, 267, 300, 333
Muetterties, E.L. 75, 285
Mugford, R.A. 141, 183, 184, 265, 266
Muller, J. 114
Murphy, M.R. 252, 256
Murphy, R.X. 141
Murray, K.E. 275-277, 280, 284, 285
Mutiso, F.M. 94
Mykytowycz, R. 259, 262, 275-277, 280, 283-285, 287, 300, 339
Myllymäki, A. 226
Nadi, N.S. 108
Naftolin, F. 165
Nalbandov, A.V. 135
Nanjundiah, V. 160
Nash, D. 317, 318
Naulleau, G. 17
Nault, L.R. 310
Negus, V. 27, 29
Neill, J.D. 162
Nelson, A.K. 259
Nequin, L.G. 162, 165
Nevo, E. 196
Newman, A. 120, 124, 130
Nielsen, J.T. 114
Nikitovitch-Winer, M.E. 168
Noble, G.K. 34, 57, 63
Nodelman, J.L. 113
Noonan, K.M. 265
Norvell, J.E. 249
Nottebohm, F. 124, 126, 129, 131
Novotny, M. 241, 242, 285
Nowell, N.W. 141, 183, 184
Nunez, A.A. 180
Nyby, J. 73, 75, 128, 180, 183-186
O'Connell, B. 1, 6, 7, 9, 13, 15, 17
O'Connell, R.J. 67, 73, 79, 91, 92, 180, 186, 187, 253
O'Farrell, M.J. 322
Olson, E.C. 35
O'Malley, B.W. 127
Ono, T. 111
Oomura, Y. 111
Orsulak, P.J. 181, 182
Ortega, S.N. 110
Owen, J.W. 265
Owen, K. 130, 222, 303
Palade, G.E. 127
Palay, S.L. 127
Palner, T. 120
Papez, J.W. 105
Parkes, A.S. 165, 341
Parrott, R.F. 96, 145
Parsons, T.S. 26, 27, 35
Passman, R.H. 262
Paulus, K. 291, 293
Payman, R.C. 168
Payne, A.P. 249, 250, 296
Pazo, J.H. 114
Pendergrass, M. 296, 297
Perkel, D.H. 161
Perkins, J. 294
Perrigo, G. 197, 198, 204-206
Perry, G.C. 287
Persoon, S.C.J. 237
Peters, R.P. 343
Peterson, D.M. 89
Peterson, E.H. 67
Pettengill, O.S. 141
Petterborg, L.J. 113
Pettijohn, T.F. 300
Pevet, P. 113
Pfaff, D.W. 181
Pfaffman, C. 181, 186, 187
Pfeiffer, W. 310
Pfister, A. 114
Phelan, R.L. 6, 28, 63, 64
Pietras, R.J. 168
Planel, H. 75
Plewig 250

Poduschka, W. 27, 29
Pollack, E. 265
Pond, C.M. 26
Poole, T. 1, 7, 17, 18
Porter, R.H. 34, 261
Potter, M. 141
Powell, T.P.S. 108, 110, 111
Powers, J.B. 29, 67, 73, 79, 90, 252, 253, 255
Pratt, H.P.M. 73, 252
Pratt, K.C. 259
Pratt, J.J. 181
Preti, G. 265
Pribram, K.H. 105
Price, E.O. 98, 99
Price, J.L. 108, 110
Priode, B.M. 155
Pruitt, G.H. 45
Quadagno, D.M. 267
Quay, W.B. 31, 113, 239
Rabb, G.B. 1, 18, 19
Radcliffe, C.W. 1-3, 6-9, 13, 15, 17, 18
Raisman, G. 110, 121, 126
Ralls, K. 262, 300
Randall, J.A. 321-324
Randall, W. 298
Rasa, O.A.E. 245, 249
Ravizza, R. 171
Ray, W.J. 171
Regnier, F.E. 119, 296
Reinhardt, A. 94
Reinhardt, V. 94
Reisch, A.M. 291
Reiter, R.J. 113
Reynolds, J. 67, 73
Rice, M. 300
Richards, S.A. 299
Richmond, M.G. 318, 339
Riss, W. 67
Ritter, F. 237, 238
Roberts, J.C. 300
Robertson, S.M. 142, 184
Robinson, P.F. 300
Roelofs, W.L. 195
Rogel, M.J. 263, 268
Rogers, Jr., J.G. 145
Romer, A.S. 25, 26, 35
Rønnekleiv, O.K. 114
Rood, J.P. 171
Rose, E.A. 138, 185
Rosenberg, L.E. 329
Rosenblatt, J.S. 262
Rosenblum, L. 267
Rosene, D.L. 67
Roser, S. 165
Ross, Jr., P. 32
Roth, E.C. 26
Roth, J.J. 26
Rowe, F.P. 229, 237
Rowell, T.E. 267
Ruben, J. 25
Ruddy, L.L. 187
Rudran, R. 347
Rümke, P. 141
Runner, C.C. 141
Russell, J.A. 137
Russell, M.J. 259, 260, 262, 263, 267, 268
Ryan, K.D. 137
Ryan, S.M. 129
Sachs, B.D. 89
Sade, D.S. 267
Saltiel, A. 128
Sapolsky, R.M. 252
Sastry, S.D. 265
Satli, M.A. 266
Scalia, F. 67, 89, 110
Schaal, B. 260, 261, 263
Schadé, J.P. 105
Schilling, A. 75
Schleidt, M. 261, 263, 266
Schmidt-Nielsen, K. 323
Schmidt, T. 75
Schneider, J. 180, 183-186
Schneider, K.M. 100
Schneider, S.P. 67
Schulman, J.H. 161
Schultz, C. 262
Schultze-Westrum, T.G. 300
Schwanzel-Fukuda, M. 107
Schwartz, N.B. 137, 162, 164, 165, 186
Schwarz, E. 197
Schwarz, H.K. 197
Scouten, C.W. 120
Scudder, K.M. 2, 3, 6-8, 13, 17
Segundo, J.P. 161
Sekulic, R. 347, 349
Shair, H. 35

Shaw, K.S. 275-277, 280, 284, 285
Shedlack, K.J. 252
Sheffield, L.P. 63
Shehadah 264
Sheik, Y.M. 310
Shelly, W.B. 264
Sherry, C.J. 91
Shillito, E.E. 223, 229
Shine, R. 1, 17, 18
Shorey, H.H. 26, 31, 259
Shoushen, Hao 215
Short, R.E. 155
Short, R.V. 168
Shryne, J.E. 121, 126, 131
Siegel, A. 111
Shumake, S.A. 241, 333
Shuveita, H.E. 267
Signoret, J.P. 99
Silverman, A.J. 107
Simon, C.A. 29, 32-34
Simonsen, L. 17
Simpson, G.G. 25, 28, 30
Singer, A.G. 180, 187, 192, 285
Singh, P.J. 34, 35
Sis, R.F. 91
Skeen, L.C. 67, 110
Slimp, J.C. 89
Smith, III A.B. 67, 75, 79, 80, 90, 196
Smith, A.T. 329
Smith, E.R. 164
Smith, H.M. 1, 6-8, 17, 18, 34
Smith, J.C. 27
Smith, M.S. 162
Smith, R.J.F. 310
Snyder, H.A. 310
Snyder, N.F.R. 310
Southam, A.M. 121, 126, 131
Spitz, C.J. 265
Sprott, R.L. 259
Stacewicz-Sapuntzakis, M. 181, 182
Stagge, B. 236
Stanley, G. 275-277, 280, 284, 285
Staudt, J. 121
Steffen, J.M. 300
Stehn, R.A. 318, 339
Steiner, A.L. 300
Stenzler, D. 310, 311
Stephens, D.R. 124
Stern, F.L. 67, 73, 92
Stern, J.J. 180, 181
Stevens, D. 32
Stimac, K. 1, 17, 18
Stoddart, D.M. 26, 31-34, 196, 224, 229-231, 238, 262, 266, 291
Stone, H. 108
Strada, S.J. 114
Strassman, B.I. 265
Sugowdz, G. 275-277, 280, 284, 285
Summerlin, C.T. 229, 237
Sun, K.S. 259
Swanson, H.H. 168, 300
Switz, G.M. 267, 268
Takahashi, F.T. 311
Tambosso, L. 73, 252
Taub, A. 91
Taylor, G.T. 183, 184
Taylor, S.B. 7
Teichmann 313
Teicher, M.H. 34, 35
Terman, C.R. 341
Thaler, H.T. 207
Thiessen, D.D. 120, 126, 128, 130, 222, 250, 262, 291, 294-298, 300, 303, 304, 306
Thomas, L. 207
Thompson, G. 249
Thompson, K. 267, 268
Thornhill, R. 265
Thung, P.J. 141
Tokizane, T. 105
Townes, B.D. 262, 263
Trakulrungsi, C. 113
Trakulrungsi, W.K. 113
Troiano, R. 111
Trupiano, J. 8, 34
Tucker, D. 27, 108
Tuite, M. 107, 111, 113
Turner, B.R. 33
Turner, J.W. 120, 300
Tutin, C.E.G. 318
Ulberg, L.C. 156
Ulibarri 124, 130
Ulinski, P.S. 67
Vaccarino, F. 73, 252

Vagvolgyi, A. 29, 45, 57, 63
Vandenbergh, J.G. 135, 136, 139, 140-142, 144, 145, 153-156, 204, 267, 285, 300
van der Lee, S. 137, 171
Van Valen, L. 25, 32, 33, 35
Vendrely, E. 114
Verweil, P. 237, 238
Vessey, S. 267
Voith, V.L. 120
Volpe, L.C. 32, 293
von Holst, E. 160
von Meyenburg, H.K. 160
Wada, M. 120
Wagner, U. 114
Wales, N.A.M. 283
Walker, T.J. 313
Wallace, P. 261, 263, 264, 303
Walton, A.H. 264
Wang, R.T. 27
Watson, D.M.S. 25, 33
Weick, R.F. 164
Weisberg, P. 262
Weise, R.E. 181
Weldon, P.J. 28, 31, 32, 311
Wellington, J.L. 29, 67, 75, 76, 79, 80, 90, 91, 142
Wenzel, B.M. 105, 110
Weston, J.S. 156
Wheeler, J.M. 120
White, Jr., L.E. 110
Whitney, G. 73, 180, 183-186
Whitsett, J.M. 140, 141, 155
Whitten, W.K. 67, 73, 137, 140-142, 144, 153, 160, 170, 241, 242, 252, 262, 267, 285
Wiedemann, E. 17
Wiener, H. 264
Wilde, W.S. 45, 47, 64
Williams, B. 108
Williams, R.J. 249
Williams, R.M. 238
Wilson, E.O. 32, 24, 264
Wilson, J.R. 246
Wilson, M.C. 67, 73, 241, 242, 252, 285
Wilson, M.I. 229, 237
Wilson, R. 171
Wilson, S.R. 241, 242, 285
Winans, S.S. 29, 67, 73, 79, 89, 90, 110, 252, 253, 255
Winfree, A.T. 159, 160, 161,171
Wolfe, J.L. 229, 237
Wood, T.K. 310
Woodside, B. 171
Wright, J.W. 108
Wright, W.W. 139
Wuensch, K.L. 238
Wutke, W. 114
Wysocki, C.J. 29, 67, 73, 75, 76, 79, 80, 90, 91, 142, 180, 183-186, 189, 251, 252, 254
Yahr, P. 119, 120, 124, 126, 128, 130, 291, 300
Yamazaki, K. 207, 285
Yingman, Liang 215
Yong, Zhao 215
Yutzey, D.A. 105
Youngs, W.M. 67, 108
Zakeski, D. 75
Zarrow, M.X. 135, 144
Zayas, Z.A. 207
Zbrozyna, A.W. 111
Zhiqin, Fan 215
Ziegler, H.P. 110
Zimbelman, R.G. 156
Zinn, T.L. 243
Zucker, I. 165
Zumpe, D. 264

SUBJECT INDEX

Accessory olfactory bulb, 75, 107
  neural projections, 109
Accessory olfactory system, 73-84, 89
Acetylcholinesterase, 128-129
Aggression, male-female, house mouse, deermice, 195-210
Agkistrodon bilineatus, 17
A. piscivorus, 17
Alarm and defense, 309-319
Alarm pheromone
  amphibians, 310
  ants, 310-319
  birds, 310
  echinoderms, 310
  evolution, 309-319
  fish, 310
  gastropods, 310
  rodents, 310
  sea anemones, 310
  treehoppers, 310
Alouatta seniculus, throat rubbing, 347-350
Amygdala and reproduction, 112
Anal gland
  rabbit, 275-290
  yellow vole, 211-222
5-α-Androstenone, 155
Anesthesia, 57
Anogenital odor, guinea pigs, 187-188
Apodemus flavicollis, 223-233
Apodemus sylvaticus, 223-233
Apodemus, trap entry, 237
Artificial insemination, 156
Attractants, eel, 313-319
Attractants, sex, humans, 264-266
Attraction to bait, rabbit, 333-339
Autogroom, gerbil, 291-308
Aversive odors, humans, 266

Baboon, synchronous menarche, 17
Bait response, rabbit, 333-339
Bats, 27
Beaver, free-ranging, 236-238
Birds, 27
Birdsong control nuclei, 124, 126, 129
Birth cycle odors, rat, 168-171
Bison, 99
Bitis arietans, 17
Bitis gabonica, 17
Black-tailed deer
  Flehmen, 96-98
  response to odors, 238-241
Body temperature, gerbil, 291-308
Bothrops schegeli, 17
Breeding, pika, 329-332
Brood patch, 36
Bruce effect, 140, 145, 318
  mouse, vole, 339
  rabbit, 339-342
Bulbectomy, 106, 108-110
  behavioral effects, 111
  complex effects, 111
  endocrine effects, 112-113
  Flehmen, 97
  pig, 156
  pineal potentiation, 113-114
  puberty acceleration, 156

Callitrichids, sternal marking, 349

Calloselasma rhodostoma, 17
Canary, sexually dimorphic brain areas, 124, 126
Canis familiaris, urine, discrimination, 343-346
Canis latrans, response to odors, 241
Canis lupus, response to odors, 241
urine, discrimination, 343-346
Carnivores, response to odors, 241, 242
Carrion feeding, snakes, 7
Castor canadensis, free-ranging, 236-238
Castor extract (beaver), effect on white-tailed deer, 240
Cat
Flehmen, 88
vomeronasal organ, 110
urine marking, 120
Cattle, estrus synchronization, 155-157
Flehmen, 88, 91, 94
pheromonal estrus synchronization, 156
puberty-accelerating male urine, 153-155
Cavia aspera (guinea pig), 77
Cell division synchrony, 160, 167
Cetaceans, 27
Cheek gland, pika, 329-332
Chemosensory searching, strike-induced, 2-20
Chemosignals,
hormonal regulation, 179-193
non-urinary, mice, 186
rodents, classification, 188-190
Chin gland, rabbit, 283-286
Cicada, emergence cycle, 161
Clethrionomys glareolus, 135, 223-233
Clethrionomys, trap entry, 237
Communication disparities in deermice, 195-210
Communication, sexual dimorphism, 120
Competition, male-male, 145-146, 324
Coolidge effect, 246, 253
Cotylosaurs, 25, 26
Courtship
mice, 180
snakes, 46, 57
Coyotes, response to odors, 241
Crocodilians, 27
Crotalus adamanteus, 17
atrox, 6, 7, 17
durissus, 6, 8, 9, 17
enyo, 6, 7, 9, 17
horridus, 9, 17
lepidus, 2, 3, 17
molossus, 17
pricei, 17
scutulatus, 17
ruber, 2, 6, 9, 17
triseriatus, 3-5, 17
viridis, 6-9, 17
Cynodonts, 26, 30
Cynognathus, 33

Deermice,
communication disparities, 195-210
pregnancy block, 317-324
Degeneration, transneural, 111
Diademodon, 33
Dicrostonyx groenlandicus, 135
Dicynodon, 30
Dicynodonts, 26, 30
Dinocephalians, 26
Dipodomys merriam, olfactory communication, 321-324
Dipodomys, trap entry, 237
Discrimination
individual, hamsters, 245-258
individual, mammals, 245
urine, dog, wolf, 343-346
Dog
urine discrimination, 343-346
urine marking, 120
Dominants, entraining estrus cycles, 171
Dorsal gland, kangaroo rat, 322
Dung,
coyote, effect on white-tailed deer, 239-241

Dung (continued)
wolf, effect on white-tailed deer, 239-241

Earthworm extracts, 47, 56, 64
Echis carinatus, 17
Eel attractants, 313-316
Elapids, 1-20
Electrophoresis, chin gland, rabbit, 285
Entrainment of coupled oscillators, 160
Envenomation, 1-20
Equids, 27
Eristocophis macmahoni, 17
Estemmenosuchus, 32
Estradiol effect on scent marking, 120
Estrogen stimulating Flehmen, 96
Estrus cycle odors, rat, 161-165
Estrus detection, 96-100
synchronization, cattle, 156
mice, 140
pheromonal, 140, 156, 159-173
rat, 159-173
Eumeces fasciatus, 34
Evolution
alarm pheromones, 309-319
pheromones, 31, 32
preying by snakes, 18, 19
Excrements, 32

Fecal pellets
rabbit, 275-290
voles, 214, 215
Feces, 32
lactating female rats, 325-328
Female urine
effects on male mice, 139
Fenestra vomeronasalis, 45
Film records, snake behavior, 48
Fireflies, synchronous flashing, 160
Flehmen, 29, 73, 87-100
black-tailed deer, 96-98
bulbectomy, 97
bulls, 91
castrates, 94, 95
diestrous urine, 98
estrus detection, 96-100
Flehmen (continued)
females, 94
freemartins, 96
hormonal control, 93-96
mechanisms, 91
mountain sheep, 99
post-copulatory, 99, 100
releasing stimuli, 96, 97
sheep, 95, 96, 99
vomeronasal organ, 87-100, 92
Follicular odor, rat, 163-166
Food attraction, effect of urine, 333-339
Food searching, 1
Fox, red, response to odor, 241
Freemartins, 96
Frogs, vocalizations, 120
FSH, 162, 165

Gas chromatography effluent, rabbit, 275-290
Genetics
pheromone production, 143-144
pheromone sensitivity, 143-144
Gerbil
hormone effects on brain, 124-128
pheromone communication, 291-308
scent marking, 120-131
social fertility regulation, 168
Gland
dorsal, kangaroo rat, 322
ear, flank, hamster, 250-258
gerbil, 291-398
Harderian, hamster, 249-258
skin, rabbit, 275-290
Goat
Flehmen, 74, 88, 89
vomeronasal organ, 90
Grooming, gerbil, 291-308
Guinea pig
29, 74-84, 90-91
anogenital odors, 187-188
response to urine, 75-84, 90-91

Hamsters, 29
glands, 250-258

Hamsters (continued)
individual discrimination, 245-258
sexual behavior, 74
VNO, 74
social cycle regulation, 121
vaginal chemosignals, 186-187
Harderian gland
gerbil, 291-308
hamster, 249-258
Hartebeest, lack of Flehmen and VNO, 92
Headbobbing, guinea pig, 75
Heart, pacemaker, 161
Heart rate, rabbit, 275-290
Hedgehog, 29, 100
Hedonic sensory system, 84
Helogale parvula, 171
Hormone regulation of chemosignals in rodents, 179-193
effects on bird brain, 124, 126
effects on mammalian brain, 124-128
scent marking, 120
Horse, Flehmen, 88, 97
Hostility, monkey, 348
Human olfactory communcation, 259-273
Human sexual behavior and odors, 263-266

Incisive foramina, 29
Individual discrimination
dog, wolf, 343-346
hamsters, 245-258
humans, 259-273
Infant recognition of mother, humans, 259-262
Infanticide, 146
Inguinal gland, rabbit, 286-290
Imprinting, primates, 266
Insectivores, 29

Jacobson's organ, see vomeronasal organ

Kangaroo rat
olfactory communication, 321-324
Kangaroo rat (continued)
sandbathing, urine marks, 321-324
Kairomones, 311
Killing of female by male deer-mice, 195-210
Kin selection, 311

Lactating female odor, rat, 170
Lactation, gerbil, 303-308
Lagurus lagurus, effect of anal gland secretion of *L. luteus*, 218-220
Lagurus luteus, anal gland, 211-222
Lateral commissure, 121
Lee-Boot effect, 137
Lemming, 146
Lemur, vomeronsal organ, 100
LH, 162
response to urine odor, 139
Limbic system, 105-112
Lion, 146
Lip-curl, *see* Flehmen
Lizard
courtship and agression, 120
spiny, 29, 32
Western fence, 28
Lordosis, rat, 167, 168
Lutra canadensis, response to odors, 241, 242
Lystrosaurus, 30, 31

Magicicada (13-year cicada), 161
Mammals, free-ranging, 235-244
mesozoic, 30
Mammary gland, derivation, 35
Marking, throat rubbing, monkey, 347-350
Maternal odors, 35
rat, 325-328
Maternal pheromone 170, 183
prolactin regulation, 1-15, 179-193
rats, 180
Mating competition, kangaroo rat, 324
Mating, pregnancy block, deer-mice, 317-324

Maturation-delaying factor in mice, 185
Maturation acceleration, male mice, 139
Medial preoptic area
  control of scent marking, 120
  sex differences, 121-124
  hormone effects, 124-128
  lesions, 130
Melatonin, 113
Menstrual synchrony, 267-273
Meriones unguiculatus, pheromone communication, 291-308
Metatarsal secretion, black-tailed deer, 239-241
Mice, 29
Microtus agrestis, 135, 145, 223-233
Microtus agrestis, trap entry, 238
Microtus ochrogaster, Bruce effect, 135, 318
Microtus pennsylvanicus, 135, 14.
Microtus townsendii, trap entry, 237
Migratory behavior
  alewives, 313, 314
  eels, 313
  salmon, 313
Mink, Everglades, response to odor, 241, 242
Modal action pattern, 8, 9
Mongoose, social fertility suppression, 171
Monkey, red howler, throat rubbing, 347-350
Mountain sheep, Flehmen, 99
Mouse, Bruce effect, 339
Mouse (house), genetic divergence, 195-210
Multiple-male copulation, deermice, 317-324
Mus musculus, 2, 172
  genetic divergence, 195-210
  priming pheromones, 135-146
  trap entry, 237
Musk, muskrat, effect on white-tailed deer, 238
Muskrat, trap entry, 237
Mustela vison evergladensis, response to odors, 241, 242

Naja mossambica, 17, 18
Naja naja, 17, 18
Nasal mucus and Flehmen, 97
Nasopalatine duct and Flehmen, 97
Natural selection, snakes, 19
Nervus terminalis, 107
Nervus terminalis, neural projections, 107

Ochotona princeps, scent marking, breeding, 324-332
Odocoileus hemionus, response to odors, 238-241
Odocoileus virginianus, response to odor, 239-243
Odor, effect on trap entry, 223-233
  of individuals, 245-258
  preference, rat, 325-328
Olfactory bulbectomy, *see* Bulbectomy
Olfactory bulb, 107
  neural projections, 109, 110
Olfactory communication
  dog, 343-346
  humans, 259-273
  kangaroo rat, 321-324
  monkey, 347-350
  wolf, 343-346
Olfactory mucosa, destruction, 255-258
Olfactory nerve, 107
Olfactory nerve transection, snakes, 49, 54-55
Olfactory neural projections, 89
Olfactory system, hamster, 250-258
Olfactory vs. vomeronasal reception, 73-84, 99
Ondatra zibethicus, 237, 238
Opistoglyphs, 20
Oryctolagus cuniculus
  Bruce effect, 339-342
  response to urine, 333-339
  skin glands, 275-290

Oscillators, olfactory coupling, 159
Otters, river, response to odors, 241, 242
Ovariectomy, effect on chemosignals, 1-15, 179-193
Ovulation, timing in rat, 159-168
Oxyuranus microlepidotus, 17
  scutellatus, 17

Paleoethology, 34-37
Papio cynocephalus, 171
Pelycosaurs, 25, 26, 30
Perognathus, trap entry, 237
Peromyscus leucopsis, trap entry, 237
Peromyscus maniculatus
  genetic divergence, 195-210
  pregnancy block, 135, 317-324
  trap entry, 237
Phase-response curve, 161, 168
Pheromone, criteria, 172-173
Pig, puberty acceleration, 155-156
Pika, scent marking, breeding, 329-332
Pineal gland, 112
Pinealectomy, 112
Plasticity, CNS, 108-109, 131
Population density and priming pheromones, 144-145
Populations of the same species of deermice, 195-210
Prairie vole, Bruce effect, 318
Pregnancy block, 140, 145
  deermice, 317-324
  rabbit, 339-342
  strain differences, 142-143
Pregnant female odor, rat, 120
Preputial gland, mouse, 141
Prey odors, 46-47, 63
Preying, 1-20
Primates, 27, 146
Primates, sex pheromones, 265, 266
Priming compatibility, house mouse, deermice, 203-205
Priming pheromones
  adaptive functions, 144
  applications, 153
Priming pheromones (continued)
  chemistry, 139-142
  mice, 135-146
  population density, 144-145
Probainognathus, 26, 30
Probelesodon, 30
Progestagens, 156
Progesterone, 161, 162, 165
Prolactin, 162
^{3}H-Proline, 46
Prostaglandin, 156
Prosynosuchus, 31
Proteroglyphs, 20
Prototheria, 26
Pteroptyx (firefly), 160
Puberty acceleration, cattle, 154-155
  male urine, 140, 153-155
  pig, 155-156
Puberty delay
  mice and voles, 145
  pheromonal, 136, 145

Rabbit
  Bruce effect, 339-342
  fecal pellets, 275-290
  response to urine, 333-339
  skin glands, 275-290
Rabbits, 29
Rat
  aggression chemosignals, 5, 6, 183, 184
  brain, sex differences, 121
  bulbectomy effects, 111-114
  estrus cycle odors, 161-165
  estrus cycle phases, 161-165
  estrus synchronization, 159-173
  feces of lactating female, 325-328
  maternal pheromone, 170, 183
  odor preference, 325-328
  sex odors, 181-183
  neuroendocrine control of urine marking, 120
  sexually dimorphic nucleus, 121
Rats, 29, 74
Rattlesnakes, 1-20, 28
Recognition, individual, humans, 259-273

Regulation of chemosignals in rodents, 179-193
Relative coordination, 160
Reproductive isolation, deermice, 195-210
Reproductive odors, 29
Reproductive status, kangaroo rat, 323
Reptiles, mammal-like, 25-37
Rhinencephalon, 105
Rhodamine, 91
Rodent prey, 1-20
Rodents, free-ranging, 236-238
Rodents
  chemosignals, hormonal regulation, 179-193
  housemouse, deermice, 195-210
  trap-entry behavior, 223-233
  voles, anal gland, 211-222
Rudolfo-Masera, organ, 107

Saliva, 32
  gerbil, 291-308
  sandbathing, gerbil, 298
  kangaroo rat, 321-323
Sceleporus occidentalis, 28
Scent marking
  humans, 262, 263
  neuroendocrine control, 119-131
  pika, 329-332
  sandbathing, kangaroo rats, 321-323
Scincella lateralis, 34
Search image, snakes, 6
Septal olfactory organ, 107
Sex attractants, humans, 264-266
Sex differences, rat brain, 121
Sexual behavior and VNO projections, 89
Sexual identifiation, humans, 263-264
Sexually dimorphic area of brain, histochemistry, 128-129
Sexually dimorphic nucleus
  gerbil, 121
  rat, 121
Sexual odor
  garter snakes, 57
Sheep
  Flehmen, 99
  vomeronasal organ, 89
Shrews, trap entry behavior, 223-233
Sigmodon hispidus, trap entry, 237
Signaling compatibility, house mouse and deermice, 195-210
Sistrurus
  catenatus, 7, 17
  miliarus, 17
Skin glands, rabbit, 275-290
Slime mold, 160
Snake, garter, 29, 45-68
Soay sheep, Flehmen, 95-96
Social fertility regulation, 168
Sodium fluorescein, 92-93
Speciation, deermice, 195-210
Steroid effect on brain cells, 126-127
Strange male, Bruce effect, 339-342
Stria terminalis, 121
Sweat, 32
Synapsida, 25-37
Synchronization, menstrual, hormonal, 267-273
Synchronized leaf movements, 160
Synchronous flashing, fireflies, 160
Synchronous menarche, baboons, 171
Synchrony, cell division, 160

Territorial marking
  rabbits, 275-290
  voles, 211-222
Testosterone effect on scent marking, 120
Testosterone and protein synthesis, 127-128
Testosterone stimulation by females, 139-140
Testudines, 29
Thamnophis, 63
  radix, 47
  sirtalis, 47
Therapsids, 25-37
Theriodonts, 26

Thermal and osmolarity properties, gerbil, 291-308
Thrinaxodon, 33
Throat rubbing, monkey, 347-350
Tongue-flicking, snakes, 2-18, 45-68
Topi, lack of Flehmen and VNO, 92
Trail discrimination, simultaneous, 10
Trail discrimination, successive, 13
Trail-following, snakes, 1-20
Trailing, snakes, 8, 47
Trap entry, rodents, 236-238
  odor, rodents, 223-233
Trigeminal nerve, 108
Tuatara, 29

Ultrasonic vocalization, mice, 180, 183
Ungulates, 29
  response to odors, 238-241
Urinary proteins, 141
Urinary sex signals in mice, 183-185
Urine
  dog, effect on white-tailed deer, 239-241
  coyote, effect on white-tailed deer, 239-241
  discrimination, dog, wolf, 343-346
  guinea pig, 76-78
  kangaroo rat, 323-324
  mouse, 179-193
  response of rabbit, 333-339
  trails, mice, 16

Vaginal chemosignals, hamsters, 8, 9, 186-187
Vaginal gland, hamster, 250-258
Vaginal secretion, hamster, 179-180
Vandenbergh effect, 140
Ventral scent gland, gerbil, 291-308
Vipera
  ammodytes, 17
  aspis, 9
  berus, 17
  latifii, 17
  xanthine, 17
Viperids, 1-20
Vole
  anal gland, 211-222
  Bruce effect, 339
  trap entry behavior, 222-233
Voles, infanticide, 146
Vomeronasal neural projections, 89
Vomeronasal organ, 4, 26, 27, 107
  anatomy, 89
  distribution, 27
  dye uptake, 92-93
  evolution, 27
  Flehmen, 87-100
  function, 87-100
  function, guinea pig, 75-84, 90
  guinea pig, 73-84
  mammal-like reptiles, 29-31
  removal, 76
  rodents, 90
  snakes, 45-68, 90
Vomeronasal pump, 91-92
Vomeronasal reception. 2. 73-84, 90
Vomeronasal system
  hamster, 250-258
  rat, 252
Vulpes vulpes, response to odor, 241

Water loss, gerbil, 298, 299
White-tailed deer, response to odor, 239
Whitten effect, 140, 160
Wolf, urine discrimination, 343-346
Wolves, response to odors, 241

Xylocaine, 57

Zeitgeber, 160
Zinc sulphate, 255

GPSR Compliance

The European Union's (EU) General Product Safety Regulation (GPSR) is a set of rules that requires consumer products to be safe and our obligations to ensure this.

If you have any concerns about our products, you can contact us on ProductSafety@springernature.com

In case Publisher is established outside the EU, the EU authorized representative is:

Springer Nature Customer Service Center GmbH
Europaplatz 3
69115 Heidelberg, Germany

Batch number: 10838292

Printed by Printforce, the Netherlands